GOALS AND ECONOMIC INSTRUMENTS FOR THE ACHIEVEMENT
OF GLOBAL WARMING MITIGATION IN EUROPE

Goals and Economic Instruments for the Achievement of Global Warming Mitigation in Europe

Proceedings of the EU Advanced Study Course held in Berlin, Germany, July 1997

Edited by

JÜRGEN HACKER
Environmental Management Consultancy UMB Hacker GmbH,
Berlin, Germany

and

ARTHUR PELCHEN
Environmental Management Consultancy UMB Hacker GmbH,
Berlin, Germany

Advanced Study Course, sponsored by DG XII/D Environment and Climate RTD Programme, European Commission

SPRINGER-SCIENCE+BUSINESS MEDIA, B.V.

A C.I.P. Catalogue record for this book is available from the Library of Congress.

ISBN 978-0-7923-5337-9   ISBN 978-94-011-4726-2 (eBook)
DOI 10.1007/978-94-011-4726-2

Printed on acid-free paper

TABLE OF CONTENTS

Foreword ix
Preface xi
Acknowledgments xiii

SECTION I. STATE OF SCIENTIFIC KNOWLEDGE IN CLIMATE CHANGE

K. HASSELMANN and U. CUBASCH
Climate and its Influencial Factors, especially the Anthropogene Enhancement of the Greenhouse Effect and its Possible Impacts - Results of the Second Assessment Report of the IPCC 3

B. LIM, P. BOILEAU and Y. BONDUKI
The IPCC/OECD/IEA Greenhouse Gas Inventories Programme: International Methods for the Estimation, Monitoring and Verification of GHG Emission Inventories 27

M. HA DUONG
Necessities and Problems of Coupling Climate and Socioeconomic Models for Integrated Assessments Studies from an Economist's Point of View 39

F. TÓTH, G. PETSCHEL-HELD and T. BRUCKNER
Climate Change and Integrated Assessment: The Tolerable Windows Approach
(The ICLIPS Project: Integrated assessment of **CLI**mate **P**rotection **S**trategies) 55

G. PETSCHEL-HELD and F. REUSSWIG
Climate Change and Global Change: The Syndrome Concept
(The QUESTIONS Project: **QU**alitativ**E** Dynamics of **S**yndromes and Transi**TION** to **S**ustainibility) 79

SECTION II. FROM SCIENTIFIC RESULTS TO THE POLITICAL AGENDA 97

G. McINNES
A Comparison of GHG Inventories and Reduction Goals for Different Countries in Europe 99

U. COLLIER
Allocating Emission Reduction Targets in the European Union - Issues and Proposals 113

C. BAIL
The EU Position in the Negotiation Process towards a Protocol on the Limitation and Reduction of Greenhouse Gas Emissions 129

SECTION III. POLITICAL AND ECONOMICAL INSTRUMENTS - AN OVERVIEW

- INTRODUCTION - 137

J. HACKER
The Range of Possible Political and Economical Instruments for the Mitigation of Global Climate Change 139

A. PELCHEN
A Framework for the Evaluation of Political and Economical Instruments for Global Warming Mitigation 145

- VOLUNTARY APPROACHES - 159

K. MÜSCHEN
Global and European Voluntary Approaches in Mitigation of Greenhouse Gas Emissions on Municipal Level and their Results 161

F. SCHAFHAUSEN
Voluntary Commitments to Mitigate Greenhouse Gas Emissions - The Example of German Industry and Trade 169

K. RENNINGS, K.L. BROCKMANN and H. BERGMANN
Voluntary Agreements in Climate Protection - Experiences in Germany and Future Perspectives 183

- TAXES - 205

R. BARON
Carbon and Energy Taxes in OECD Countries 207

S. BILL
European Commission's Experience in Designing Environmental Taxation for Energy Products 231

T. BARKER
Limits of the Tax Approach for Mitigating Global Warming 239

- JOINT IMPLEMENTATION - 253

C.J. JEPMA
Activities Implemented Jointly (AIJ) as an Instrument for the Mitigation of Global Warming 255

F. TATTENBACH
Practical Examples of Activities Implemented Jointly (AIJ) in Costa Rica 273

J. HACKER
Problems and Limitations of AIJ and the JI Potential from the Perspective of a German Project Broker 283

SECTION IV. TRADEABLE EMISSION PERMITS

- A PRACTABLE EXAMPLE: TRADEABLE EMISSION PERMITS FOR SO_2 - 293

B. McLEAN
US Experience with Tradeable SO_2 allowances 295

K.L. BROCKMANN, H. KOSCHEL and T.F.N. SCHMIDT
A European Model for Tradeable SO_2-Emission Permits 309

- DESIGN OF A CO_2 EMISSION SYSTEM - 329

J. OLIVEIRA MARTINS
Efficiency, Equity and Optimal Carbon Abatement 331

P. KOUTSTAAL
Posibilities of an EU System of CO_2 Emission Permits 339

F. T. JOSHUA
Launching a Plurilateral Greenhouse Gas Emissions Trading System
- the UNCTAD-Earth Council Initiative 357

S. VAUGHAN
Tradeable Emissions Permits and the WTO 367

- PERSPECTIVE OF A CO_2 EMISSION SYSTEM - 377

J. DEPLEDGE
Status of Discussion and Negotiation for a System of Tradeable
CO_2 Emission Permits within the UNFCCC 379

J. WOLFE
The United States Proposal for an International CO_2 Emissions
Trading System 385

- EMISSION QUOTA TRADE - A SIMULATION EXPERIMENT - . . 395

A. PELCHEN
Results of the Emission Quota Trade Simulation Experiment
by the Participants 397

SECTION V.
COMPARISON OF INSTRUMENTS AND SUMMARY 407

A. ENDRES
Assessing the Different Instruments in Climate Change Mitigation
from the Perspective of Economics 409

A. PELCHEN
Assessment of the Different Economic Instruments by the Participants and
Selected Results of the Working Groups 427

List of Authors 441
List of Participants 447
Colour Plate Section 457

FOREWORD

Climate change poses important challenges to research and policy. Within three decades, an issue that was initially confined to the attention of a few scientists became the topic of large-scale research programmes, national and European policies and an international Convention. While significant uncertainties remain on the timing and scale of the changes to be expected and of their impacts, an appreciation emerged of the high ecological, economic, political and social stakes involved and lead to governmental, business and citizens' initiatives.

After focusing on the understanding of climate processes and the possible impacts of climate change on ecosystems, European research - and international research more generally - started addressing also the social, economic and policy causes of and responses to climate change. In the meantime, local, national and European measures started being developed to mitigate carbon dioxide emissions, a European target was agreed to achieve the stabilization of carbon dioxide by 2000 at the levels of 1990, the Framework Convention on Climate Change (FCCC) was adopted and was followed by its Kyoto Protocol.

When the Advanced Study Course "Goals and Economic Instruments for the Achievement of Global Warming Mitigation in Europe" was held, the research and policy communities in Europe and worldwide were intensely debating what targets should be agreed in Kyoto, what instruments and policies could be developed to mitigate greenhouse gases' emissions, how should the responsibility and costs for such mitigation be shared across countries and socio-economic sectors. The Advanced Study Course complemented other research initiatives, such as the series of policy-research interface workshop "EU Climate Policy: Research Support for Kyoto and beyond", undertaken by DG XII (Directorate General for Science, Research and Development) on socio-economic aspects of climate change. The Course focused on the formulation and use of the so-called 'flexibility instruments' (such as joint implementation and emission trading) that were among the most intensely debated issues at the Conference of the Parties of the FCCC in Kyoto and that remain among the 'pending issues' after such Conference.

There is no doubt that the organizers of the Course identified a policy relevant issue, and an issue where the need for improved understanding and capacity building - including training of young researchers - was needed. The quality of the contributions by the academic and policy experts and the enthusiasm shown by the students made the Course a useful and challenging event. Given the fact that climate change is a long-term issue which complexity is far from being 'managed', the transmission of analytical and practical knowledge to younger generations seems an effort worth pursuing.

Angela Liberatore
Unit on socio-economic environmental research
DG XII-D5, European Commission

PREFACE

From July 20 to 26, 1997 an Advanced Study Course "Goals and Economic Instruments for the Achievement of Global Warming Mitigation in Europe" was held at the Technical University of Berlin, Germany. The Course has been carried out by UMB UmweltManagementBeratung Hacker GmbH on behalf of the European Commission, Directorate-General XII/D Environment and Climate RTD Programme.

The Course was attended by fifty-one young scientists from twelve EU and seven non-EU nations. Lectures were given by twenty-nine experts from eleven different EU universities or research institutions, seven international organizations and five governments or governmental agencies. The lectures were structured in five sections like this volume. Additionally, there was an excursion to the Potsdam Institute for Climate Impact Research near Berlin. In four working groups the participants evaluated the results of the lectures of each section and presented and discussed their evaluations in plenary sessions. In the second part of the Course the participants performed an simulation experiment for emission trading.

The aims of the Advanced Study Course were:

- to improve the communication between students and scientists on a European level as well as fostering the interdisciplinary discussion of the subject of global warming,
- to bridge the gap between the scientific knowledge about mitigation of global warming on the one hand and the practical implementation of the scientifically necessary measures for the reduction of GHG-emissions in the political process on the other hand,
- to evaluate the different political and economical instruments for the mitigation of global warming that are in today's public, political and scientific discussion in relation to their accuracy of reaching the agreed targets and their cost-efficiency.

To enhance the sustainability of these aims the proceedings are published in this book, also sponsored by the European Commission. Part of the papers presented here are the original lectures, part of them are versions modified after the Course. Since especially the contributions concerning the political implementation are very time-related, their date of completion is given at their beginning.

We hope that this book stimulates the ongoing public, political and scientific discussion and helps to find sustainable, efficient and fair solutions to the problem of global warming.

Jürgen Hacker
Dr. Arthur Pelchen

ACKNOWLEDGMENTS

We wish to acknowledge with sincere thanks all of those responsible for the funding, preparation and organization of the EU Advanced Study Course, which was the basis for editing this book.

Firstly, we are grateful to the European Commission, DG XII/D Environment and Climate RTD Programme, for funding the Advanced Study Course and this book. Special thanks is given to Dr. Angela Liberatore, the scientific responsible in the Commission, for her help to cope with all needs of the Commission's administration.

Secondly, we wish to thank the members of the Scientific Panel of the Course, Prof. Dr. Jürgen Starnick (Technical University of Berlin), Dr. Gerhard Petschel-Held (Potsdam-Institute for Climate Impact Research), Prof. Dr. Alfred Endres (Distance Teaching University Hagen), Joaquim Oliveira Martins (Organisation for Economic Co-operation and Development) and again Dr. Angela Liberatore for their advice and comments on the programme of the Course and the selection of the lecturers. Especially we thank the Chairman, Prof. Dr. Starnick, for the hospitality of the Technical University of Berlin and the Vice-Chairman, Dr. Petschel-Held, for his efforts in organizing the excursion to the Potsdam-Institute for Climate Impact Research.

We also thank Prof. Peter Bohm of Stockholm University for allowing us to rerun his Emission Quota Trade Experiment among four Nordic countries with the participants of the Course despite his doubts of its suitability for this purpose.

Naturally we wish to acknowledge with thanks all of the lecturers not only for their presentations to the Advanced Study Course and their readiness and engagement in discussing their lectures with the participants but also for their contributions in this book.

We are grateful to Prof. Dr. Dietmar Winje, chairman of the board of the Berlin Electricity Utility Bewag for sponsoring the farewell dinner in the Spandau Citadel.

And finally we also thank Astrid Zandee of Kluwer Academic Publishers for her publication assistance.

Jürgen Hacker
Dr. Arthur Pelchen

ACKNOWLEDGEMENTS

SECTION I

STATE OF SCIENTIFIC KNOWLEDGE IN CLIMATE CHANGE

CLIMATE AND ITS INFLUENTIAL FACTORS, ESPECIALLY THE ANTHROPOGENIC ENHANCEMENT OF THE GREENHOUSE EFFECT AND ITS POSSIBLE IMPACTS - RESULTS OF THE SECOND ASSESSMENT REPORT OF THE IPCC

Prof. Dr. K. HASSELMANN
Max-Planck-Institute for Meteorology
Bundesstr. 55, D-20146 Hamburg, Germany

Dr. U. CUBASCH
Deutsches Klimarechenzentrum GmbH
Bundesstr. 55, D-20146 Hamburg, Germany

1. The climate system

The climate system is a large heat engine driven by the radiation from the sun (FIGURE 1). About 30% of the energy of the sun is reflected, 70% is absorbed. If the earth were a black body, the equilibrium temperature needed to re-emit the absorbed radiation would be -20°C. However, the earth is surrounded by the natural greenhouse gases (in order of importance: the water vapour, CO_2 and methane). These are essential for a liveable climate. The solar radiation passes unhindered to the earth surface, but the infrared (thermal) radiation is partially absorbed and re-emitted to the earth's surface. This results in an increase of the atmospheric temperature by about 35 °C yielding a global mean average of 15 °C.

The tropical regions receive more radiation than the higher latitudes. This results in a temperature contrast which drives the atmospheric circulation, redistributing the heat from the equator to the poles. About the same amount of heat is transported through ocean currents like the Gulf stream and the Kuroshio. The net effect of the heat redistribution through the atmospheric and oceanic circulation is a reduction of the temperature contrast between the low and high latitudes of about 10-20 °C. The details of the actual processes governing the circulation systems and the energy transfer in the climate system are highly complex.

J. Hacker and A. Pelchen (eds.), Goals and Economic Instruments for the Achievement of Global Warming Mitigation in Europe, 3–26.

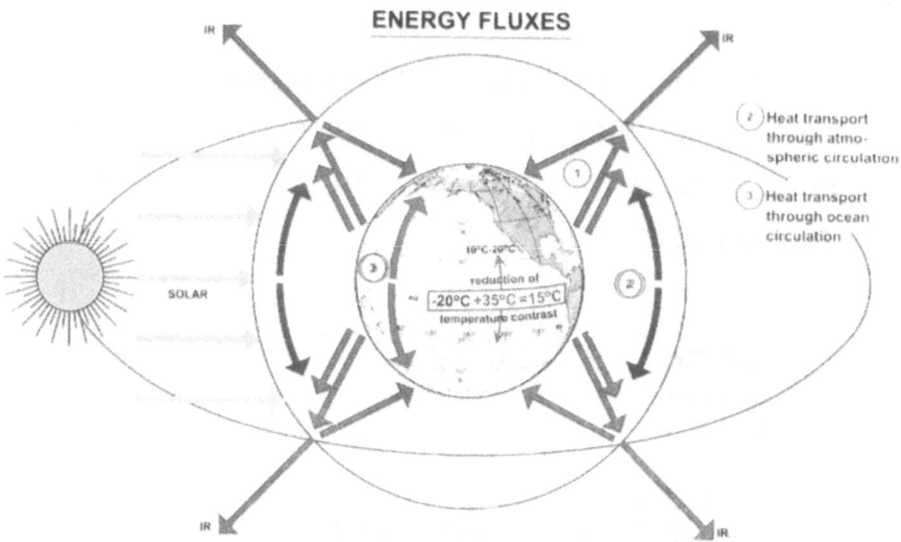

FIGURE 1 (A). The energy fluxes in the earth-climate system.

Figure 1, 9a, 9b, 13, 16, 17, 20, 21 of this paper are reproduced additionally in the colour section at the end of the book. The letters in paranthesis indicate their number there.

2. The Greenhouse Gases

The climate problem mankind is now facing is a continued increase in the greenhouse gas emissions through man's activities, in particular CO_2, which is responsible for about 60% of the increase in the greenhouse gas forcing (see the 1995 report of the UN Intergovernmental Panel for Climate Change (IPCC)).

FIGURE 2 shows the observed CO_2 concentrations in the atmosphere. From 1750 to 1958, the CO_2 data have been extracted from air-bubbles in ice cores. Since about 1957 onwards direct measurements have been performed, first at the Mauna Loa Observatory in Hawaii, later at various other locations. One sees a strong increase in the CO_2 concentrations during the last decades. The growth rate of the CO_2 concentrations is currently about 1.0% per year, caused mainly by fossil fuel burning. CO_2 is not the only greenhouse gas that is increasing. One finds also increases in the methane concentration and the CFCs, particularly CFC-11. However, according to the Montreal Protocol, the CFCs will be phased out (not because of their greenhouse gas effect, but for their role in destroying the ozone layer).

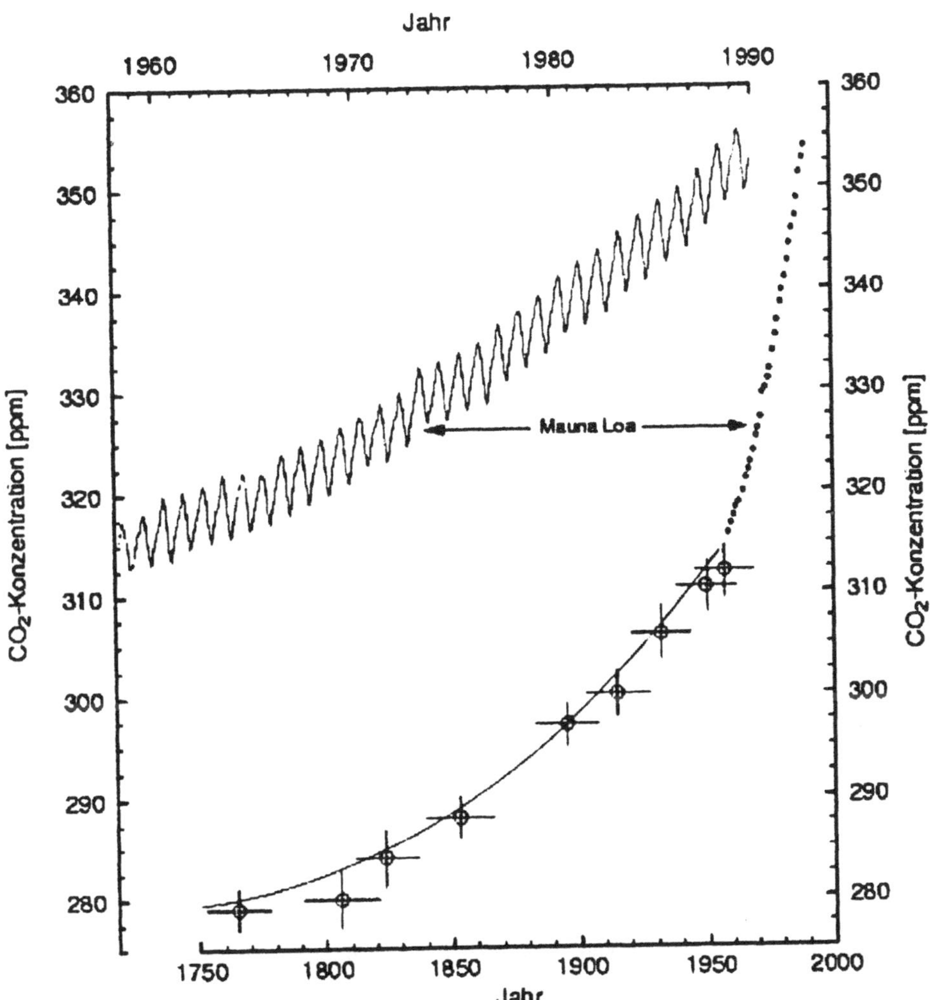

FIGURE 2. The CO_2 concentration over the past 250 years (lower curve) from ice core records (crosses) and since 1958 from Mauna Loa, Hawaii (upper curve) in ppmv/year (after IPCC, 1996).

FIGURE 3 shows a summary the principal global-mean radiative forcings caused by the greenhouse gases like CO_2 and other sources such as sulphate aerosols and changes in solar radiation. The forcings can be computed with radiative transport models. Sulphate aerosols are formed from SO_2, which is emitted during the burning of fossil fuels. They have a cooling effect in the atmosphere. Soot and biomass burning is also a source of cooling aerosols. Finally, there is much discussion on whether variations in the solar radiation contribute measurably to the observed climate change.

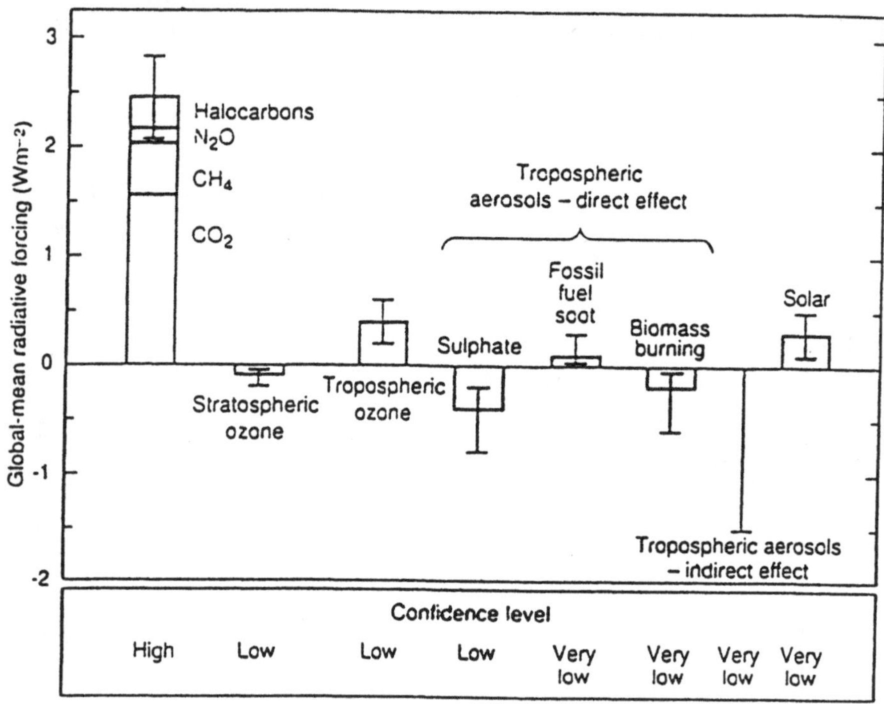

FIGURE 3. Estimates of the globally and annually averaged anthropogenic radiative forcing (in W/m^2) due to changes in concentrations of greenhouse gases and aerosols from pre-industrial times to the present (1992) and to natural changes in solar output from 1850 to present. The height of the rectangular bar indicates a mid-range estimate of the forcing whilst the error bars show an estimate of the uncertainty range based largely on the spread of published values; The "confidence level" indicates the IPCC's confidence that the actual forcing lies within this error bar (after IPCC, 1996).

3. Modelling the Climate System

The climate system (FIGURE 4) consists of various subsystems: the atmosphere, the oceans, the land surface, the biosphere, the biogeochemical cycles and the cryosphere.

For all the subsystems models have been developed (FIGURE 5), for example for the biosphere, ice-sheets, trace gases, the CO_2 carbon cycle and the general circulation of the atmosphere and the ocean. The theoretical goal is to couple the various subsystem models together to obtain a model of the complete climate system. However, this can be only partially realized due to finite computing resources.

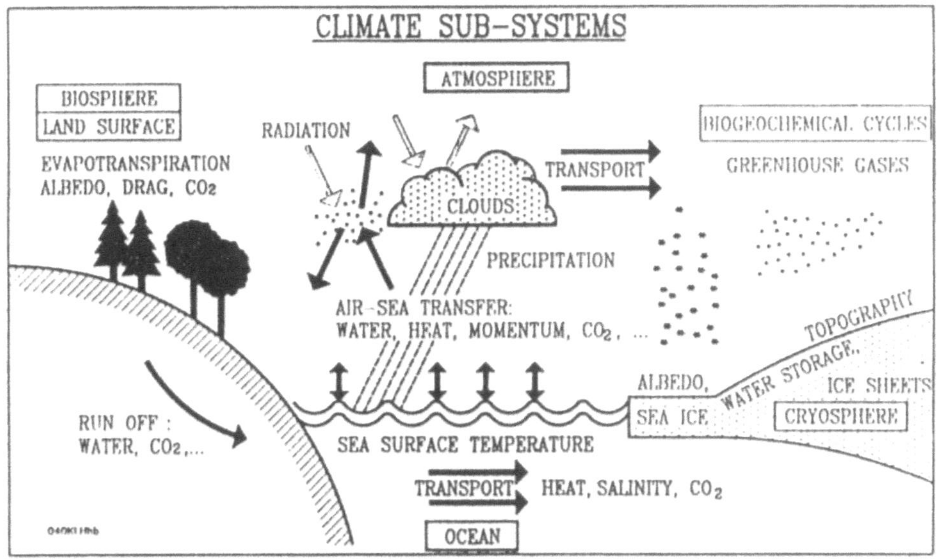

FIGURE 4. Schematic view of the components of the global climate system, their processes and interactions.

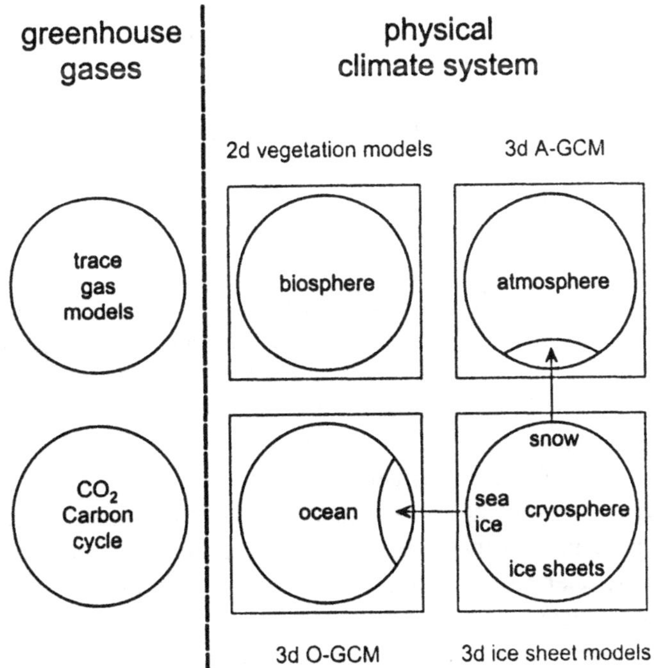

FIGURE 5: Schematic diagram of the models of the climate subsystems.

Models consist of a mathematical/physical description of the processes acting in the different subsystem. The equations describing the system are solved on a finite grid (FIGURE 6). For a typical climate model a horizontal resolution of 500 km (T21) to 250 km (T42) and a subdivision in the vertical in 10-30 layers is currently state-of the-art.

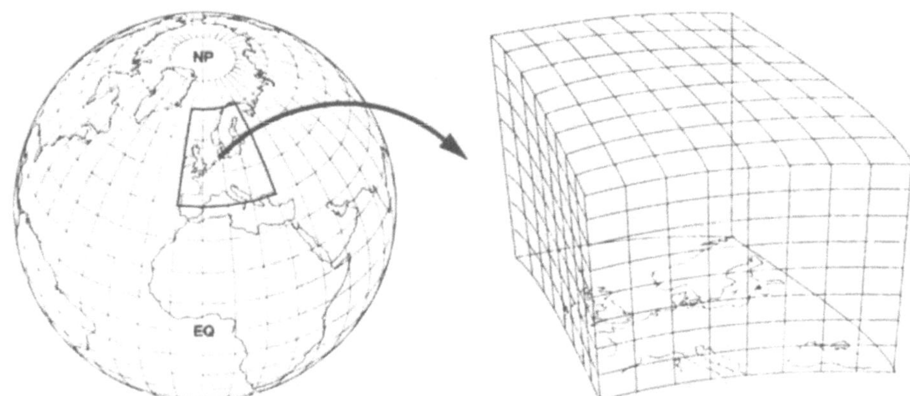

FIGURE 6. Schematic representation of the grid structure used in climate models to solve the equations describing the climate system.

FIGURE 7 shows Europe at different horizontal resolutions. T21 corresponds to a resolution which was common about 5 years ago, T42 is the resolution currently achieved for century-long integrations. Both resolutions give a rather coarse picture on the regional scale. For climate impact studies one would like resolve down to regional scales of 50 km or less. This can be achieved only by applying regional models or statistical (downscaling) methods.

A basic problem is the different response time of the climate subsystems: The atmosphere is a fast system. It responds to forcing from the outside on time scales from minutes up to a month. The ocean is a much slower system. It has about 1000 times the inertia or density of the atmosphere, and therefore responds on much longer time scales of up to a thousand years. The ice sheets have response times up to a million years. The biosphere and the biochemical cycles have response times between minutes and centuries (FIGURE 8). The extent to which the different subsystems can be coupled together depends on the available computing resources.

With present-day supercomputers it is possible to carry out simulations, in which the climate subsystem atmosphere is computed by a circulation model that follows the details of weather patterns from day to day or hour to hour, and which are sufficiently long to realistically represent the oceanic circulation over periods up to thousand years. These models can be used for transient computations over several centuries to determine the response of the climate system to an increase of greenhouse gas concentrations and/or other anthropogenic forcing.

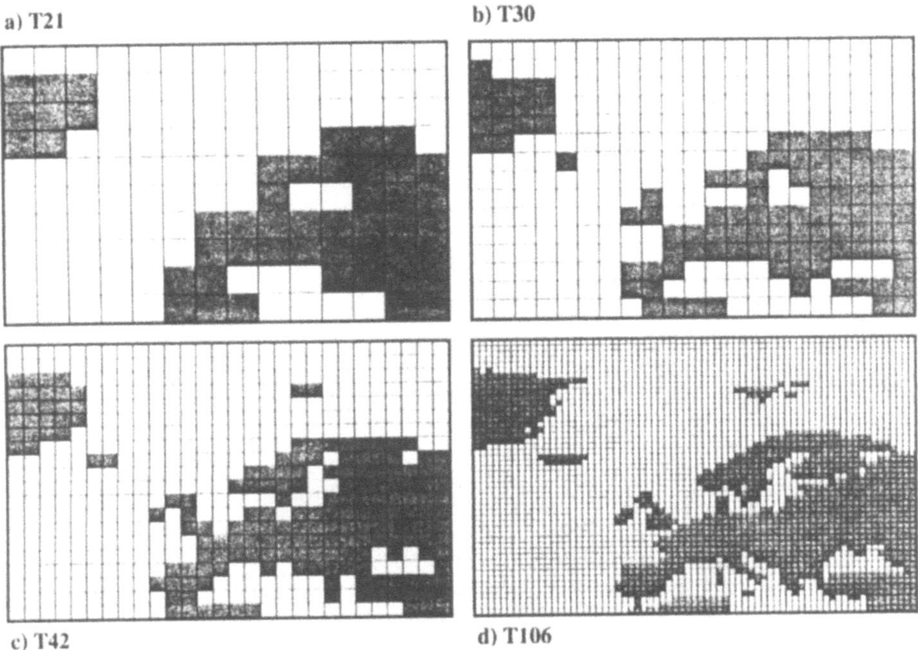

FIGURE 7. The horizontal grid over Europe for different horizontal resolutions (after Cubasch et al., 1995).

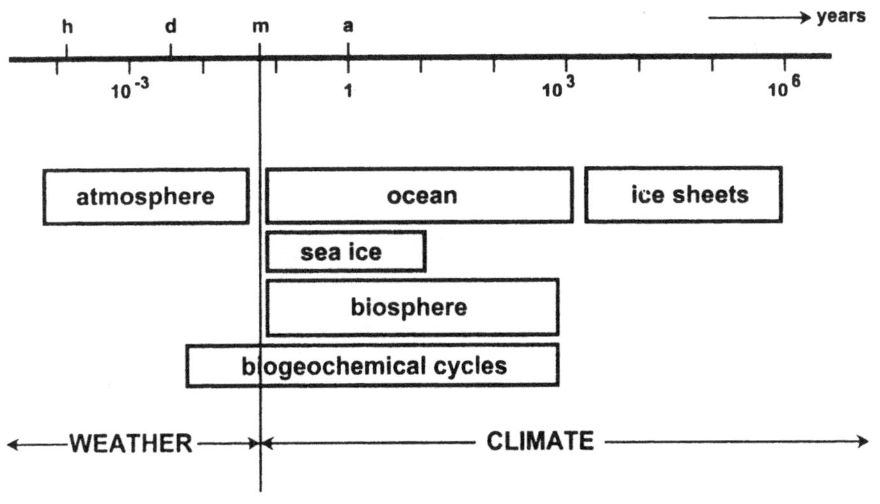

FIGURE 8: The characteristic timescales of the climate subsystems.

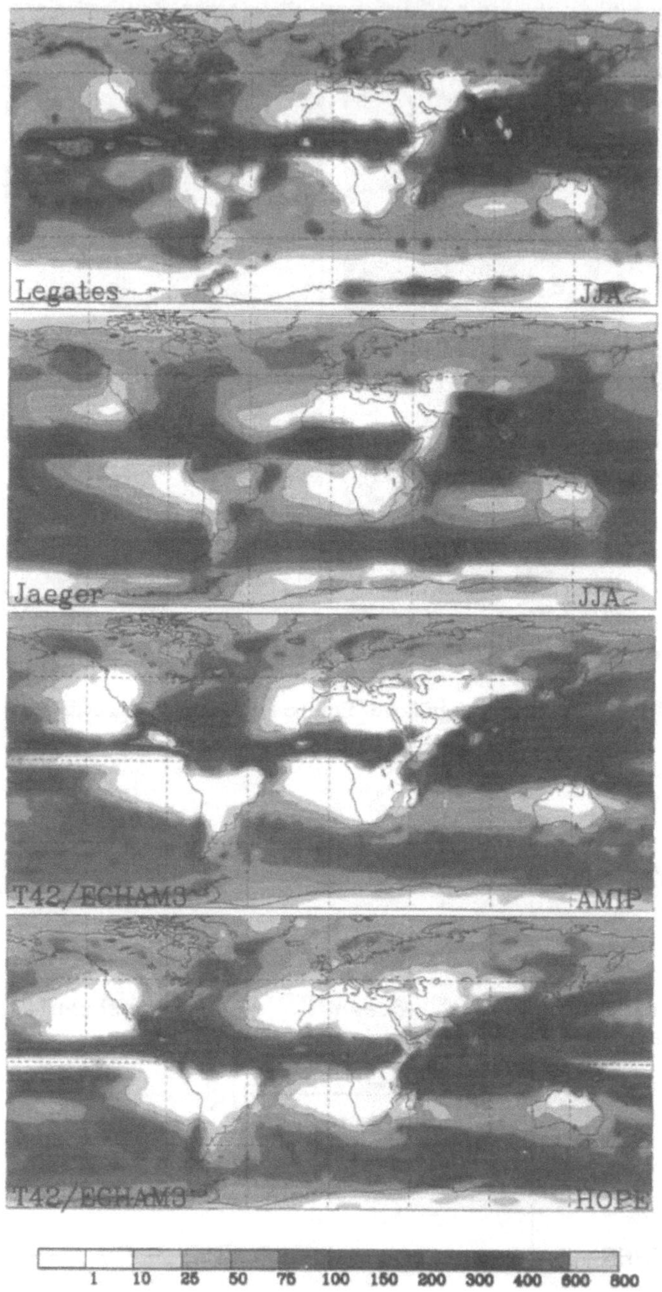

FIGURE 9a (C). The observed and the modelled precipitation for summer (JJA). The observations (top two panels) have been compiled by Legates and Willmott (1990), one model run was forced with observed sea-surface temperatures (third panel) and one model run has been performed with a fully coupled ocean-atmosphere climate model (bottom panel). (Arpe, pers. com)

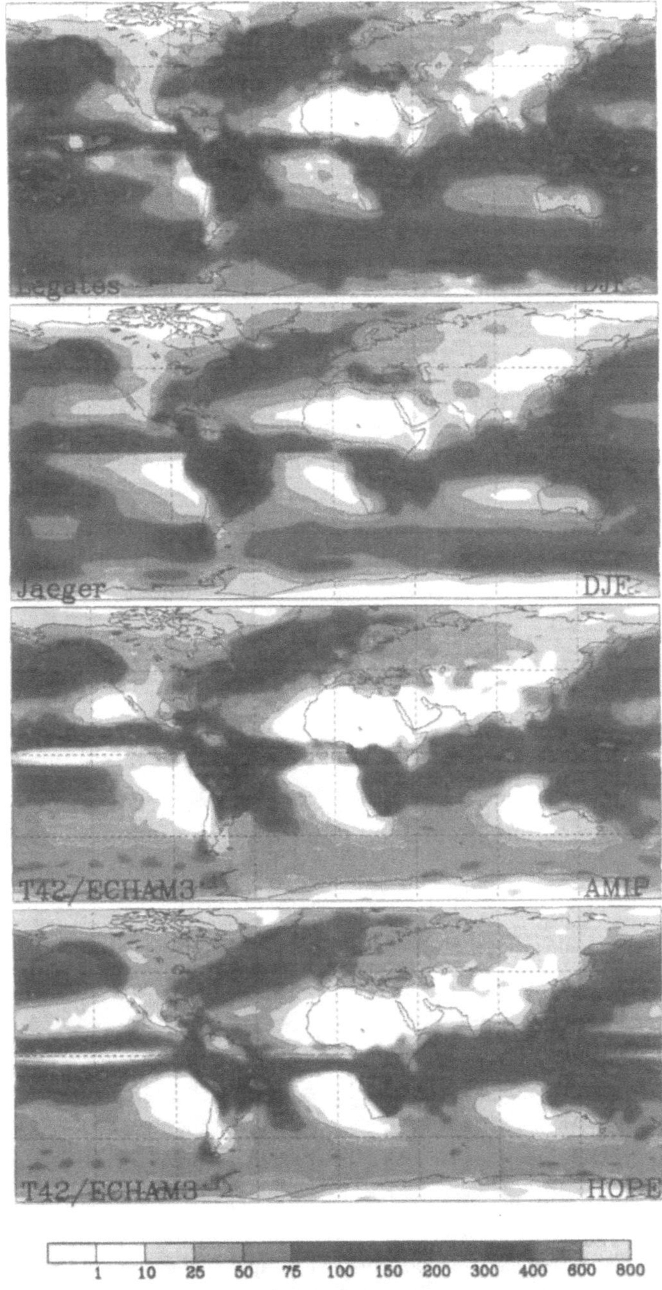

FIGURE 9b (D). The observed and the modelled precipitation for winter (DJF). The observations (top two panels) have been compiled by Legates and Willmott (1990), one model run was forced with observed sea-surface temperatures (third panel) and one model run has been performed with a fully coupled ocean-atmosphere climate model(bottom panel). (Arpe, pers. com)

The models are verified and calibrated against present-day climate data. Historic data contain too many uncertainties for useful validation. FIGURE 9 a, b show as an example a comparison of the observed and simulated precipitation for summer and for winter. Two different analyses of the observed data are compared with two sets of data generated by two different model experiments. One data set represents a simulation with an atmosphere model only that was driven by observed sea surface temperatures, the second one a coupled ocean-atmosphere model simulation. In both cases the general patterns of the precipitation are simulated to the same level of accuracy as in the observational data. The models of the other climate subsystems are tested in a similar fashion.

FIGURE 10 shows the $\delta^{18}O$ data from an ocean sediment core that represents a proxy for the volume of the ice sheets, or equivalently, sea level. A second $\delta^{18}O$ curve from an ice core in Greenland, represents a proxy for the temperature, while a third curve shows pollen data, which serve as an indicator of the local climate change.

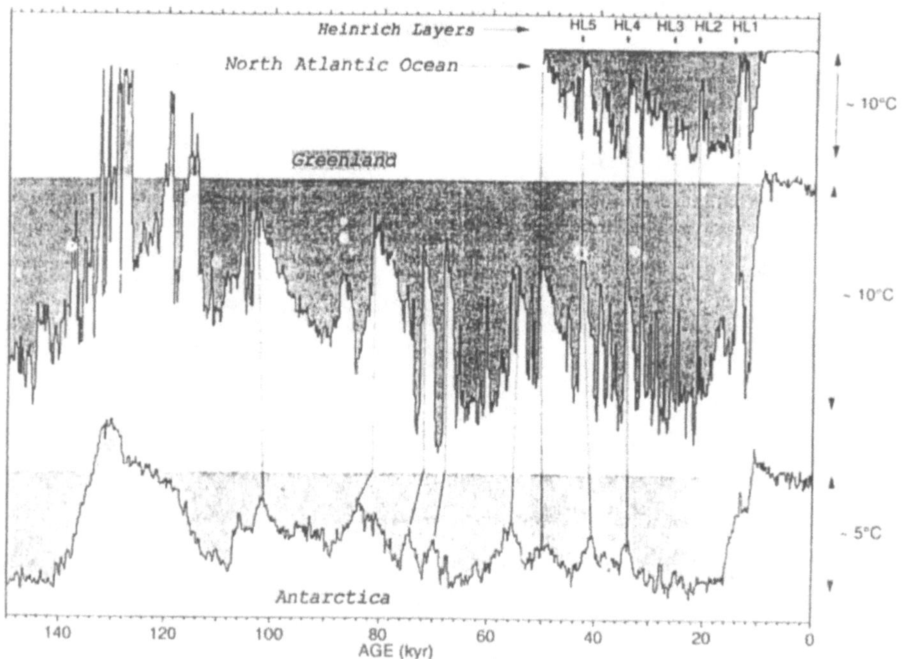

FIGURE 10. Reconstructed climate records showing rapid changes in the North Atlantic and in Greenland; the corresponding events (indicated by thin dashed vertical lines) are damped in the Antarctic record. Temperature changes are estimated from the isotopic content of ice (Greenland and Antarctica) and from faunal counts (North Atlantic). HL1 to HL5 indicate sedimentary "Heinrich" layers (after IPCC, 1996).

All of these records show very rapid changes on time scales of a few hundred years to a few thousand years and sometimes even decades. The cause of these rapid fluctuations is still not understood. These variations become smaller in the Holocene (last 10.000 years).

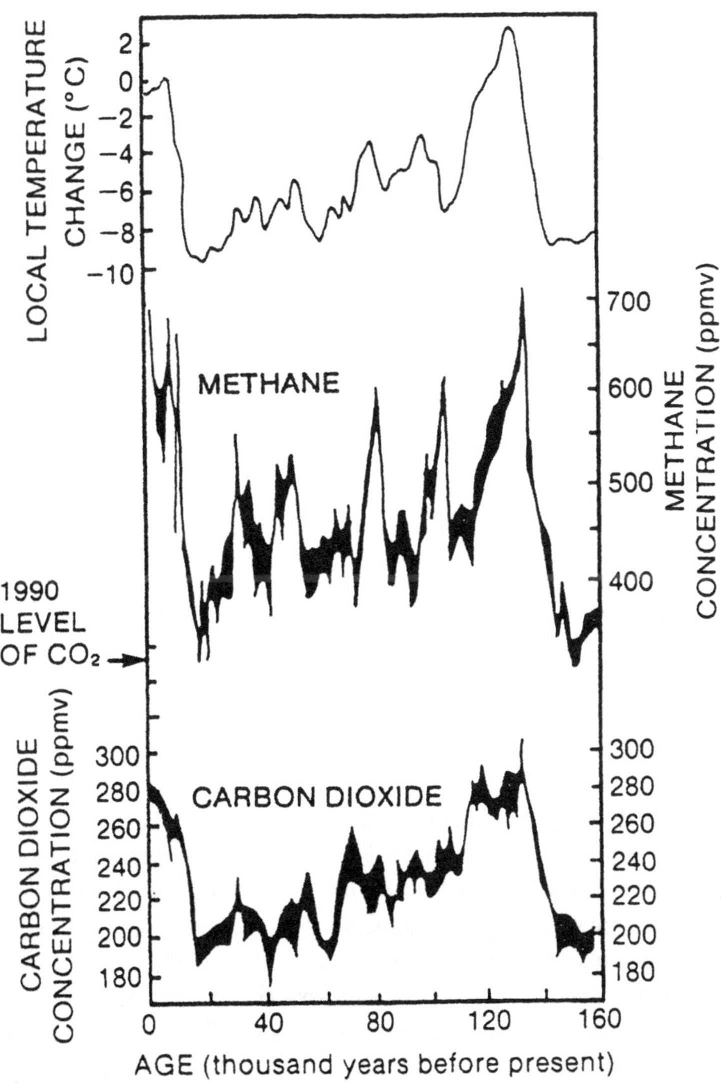

FIGURE 11. Analysis of air trapped in Antarctic ice cores shows that methane and carbon dioxide concentrations were closely correlated with the local temperature over the last 160 000 years. Present day concentrations of carbon dioxide are indicated (after IPCC, 1990).

Another example (FIGURE 11) shows the simultaneous measurement in an ice core in Antarctica of CO_2 concentrations, going back from the present to the last 200.000 years. The decrease of the CO_2 concentrations towards the last ice age and an increase again to the present day levels in the Eem a hundred thirty thousand years ago is highly correlated with the temperature and the methane concentrations. However, it is not clear yet whether the CO_2 drives the temperature change or the temperature change drives the CO_2 or the methane concentrations, or (which is probably the most appropriate view) whether all records are mutually coupled through climate-carbon cycle feedbacks.

A certain percentage of the observed natural climate variations at specific periods can be explained through the changes in the earth orbit (Milankovic theory). It is speculated that some of the changes are caused by instabilities in the ocean circulation, which can switch between different modes (but this has not appeared to have happened in the last 10.000 years). Unfortunately, it is not yet possible to simulate such extensive timeperiods with realistic climate models due to lack of computing resources. However, most of these changes occur on timescales that are large compared with the timescale of anthropogenic forcing and can therefore be disregarded in this context.

4. Simulations of Climate Change

The IPCC compared the impact of various scenarios of the future evolution of the emissions of CO_2, methane and other GHG on the climate system. The most widely used scenario is 'IS92a' ("business-as-usual"), which assumes that no regulatory mechanisms are imposed. A wide variety of other scenarios have also been investigated, some more pessimistic, for which the increase is more rapid (IS92 e + f), some more optimistic (IS92 b - d), assuming the emissions are reduced through regulatory instruments. FIGURE 12 shows the corresponding CO_2 concentrations computed with a carbon cycle model for these emissions. For the scenario 'IS92a' there is an increase by about a factor of two between the atmospheric CO_2 concentration of 360 ppm today to over 700 ppm in the year 2100.

FIGURE 13 shows the response simulated by a comprehensive climate model for the "business-as-usual" scenario (EIN), and for an "accelerated policies" scenario (SZD), in which a stabilization of the atmospheric CO_2 concentration is assumed. Although the global mean temperature increase is 2.5 to 3 °C in the scenario "business-as-usual", the temperature increase over the continents is considerably higher. The climate change in this scenario is comparable in magnitudes, but occurs far more rapidly than the major climate changes in the past in going from an ice age to an interglacial period. In the scenario "accelerated policies" the climate change is restrained during the next 100 years to an acceptable level.

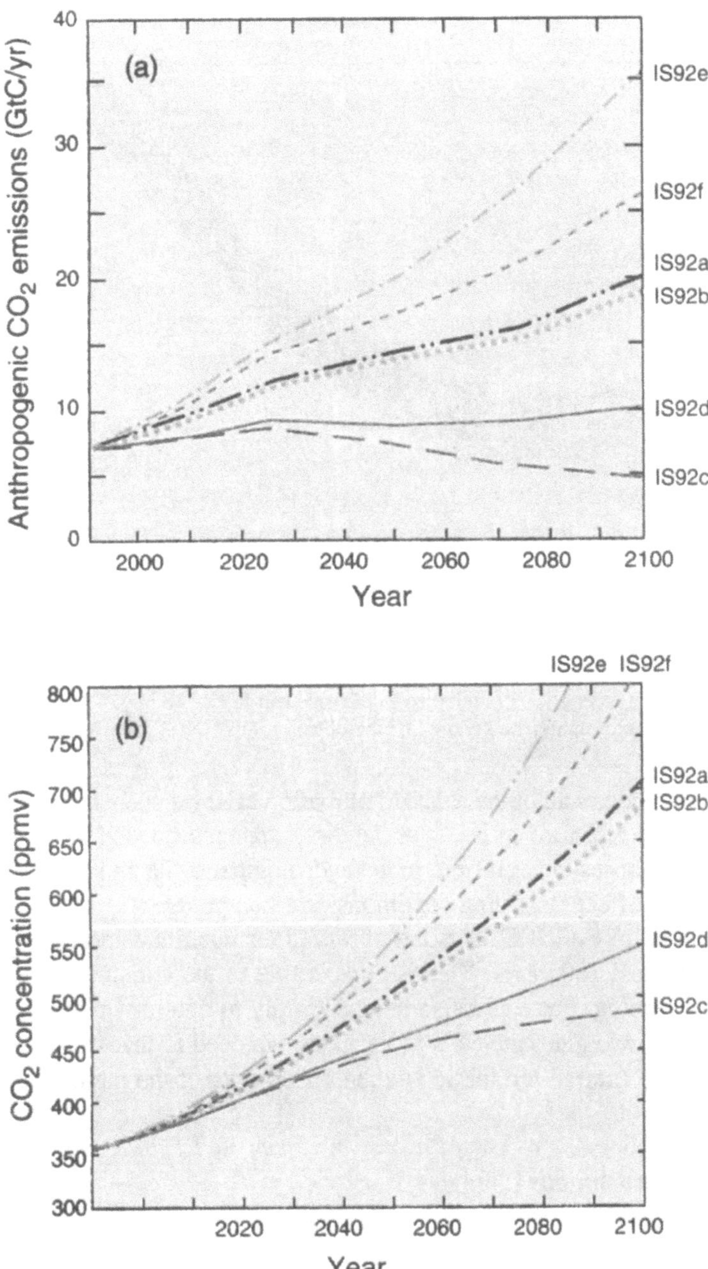

FIGURE 12a+b. (a) Total anthropogenic CO_2 emissions under the IS92 emission scenarios and (b) the resulting atmospheric CO_2 concentrations using the "Bern" carbon cycle model (IPCC, 1996).

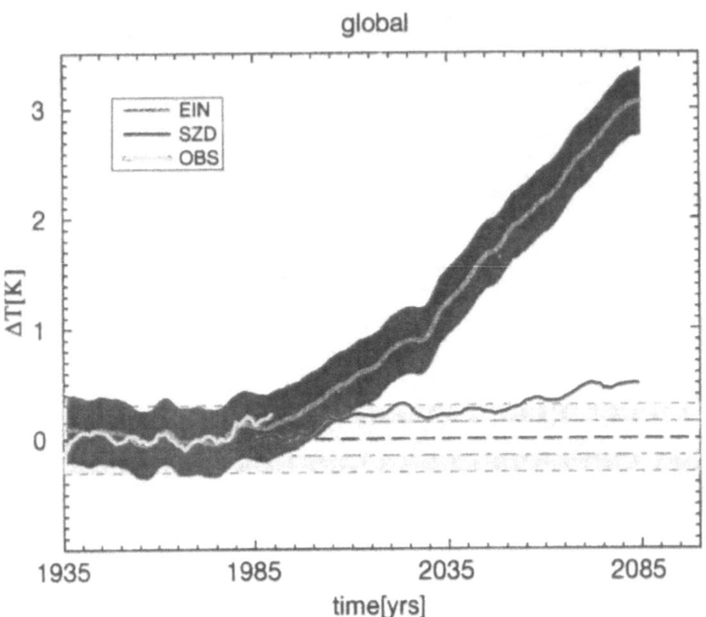

FIGURE 13 (B). The global mean temperature change for scenario A (similar to IPCC scenario IS92a) (EIN) and of scenario D (similar to IPCC scenario IS92d) (SZD). The gray and blue belts indicate the model uncertainty, the yellow line the observations (OBS) (after Cubasch et al., 1995).

With these models all of the standard climatic variables such as precipitation, the sea-level, soil moisture as well as factors, that are needed to determine, for example, the changes in vegetation, in the hydrological cycle and other environmental factors that effect our living conditions, are computed.

Also shown in FIGURE 13 is the observed termperature increase of about 0.5 K during the last 100 years. This is comparable to the computed increase due to greenhouse forcing, but could also be due simply to natural variability. To decide which of these two alternatives is more likely we need to investigate more closely the form of the natural variability and the uncertainty of the model prediction.

5. Detection of Climate Change

FIGURE 14 depicts in more detail the observed increase in global mean near-surface temperature from the beginning of the industrialization up to today. The inherent inaccuracy of the simulations is indicated by a set of computations with different models (FIGURE 15). The top two curves show the response to GHG increases

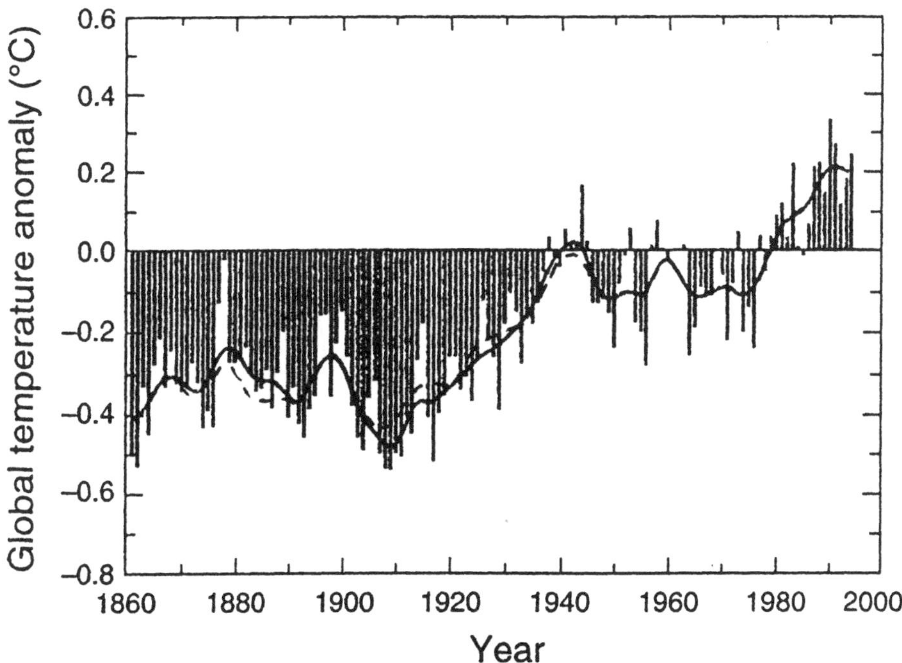

FIGURE 14. Combined land-surface air and sea-surface temperature changes (°C) 1861 to 1994, relative to 1961 to 1990. The solid curve represents smoothing of the annual values shown by the bars to supress sub-decadal time-scale variations. The dashed smoothed curve is the corresponding result from IPCC (1992). (after IPCC, 1996).

(CO_2 increase and methane) simulated with two different models, the bottom two curves similar simulations including also the impact of the direct radiative effect of sulphate aerosols. The indirect effect on the clouds, which is more difficult to compute, has probably the same magnitude. The direct aerosols effect reduces the warming by nearly 1 °C.

To determine whether the slow increase found in the observations is just a slow natural variation, inherent in the coupled ocean-atmosphere system or is forced by the GHG from anthropogenic emissions, we need to consider not only to the global mean temperature, but the details of the signal and natural variability patterns.

FIGURE 16 shows the spatial temperature change pattern for the "business-as-usual" experiment and the "business-as-usual + aerosols" experiment. The patterns differ in a smaller increase of temperatures due to the aerosol cooling over the middle latitudes of the Northern hemisphere. These two signal patterns are compared with the observed temperature change patterns.

Using a "finger-print" technique, from the signal patterns a detection variable is computed that suppresses regions with the highest natural variability. From an analysis of the natrual variability one can estimate the range of values the detection variable would have in the absence of an anthropogenic signal. If the observed

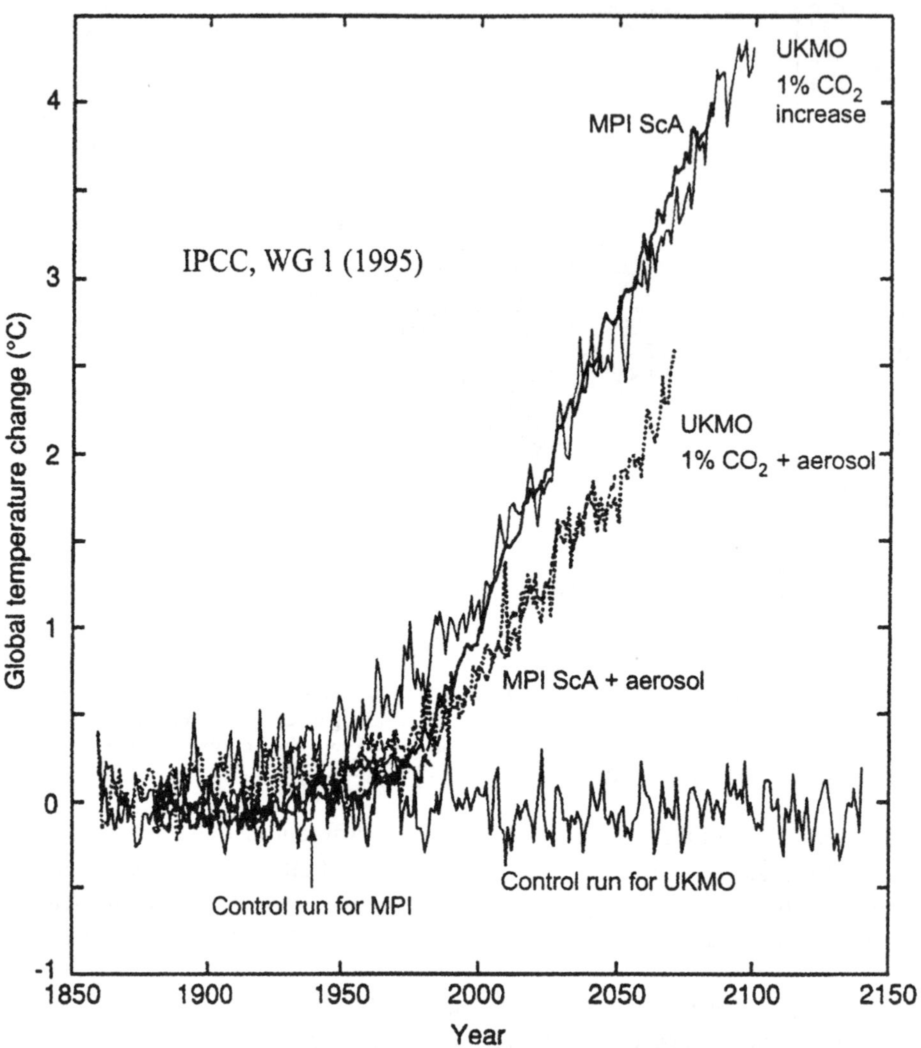

FIGURE 15. The simulated global annual mean warming from 1880 in two simulations with greenhouse gas forcing only, MPI (ScA) and UKMO (1% CO_2 increase), and two simulations which include both greenhouse gas and direct sulphate aerosol forcing, MPI (ScA + aerosol) and UKMO (1% CO_2 + aerosol). The control runs for each model are also shown (after IPCC, 1996).

detection variable is above a certain significance level, for example 95%, this implies that there is only a 5% probability level that this signal can be explained by natural variability. If the detection variable is below this significance level, it is regarded as part of the natural variability.

FIGURE 17 shows that the observed signal (Observation) exceeds the natural variability already in the year 1995, implying that one can now already see a climate change signal in the data. However, this result is still controversial. The IPCC

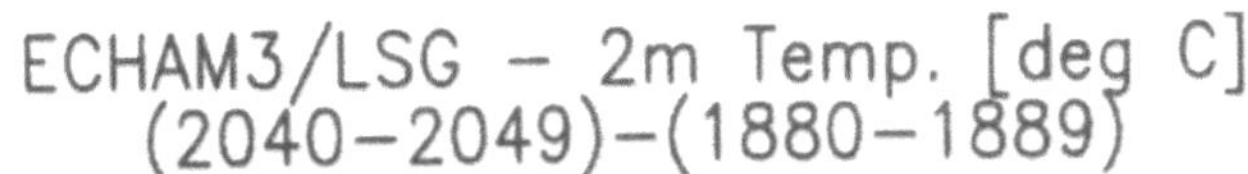

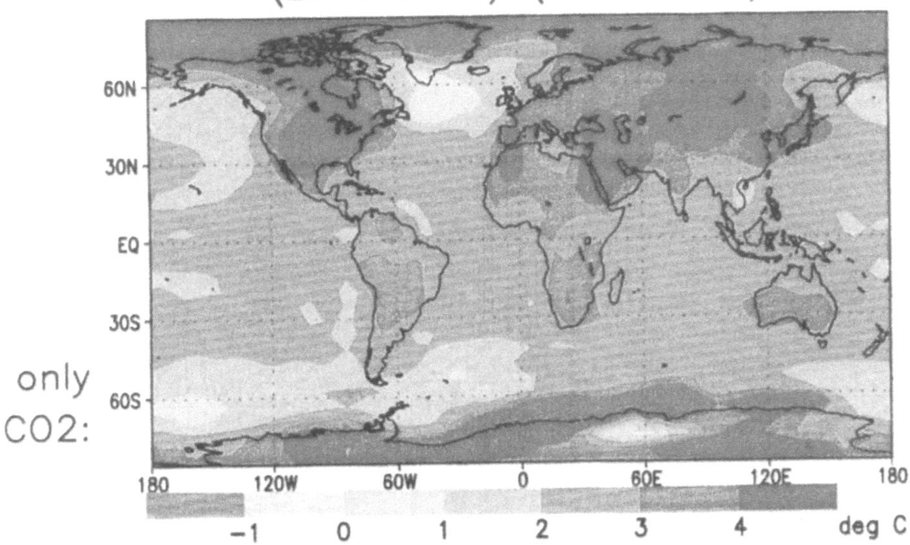

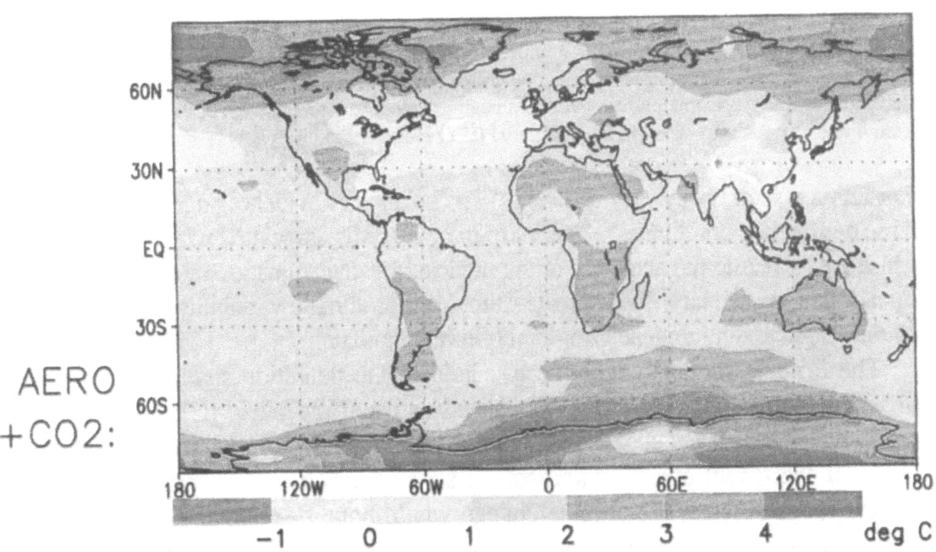

FIGURE 16 (E). Change of annual mean near surface temperature (in °C) in the mean over the decade 2041-2049 relative to the initial decade of the simulations (1880-1889) for the greenhouse gas only experiment (top) and the average of two experiments with greenhouse gas-plus-aerosol forcing (bottom) (after Hegerl et al., 1997).

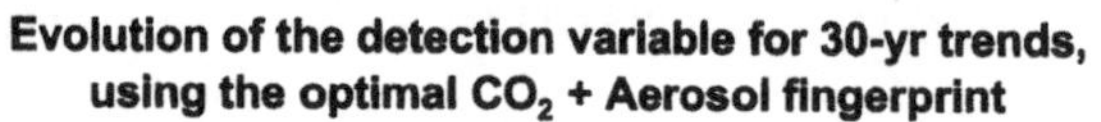

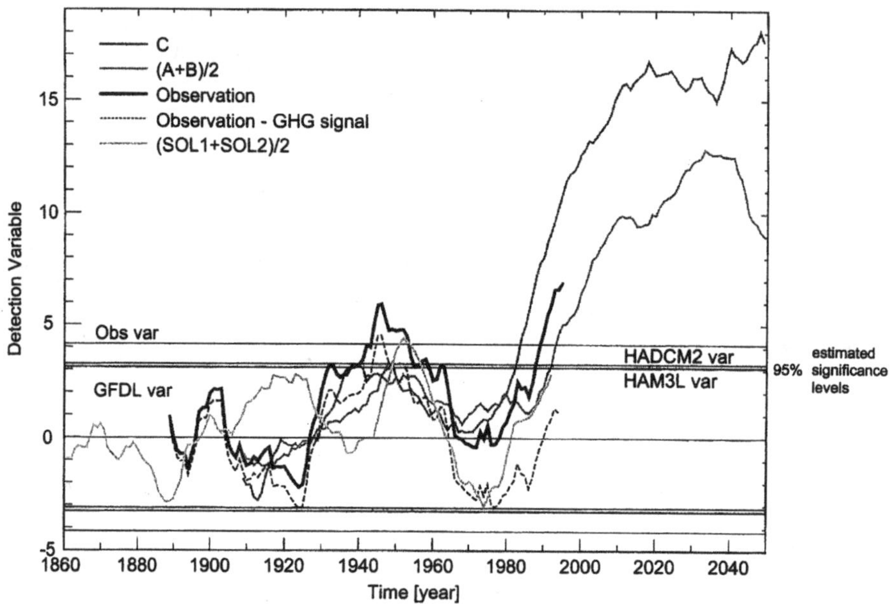

FIGURE 17 (H). Evolution of the detection variable computed with an optimal fingerprint for 30-yr. trend patterns from the observations (black), the aerosol experiment ((A + B)/2; blue), the greenhouse gas-only experiment (C; red), an experiment with prescribed solar variability as well as the observations, where the trend induced by the greenhouse gases has been removed (dashed, black). 95% confidence intervals from 4 sets of variability data are also indicated (after Hegerl et al., 1997).

conclusion is accordingly very cautious: "The balance of evidence suggests a discernible human influence on climate". There are many caveats associated with this assessment, which are related to the accuracy of the models, the determination of the anthropogenic signals, and, particularly the estimate of the natural climate variability, which is limited by the length and homogeneity of the observational data.

The climate models enable us to calculate the detection significance level for different scenarios: Assuming a "business as usual" scenario (C), climate change would have been detected at the 95% significance level (horizontal lines) already in 1980. If the direct effect of aerosols is superimposed on the "business as usual" (scenario (A+B)/2), the climate change would have been detected in 1990. If no anthropogenic greenhouse forcing existed, and only the increase in solar radiation is taken into consideration (Scenario (SOL1 + SOL2)/2), the climate change is inseparable from the noise. If an estimate of the greenhouse signal is substracted from the observed temperature change (Observation - GHG-Signal), the climate change would also remain below the noise level as expected.

6. Integrated Assessment Studies

How can one apply the results of climate change predictions to develop climate protection strategies? For this purpose one needs to combine climate models with socio-economic models that include both the impact of climate change on human living conditions and the interactions between emission control measures, the economy and the climate system. FIGURE 18 shows schematically the structure of such a couple model. Decision-makers have the task of optimizing the complete system according to some agreed upon concept of global human welfare.

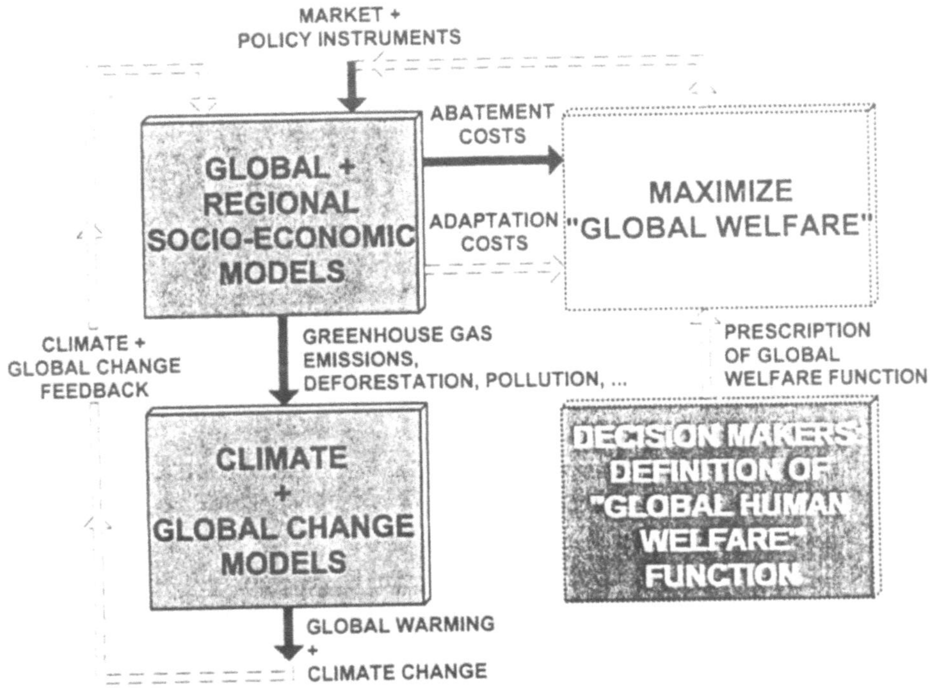

FIGURE 18. Interactions and sub-systems of an integrated Global Environment and Society (GES) model (after Hasselmann, 1991).

Assuming that an agreement on the definition of global welfare has indeed been achieved, the optimization problem reduces to a single-actor problem. The actor has the task of applying appropriate policy and market policy instruments to control the time evolution of future greenhouse gas emissions such that the global welfare, integrated over time, is maximized. The global welfare will generally be a function of the abatement costs, that arise through the measures introduced to reduce the emissions, and the climate damage costs, including both direct damages and the economic costs of adjusting to a climatic change. The sum of these two costs determine the global welfare. The task of the decision maker is therefore to determine the optimal greenhouse gas emission path that maximizes the global welfare, or minimizes the sum of the abatement and climate damage costs. Unfortunately, the climate models presented so far are far too costly to be used in a numerical optimization study, which typically requires many integrations before the emissions path is adjusted to the optimal solution. Therefore a much simpler climate model needs to be developed.

One such model simulates the response of the climate system to a change in the anthropognic emissions of CO_2 by an impulse response function. The model works as follows (FIGURE 19): Consider one ton of CO_2 that is introduced (in the form of a delta function) at time to into the atmosphere. This immediately produces a step-function increase in the CO_2 concentration at time t_o.The concentration then gradually falls off as the CO_2 content in the atmosphere is absorbed in the ocean over a time scale of about 300 years. The next step is to analyse the effect of the change in the CO_2 concentration on the climate system. For a sudden step function change in the concentration the climate model responds with a gradual increase in

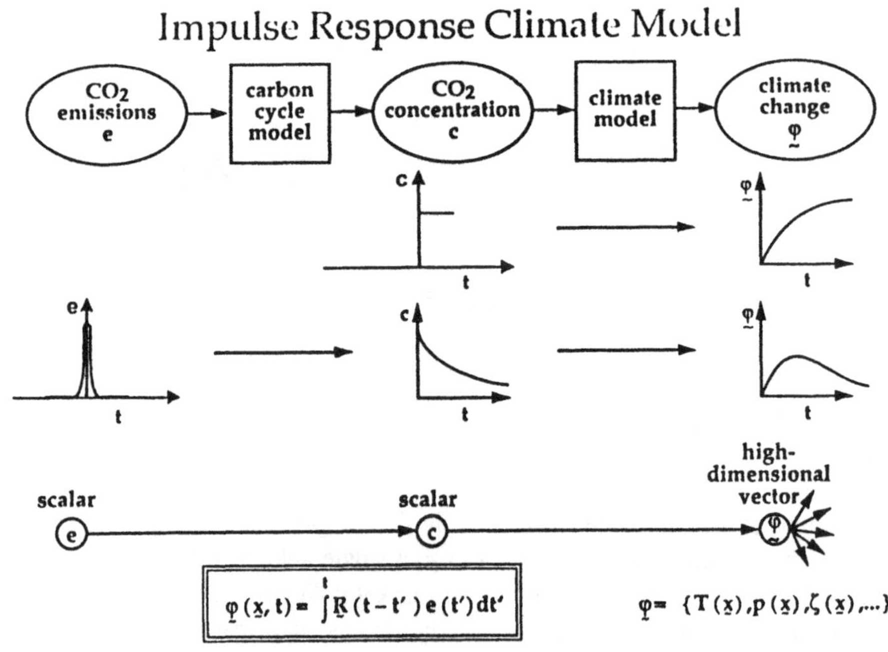

FIGURE 19. The impulse-response climate model.

the temperature as the oceans warm up. An equilibrium temperature is obtained after about a thousand years. The net response of both models acting in succession to a delta-function input of CO_2 is an initial warming with a subsequent cooling as the CO_2 concentration gradually decreases. The response of this simple model can be calibrated against the coupled atmosphere-ocean model described previously.

The climate system has a very long memory, far exceeding the planning horizons of bussiness or decision makers. Thus to assess the climate impact of fossil fuel emissions one needs to compute the climate change over many hundred years. FIGURE 20 shows the concentration of the CO_2 in the atmosphere and the global mean temperatures computed until the year 3000 for four given-emission scenarios: a hypothetical "business-as-usual scenario" (SA, which is unrealistic, however, as the fossil fuels would be exhausted before the end of the period); a more realistic modification of this path (SB) in which it is assumed that the fossil fuel emissions decline from the year 2200 onwards; a path in which the emissions are frozen at the 1990 levels (SF); and a fourth path (SG) assuming an emission reduction of 20% compared with the 1990 levels.

The temperature change in the "business-as-usual" scenario until the year 2100 is of the order of 2-3 °C, but after about 2300, 2400, the temperture increases to 8° or 9 °C. This is more than has ever been observed in the past. For the paths with

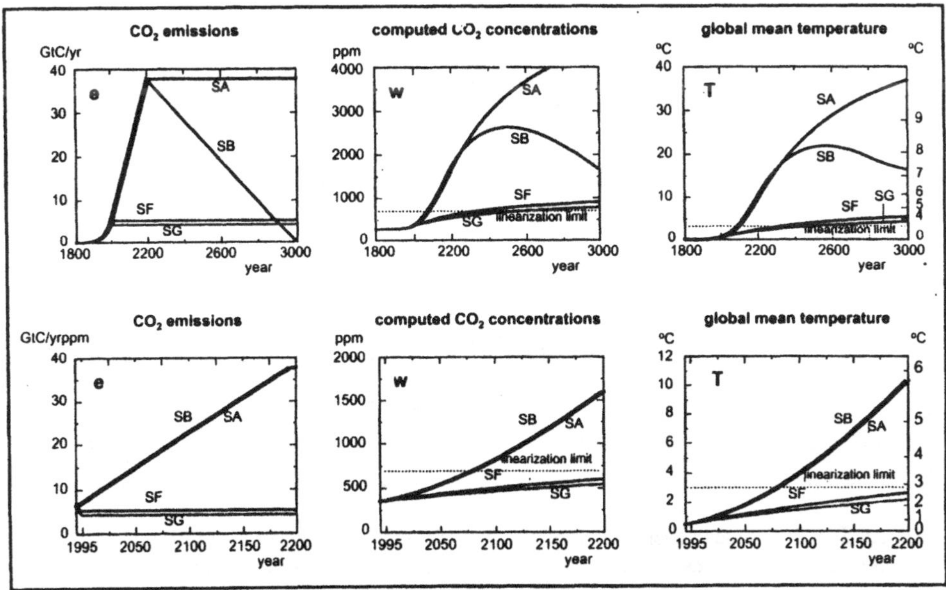

FIGURE 20 (F). CO_2 emissions, computed CO_2 concentrations and global warming (from left to right) for the time periods 1800-3000 (top) and 1995-2200 (bottom) for the "business-as-usual" (BAU) scenario (SA), for the modified BAU scenario (SB), frozen emissions at 1990 levels after the year 2000 (SF) and 20% reduced emissions relative to the 1990 levels after the year 2000 (SG). The linear model is not applicable above the based levels. (after Hasselmann et al., 1996).

frozen emissions a major climate change also occurs, but over a still longer time span. These temperature changes lie beyond the range of calibration of the models, however, and can be regarded only as order of magnitude estimates.

FIGURE 21 shows in contrast the optimal emission paths computed by minimizing the sum of the abatement and climate damage costs. Four different cases are considered: a baseline reduced-emission scenario S0; two scenarios with reduced (S1a) and zero (S1b) economic inertia terms in the abatement cost function; and a modified baseline experiment (S2) in which the climate damage costs are assumed to depend only at the rate of the temperature change. Also shown in the lower panel are the exponential abatement and damage cost discount factors Da and Dd (after Hasselmann et al., 1996). In the baseline case the optimal emission path continues to rise beyond the present day level due to the inertia of the economic system and then gradually decreases after the year 2200. The long-term temperature change remains acceptable. In all sceanarios, limited temperature changes can be achieved only by drawing down the emissions in the long term to essentially zero, but the decrease can occur over a period of several hundred years.

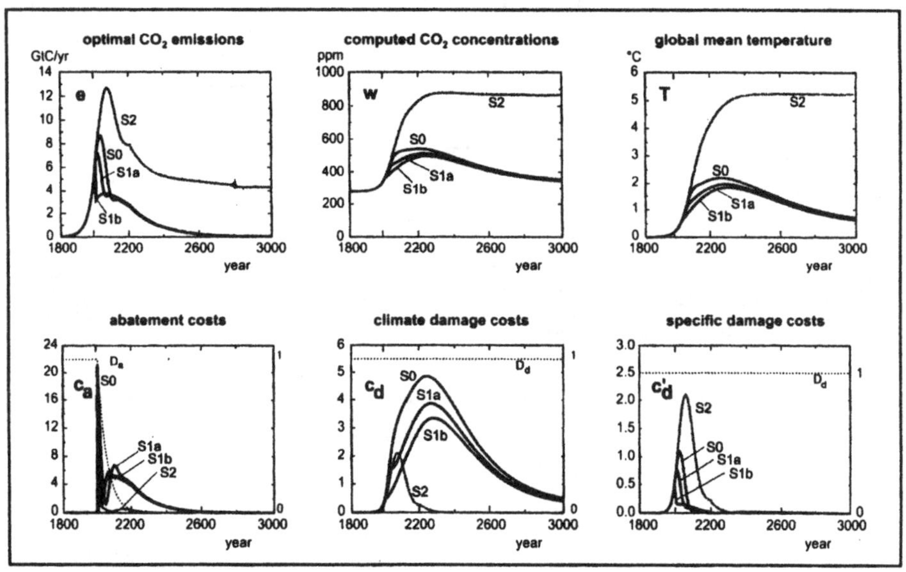

FIGURE 21 (G). Evolution over the period 1800 - 3000 of: (top, left to right) CO_2 emissions, CO_2 concentrations, global mean temperature and (bottom, left to right) specific abatement costs (c_a), specific damage costs (c_d) and the contribution to the specific damage cost (c'_d) from the rate of change of temperature for: the baseline reduced-emission scenario S0, with reduced (S1a) and zero initial terms in the abatement cost function (S1b) and a modified baseline experiment in which the climate damage costs are assumed to depend only at the rate of the temperature change (S2). Also shown in the lower panel are the exponential abatement and damage cost discount factors D_a and D_d (after Hasselmann et al., 1996).

7. Summary

The principal conclusions of the Second Assessment Report of the IPCC and other recent research may be summarized as follows:

In the past 200,000 years, the climate system has exhibited significant variations, with a strong correlation between changes in the temperature and the concentrations of methane and CO_2. With the exception of the variations in the narrow frequency bands associated with modifications in the orbit of the earth, the origin of these variations is not well understood.

There has been a measured increase in the atmospheric concentrations of the major greenhouse gases, particularly of CO_2, in this century. Parallel with this increase there has been an observed rise in the global mean near-surface temperature of about 0.5 K. Models predict a further temperature rise over the next 100 years between 1.0 and 3.5 K for the scenario "business as usual".

With the optimal fingerprint method it is now possible to detect a climate change signal in the observations exceeding the natural climate variability level, leading to the cautious IPCC statement that "the balance of evidence suggests a discernible human influence on climate".

Global climate change affects human welfare both through direct climate damages, the costs of adapting to climate change, and the costs incurred through measures to abate emissions or otherwise mitigate climate change.The optimal trade-off between the costs of emission reduction and the costs of adjusting to climate change can be calculated using Integrated Assessment Models that combine socio-economic models with climate models.

The computed optimal CO_2 emissions path including economic inertia allows some initial increase in the emissions, followed by a gradual decrease to essentially zero thereafter. This path avoids high initial abatement costs and keeps the temperate increase below a tolerable limit of 2.0 K, at the same time permitting the necessary change to fossil-free energy to be carried out over a period of a century or longer.

8. References

Cubasch, U., B. D. Santer and G. Hegerl (1995) Klimamodelle - wo stehen wir? *Phys. Bl.*, **51**, 269-276.

Cubasch, U., G. C. Hegerl, R. Voss, J. Waszkewitz and T. C. Crowley (1997) Simulation with an O-AGCM of the influence of variations of the solar constant on the global climate. *Climate Dynamics*, **13**, 757-767.

Hasselmann, K.., S. Hasselmann, R. Giering, V. Ocana and H. von Storch (1996) Optimization of CO_2 emissions using coupled integral climate response and simplified cost models. MPI Report No. 192, Max-Planck-Institut für Meteorologie, Hamburg.

Hasselmann, K. (1991) How well can we predict the climate crisis? Conf. on Environm. Scarcity: the international dimension. Symposiumbände des Institutes für Weltwirtschaft an der Universität Kiel, J.C.B. Mohr, Tübingen, 165-183.

Hegerl, G. C., K. Hasselmann, U. Cubasch, J. F. B. Mitchell, E. Roeckner, R. Voss and J. Waszkewitz (1997) Multi-fingerprint detection and attribution analysis of greenhouse gas, greenhouse gas-plus-aerosol and solar forced climate change. *Climate Dynamics*, **13**, 613-634.
IPCC (1990) Climate change: The IPCC scientific assessment. Eds. J. Houghton, G. J. Jenkins and J. J. Ephraums, Cambridge University Press, 364 pp.
IPCC (1992) Climate change: The supplementary report to the IPCC scientific assessment. Eds. J. Houghton, B. A. Callendar and S. K. Varney, Cambridge University Press, 198pp.
IPCC (1996) Climate change 1995 - The science of climate change: Eds. J. Houghton, L. Meira Filho, B. A. Callendar, N. Harris, A. Kattenberg and K. Maskell, Cambridge University Press, 572pp.
Legates, D. R., and C. J. Willmott (1990) Mean seasonal and spatial variability in gauge corrected global precipitation. *J. Climatology*, pp. **10**, 111-127.

THE IPCC/OECD/IEA GREENHOUSE GAS INVENTORIES PROGRAMME: INTERNATIONAL METHODS FOR THE ESTIMATION, MONITORING AND VERIFICATION OF GHG EMISSION INVENTORIES

Dr. BO LIM*
IPCC/OECD/IEA Programme for National Greenhouse Gas Inventories, PPCD, Environment Directorate
2 Rue Andre Pascal, F-75016 Paris, France

PIERRE BOILEAU*
International Energy Agency
9 Rue de la Federation, F-75739 Paris, France

YAMIL BONDUKI*
IPCC/OECD/IEA Programme for National Greenhouse Gas Inventories, PPCD, Environment Directorate
2 Rue Andre Pascal, F-75016 Paris, France

* The views expressed in this paper are those of the authors and do not necessarily reflect the views of the organisations in which they work.

Aims. The aims of this paper are to summarise the current status in international methods for the estimation of GHG inventories and the relevance of this work to meet the needs of the scientific and policy communities. It also describes the possible features of a future protocol under the Framework Convention on Climate Change and its implications for national greenhouse gas inventories, including the possible direction of future work under the Intergovernmental Panel on Climate Change.

1. Objectives

The current objectives of the IPCC/OECD/IEA Programme on National Greenhouse Gas (GHG) Inventories are:

- to further develop and evaluate the inter-governmentally agreed IPCC (Inter-

J. Hacker and A. Pelchen (eds.), Goals and Economic Instruments for the Achievement of Global Warming Mitigation in Europe, 27–39.

governmental Panel on Climate Change) Methodology for the estimation of national greenhouse gas (GHG) emissions and removals for all anthropogenic sources and sinks.

- to ensure that the Revised 1996 IPCC Guidelines for National GHG Inventories (Revised Guidelines) enable Parties under the United Nations Framework Convention for Climate Change (UNFCCC) to prepare and periodically update national inventories that are accurate, complete, comparable and transparent.

2. Introduction

The ultimate objective of the UN Framework Convention on Climate Change (UNFCCC) is to achieve stabilisation of greenhouse gas (GHG) emissions in the atmosphere in order to prevent dangerous anthropogenic influence on the climate system. To help achieve this aim, Parties are required to periodically update and report their national GHG inventories to the UNFCCC under Articles 4.1 (a) and 12 of the Convention. Annex-1 Parties have submitted national GHG inventories in their initial communications for 1990, the base year, and are in the process of submitting their second national communications for 1995 inventory data. About one hundred non-Annex 1 Parties are now in the process of preparing their initial communications for submission.

To facilitate the preparation of national communications, a set of internationally-agreed methodologies for the estimation and reporting national GHG inventories was required. The Intergovernmental Panel on Climate Change (IPCC), in co-operation with the Organisation for Economic Co-operation and Development (OECD) and the International Energy Agency (IEA) developed the *IPCC Guidelines for National Greenhouse Gas Inventories* (*IPCC Guidelines*) (1995). This period of methods development took place over a four-year period (1990-1994). The *IPCC Guidelines* were then approved by the IPCC in 1994, and later by the Conference of the Parties (COP-1) in 1995 for the preparation of national communications.

3. The IPCC Methodology

3.1. THE IPCC GUIDELINES

The *IPCC Guidelines* are intended for use by all Parties under the Convention, each having very different resources and technical capabilities. To ensure that the *IPCC Guidelines* can be widely applied, a balance must be maintained between flexibility, simplicity and comprehensiveness. The *IPCC Guidelines* contain a range of GHG estimation methods, from the simplified *Reference Approach* for calculating CO_2

from fossil fuel combustion, to a highly detailed bottom-up sectoral approach. This flexibility allows all countries to use the IPCC Methodology to a level of appropriate detail.

The *IPCC Guidelines* comprise of three volumes:

- REPORTING INSTRUCTIONS
- WORKBOOK
- REFERENCE MANUAL

The *Reporting Instructions* (Volume 1) provide step-by-step directions for reporting the completed national inventory. The *Workbook* (Volume 2) contains step-by-step simplified worksheets for calculating emissions of carbon dioxide (CO_2) and methane (CH_4), as well as other trace gases, from six major source/sink categories. The *Reference Manual* (Volume 3) provides a range of possible methods for estimation of emissions for a broader range of GHG sources/sinks by gas.

The IPCC source/sink sectors are defined as follows:

Energy
- Combustion-Related Emissions
- Fugitive Emissions

Industrial Processes
- CO_2 from Cement Production

Solvent and Other Product Use

Agriculture
- Domestic Livestock
- Rice Cultivation
- Prescribed Burning of Savannas
- Field Burning of Agricultural Residues

Land-Use Change and Forestry
- Changes in Forest and Other Woody Biomass Stocks
- CO_2 Emissions from Forest and Grassland Conversion
- On-Site Burning of Forests: Emissions of Non-CO_2 Trace Gases
- Abandonment of Managed Lands

Waste
- Land Disposal of Solid Waste
- Methane Emissions from Wastewater Treatment

3.2. THE REVISED 1996 IPCC GUIDELINES FOR NATIONAL GREENHOUSE GAS INVENTORIES (REVISED IPCC GUIDELINES)

Additions and revisions were recently made to update the *IPCC Guidelines*. This resulted in the publication of the *Revised 1996 IPCC Guidelines* (July, 1997). The additions and revisions were accepted by the IPCC at its Twelfth Session held in Mexico City (11-13 September 1996) after acceptance by Working Group I at its Sixth Session held in Mexico City (10 September 1996).

Further development of the *IPCC Guidelines* focused on several areas:

- *Carbon dioxide and non-carbon dioxide gases in Energy;*
- *'New gases' (HFCs, PFCs and SF_6) in Industrial Processes;*
- *Nitrous Oxide from Agriculture;*
- *Carbon dioxide from soils; and*
- *Methane from waste and rice cultivation.*

Of the additions and revisions, estimation methodologies for HFCs, PFCs and SF_6 from industrial processes and carbon dioxide from soils were amongst the most important. Under the *Guidelines for National Communications*, both Annex 1 and non-Annex 1 should report 'new gases' but until publication of the *Revised Guidelines* such a methodology had not been developed. Because sinks processes are generally less well understood, the development of the soil carbon module was a breakthrough in GHG methodologies. Much progress was also made towards improved harmonisation with other inventory methods, such as the CORINAIR methodology.

4. Quality of GHG Emission Inventories

The quality and the informational use of GHG inventories depend upon four criteria:

- accuracy
- completeness
- comparability and
- transparency.

The relative importance of these criteria to meet the needs of policy makers and scientists differ highly. For scientific purposes, inventory data can be used for validation of atmospheric GHG concentrations, as input data for atmospheric models and for the construction of regional or global emission scenarios. Although the required level of spatial and temporal resolution of models often exceeds the level detail of national information, inventory data can still be used at an aggregated regional or global level. For scientific applications, the usefulness of inventory data depends critically upon the accuracy and completeness of the inventories rather than their comparability and transparency. For monitoring targets, however, the relative changes in emission rates of GHG, i.e. trend analysis, are probably more important than either accuracy or completeness. Therefore, comparability and transparency of GHG inventories are the higher priorities for policy purposes. This situation may change post-COP-3 (Kyoto, December, 1997) if legally binding targets are agreed, given that emissions budgets will rely on the accuracy of national GHG inventories.

5. Future protocol under the UNFCCC

If strengthened commitments do emerge from a future legal framework, and quantified emission limitation and reduction objectives (QELROS) are agreed, the implications for emission inventories are likely to be profound. At present, Annex-I Parties are required to provide annual updates of inventory data together with estimated uncertainties of emissions on a gas-by-gas basis. In practice, few countries do so, and those that do, use expert judgement. There are no consistent methodologies used for estimating uncertainties. Legally-binding targets would place greater emphasis on the quality of emission data and the need to use comparable methods. In the future, if annual updates are required of non-Annex 1 Parties as suggested by the U.S.A. and the European Union, then GHG inventories will also be given a higher priority in developing countries where inventory input data are currently sparse. This would necessitate considerable, additional international research efforts to improve national statistics in emission factors and activity data, as well as capacity building to develop good data management practices.

Under QELROS, current proposals for a COP-3 negotiation text are highly varied, i.e., ranging from a gas-by-gas approach to a basket of gases, with and without sinks, and for certain IPCC sectors. Different indicators have also been proposed, such as emission intensities of GDP, exports and fossil fuels. If emission trading became part of the protocol, emissions budgets would need to be established, as well as monitoring and verification schemes. All such factors are likely to provide considerable incentives for enhancement of the quality of national emission data.

The remainder of this paper focuses on methodologies of quantifying emission data and suggests the possible direction of future work in inventories.

6. Uncertainties

TABLE 1 provides a qualitative assessment of the relative uncertainties for global emission estimates for key components of the IPCC Methodology. The overall uncertainty ranges shown here are based on an interpretation of the uncertainty information presented by the IPCC [1992]. The uncertainties of emission factors and activity data are based on expert judgement. They are provided to demonstrate relative uncertainties in the basic science underlying different components of GHG inventories.

TABLE 1.
Uncertainties due to Emission factors and Activity Data

1 Gas	2 Source category	3 Emission factor U_E	4 Activity data U_A	5 Overall uncertainty U_T
CO_2	Energy	7%	7%	10%
CO_2	Industrial Processes	7%	7%	10%
CO_2	Land Use Change and Forestry	33%	50%	60%
CH_4	Biomass Burning	50%	50%	100%
CH_4	Oil and Nat. Gas Activities	55%	20%	60%
CH_4	Coal Mining and Handling Activities	55%	20%	60%
CH_4	Rice Cultivation	3/4	1/4	1
CH_4	Waste	2/3	13	1
CH_4	Animals	25%	10%	25%
CH_4	Animal waste	20%	10%	20%
N_2O	Industrial Processes	35%	35%	50%
N_2O	Agricultural Soils			2 orders of magnitude
N_2O	Biomass Burning			100%

Note: Individual uncertainties that appear to be greater than ± 60% are not shown. Instead judgement as to the relative importance of emission factor and activity data uncertainties are shown as fractions which sum to one.

The methods for estimating CO_2 emissions from fossil fuel combustion are relatively well understood and their uncertainty is rated at +/- 10% [Marland, 1997]. Sources of CH_4 from rice paddies or CO_2 from land-use change and forestry are more uncertain because of limited knowledge of agricultural practices and land-use change. Continuing effort is being made to quantify and reduce these uncertainties in future work.

The future objectives of the programme are likely to:

- to assess uncertainty estimation methodologies, e.g. stochastic modelling and other techniques, that can be used at a national level to obtain sector-specific uncertainty estimates on a gas-by-gas basis.
- to recommend uncertainty estimation methodologies for incorporation into the *Revised Guidelines* (Annex 1 of the *Reporting Instructions)* using a tiered approach consistent with other IPCC methodologies.
- to estimate uncertainties of GHG emission estimates calculated from the IPCC default methodologies for key source/sink sectors.

Future activities could include the following:

- IPCC Default Methodologies: Development of a table of uncertainty ranges for the direct GHG by sector using the IPCC Default Methodologies based upon subjective probability assessment (e.g. expert judgement).
- Review of National Inventories and Uncertainty Estimation Techniques: Review of national inventories from countries which have used statistical techniques to assess uncertainties.
- IPCC Tiered Methodologies: Methods for the estimation of uncertainty associated with inventories to be considered for their robustness and applicability to the current requirements of emissions inventories.

These will include:
- statistical error estimation;
- stochastic error estimation.

7. Verification, Transparency and Monitoring

In preparing a national inventory, it is essential to have reliable and accurate information about activities which result in anthropogenic GHG emissions or removals. Further, in tracking a country's progress towards it's GHG reduction goals it is important to perform checks and balances with other methods of producing national emission estimates. These actions translate into a system of monitoring, transparency and verification that can be routinely performed by countries as they prepare their inventories. The *IPCC Guidelines* have not yet addressed the issue of monitoring GHG inventories. The onus has been placed on the user of the Methodology to ensure uniformity throughout the national inventory. The role of the *Revised*

Guidelines is to recommend internationally-agreed methods or procedures that can be used by both Annex-I and non-Annex-I Parties.

Four criteria are required for verification:

- empirical testing;
- documentation and the reproducibility of results (transparency);
- explicit reporting of uncertainty;
- peer review of results.

The *IPCC Guidelines* recommend that countries perform verification of their own national inventories and detail the results. Suggested methods of verification are:

- Comparison with a country of similar socio-economic structure;
- Comparison with the IPCC default methods (i.e. Reference Approach etc.);
- Comparison with similar inventories produced on a global scale.

Within the context of emission inventories, two major (analytical and political) applications of the data must be considered when applying verification techniques. Some of the major terms and their importance are outlined in TABLE 2.

Under a proposal for future work, to improve monitoring, transparency and verification for the national GHG inventories, the objectives are:

- to identify and assess monitoring procedures for GHG emissions/removals for key source/sink sectors of the *Revised Guidelines*.
- to identify and assess methodologies for improving the transparency and the verification of GHG inventories for inclusion into future editions of the *Revised Guidelines*.
- to recommend monitoring, improved transparency and verification procedures for inclusion into future editions of the *Revised Guidelines.*

The specific activities are to include:

- *Review of Methods for Monitoring National GHG Inventory Data*, including:
    - monitoring sources of activity data; and
    - monitoring actual emissions and/or emission rates or factors.

- *Comparisons of Inventory and IPCC Default Emission Estimates*: National inventory data will be compared with estimates resulting from the use of the IPCC Default Methods and data.

TABLE 2. Importance of Terms in Major Applications [EEA, 1997]

TERM	ANALYTICAL APPLICATION	POLITICAL APPLICATION
Accuracy	Very important, inaccurate data can lead to erroneous scientific conclusions; can affect credibility of policy decisions and cost/benefit analyses	Accurate data is always preferred for any application, however in some political applications adhe-herence to protocol could be more important; accuracy of data should be within the constraints of the proposed control programs
Precision	Very important for comparative analy-ses and for repeatability	Numbers generated with accepted protocol should be comparable and repeatable, but for applications it is not so important
Confidence	Not important for scientific purposes if all quality control and quality assurance standards were achieved	Important in terms of following established protocol, empowers the developer and users in internatio-nal programs
Reliability	Important in that it implies usefulness and credibility	More important as a reflection of method followed than for accuracy of specific numbers
Quality Control	Very important, this is a necessary part of data gathering and processing	Important as it relates to ensuring the adherence to accepted procedu-res and methods
Quality Assurance	Very important, in the scientific arena QA defines technique and qualifies re-sults	Important as it relates to ensuring the adherence to accepted procedu-res and methods
Uncertainty	Important, the error inherent in the data-bases used to generate emissions data should be known to understand the error in final numbers	Important, uncertainty in policy applications translates directly to confidence and reliability con-cerns,
Validation	Not important for scientific applica-tions, but is needed when data or analy-ses are used in policy applications	Very important, official acceptance is the critical test in the political arena
Verification	Very important, if the scientific applica-tions of the inventory are satisfied then the effort was a success	Important only in a relative sense that data can be repeated and that two data sets can be compared
Transparency	Only important if analyses extend into neighbouring countries	Very important to demonstrate ad-herence to common protocol
Compliance	Important if no standard procedures exist, but is relatively unimportant if methods followed are trusted	Very important in terms of adhe-rence to agreed protocol or me-thods

- *Inter-country Comparisons*: The usefulness of performing inter-country comparisons will be evaluated following the completion of the inventories database.
- *Self-evaluation Procedures*: Standard procedures that allow countries to perform self-evaluation/verification of GHG inventories could be developed.

Other inventory methodologies have also recommended different comparisons as verification techniques (TABLE 3). Three types of monitoring activities are outlined in TABLE 4.

TABLE 3. Summary of Data Comparison Opportunities in Emission Inventory Applications [EEA, 1997]

DATA COMPARISON TYPE	EXAMPLES
Alternate estimation methods	• Emissions magnitudes based on raw material feed versus product • Emissions magnitudes based on alternate measures of the inherent activity
Top-down versus bottom-up methodologies	• National or regional-level estimates versus source-specific totals within source categories
Emission density comparisons	• Aggregate estimates for per-capita, per-employee or per-area compared to total emissions from all facilities in a source category • Aggregate emissions densities compared to similar estimates from other countries or regions
Emission factor comparisons	• Source-specific emission factors compared to default or average factors • Uncontrolled emission factors with average level of control to controlled emission factors • Emission factors based on alternate measures of the inherent activity
Control total comparisons	• National totals compared to sum of all source catego ries or facilities within source categories • Summed emissions totals from detailed inventories compared to national totals in trends analyses • National totals compared to national totals of nearby countries corrected for population and economic status
Completeness checks	• Comparison to earlier inventories to check that all significant sources are considered • Checks that all important source categories are considered • Checks that all important data elements are included for facility records
Consistency checks	• Internal consistency for facility data records • Consistency of methodology for source categories • Consistency of methodologies between countries in multiple country inventory development

TABLE 4. Monitoring Types, Examples and Uses for Emissions Inventories [EEA, 1997]

Monitoring Class	Examples of Monitoring Programs	Uses of the Data for Emissions Inventories
Direct Measurements	• Inprocess emissions measurements • Process operating parameters • Random sampling of process units or potential leak tests	• Comparison to estimated values • Identification of ranges of application estimates (operating parameters, emissions factors) • Specification of fugitive emissions or process leaks
Indirect Measurements	• Remote measurement systems: FTIR, UV, Gas Filter Correlation • Ambient VOC/NOx ratio studies	• Comparison of estimated emission rates with near source concentrations • Estimation of emission fac tors for sources that do not have stacks or events
Ambient Studies	• Tunnel Studies • Aircraft Studies • Upwind-downwind difference studies • Receptor Modeling	• Identification of obvious weaknesses in procedures or under-estimation of emissions • Checking of ambient impacts of sources or mixtures of sources • Identification of principal emissions sources in a region

8. Conclusions

During the first half of this decade, international effort in the area of inventories mainly focused on the development of estimation methods. This resulted in the publication of the *IPCC Guidelines*, and more recently the *Revised Guidelines*. These *Guidelines* enabled countries to prepare their national communications and national GHG inventories in a consistent and transparent manner, thus meeting one of the aims of the Convention. Many Annex 1 Parties have now developed technical capabilities to provide annual updates of national GHG inventory data, and about 100 non-Annex 1 Parties are in the process of preparing their initial communications for submission.

The possibility of strengthened commitments under a future protocol marks a turning point in the evolution GHG emission inventories. Legally-binding targets would have profound implications for inventory data by providing considerable incentives for the enhancement of inventory data quality. Given this possibility, during the next few years, greater emphasis is likely to be on the reliability of inventory and methods for their quality assessment, such as assessing uncertainties, and procedures for monitoring, transparency and verification.

References

European Environment Agency, EMEP/CORINAIR Atmospheric Emission Inventory Guidebook, Air, CD-ROM, Copenhagen, 1997.

IPCC, Intergovernmental Panel on Climate Change (1992), *Climate Change 1992: The Supplement to the IPCC Scientific Assessment*.

IPCC/OECD/IEA Inventories Programme, *IPCC Guidelines for National Greenhouse Gas Inventories*, Paris, 1997.

Gregg Marland, Carbon Dioxide Information Analysis Centre (CDIAC), Pers. Comm.

NECESSITIES AND PROBLEMS OF COUPLING CLIMATE AND SOCIOECONOMIC MODELS FOR INTEGRATED ASSESSMENTS STUDIES FROM AN ECONOMIST'S POINT OF VIEW

MINH HA-DUONG
Centre International de Recherches sur l'Environnement et le Développement CIRED
1 rue du 11 Novembre, F-92120 Montrouge, France

Abstract. This paper has two parts. The first part is more theoretical. We will examine the definition of Integrated Assessment Models of Climate Change and identify their main aims. We will also have a look at the differences between climate and socioeconomic models. The second part will be applied. We present the Dynamics of Inertia and Adaptability Model (DIAM). That model will be used to study some of the fundamentals economic parameters of IAMs: the discount rate, the technical progress rate, the flexibility of energy systems, uncertainty and the final concentration target. We won't try to be exhaustive, see Weyant[1] leading author of the corresponding IPCC chapter for that. Our goal is simply to convince the reader that these non realistic models can profitably be used to analyse real issues.

1. Why put together climate and socioeconomic models?

An Integrated Assessment Model (hereafter IAM), is a model that combine knowledge from a wide range of disciplines to provide insights that would not be observed through traditional disciplinary research. They are used to explore possible states of human and natural systems, analyse key questions related to policy formulation and help set research priorities (IPCC, 1995). This very broad class of models could also be called applied interdisciplinary models.

These are not specific to the Climate Change, and there has been past example of integrated assessment of major environmental issues, as the European acid rain study RAINS. The main component of a climate change IAM can be graphically represented as in FIGURE 1. An IAM model will cover all or part of these boxes. It doesn't have to include all of them, for example a model of concentrations arising from the use of energy would qualify as a 'partial' IAM. If a model does represent all the aspects, then we can call it a or 'full' scale model. Typically, some boxes will be more detailed than others.

J. Hacker and A. Pelchen (eds.), Goals and Economic Instruments for the Achievement of Global Warming Mitigation in Europe, 39–53.

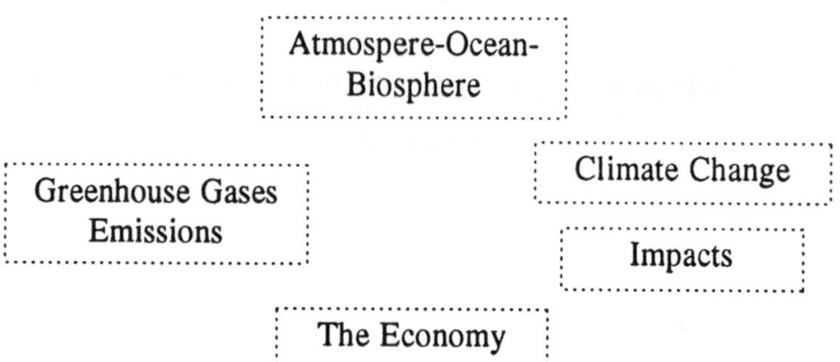

FIGURE 1. Components of an Integrated Assessment Model of Climate Change

Because IAMs are defined very broadly, it is no surprise that they can be everything from very compact to very large. So there is no "best" way of building these models and no standard interface with the policy makers or the rest of the scientific community. As TABLE 1 illustrates, an extremely wide variety of computing tools are currently used. However, models tend to be divided in two categories, as integrated assessment seems to be caught between two traditions: the tradition of natural sciences, especially physics, established for more than three centuries and the traditions of social sciences which are only around 50 years old.

TABLE 1. Some models and their programming environment.

Language	Models
GAMS	MERGE, DICE, ALICE
Mathematica	DIAM
PASCAL, FORTRAN 90	FUND, MiniCAM, PAGE
STELLA, proprietary language (M, DEMOS)	IMAGE, TARGETS, ICAM

The first tradition describes physical phenomena that can be qualified as *« pushed by the past »*. These are treated in the framework of differential equations or finite difference equations. They are accessible through models where the flow of the calculation follows the *« natural time »*: Given the state of the system at date 0, state at date 1 is computed first, then date 2 is examined..., recursively up to date T. In the second tradition, the phenomena are « driven by the future ». In economics and human affairs, there is often coordination (like general equilibrium), the existence of anticipations or a goal. Such phenomena can be best studied in the context of intertemporal optimisation where an optimal path is computed globally, taking into account initial and final conditions, as well as limited resources conditions. From a computational point of view, the solution is sought using the vector of states $V = (S_1, S_2, .. , S_T)$. The number of generic equations can be small, but multiplied by the number of time periods, probabilistic outcomes, agents, production technologies and economic goods, they can generate a large, non linear, optimisation problem.

The question of the magnitude of global warming illustrates a typical physical phenomena "pushed by the past". The question of building international co-operation to reduce polluting emissions is "driven by the future", because anticipations play the main role in it. Models about economic responses to the issue of global warming needs to integrate human activities (demography, growth, the energy sector, agriculture, forestry ...) with the evolution of the terrestrial environment (the climate, the carbon cycle...). By essence, this is at the confluence of the two traditions defined above, so the division line of TABLE 2 is no waterproof. Yet today, two types of models are built in our field:

- policy analysis models, recursively calculated and usually quite detailed, as the IMAGE model by Rotmans[7].
- policy optimisation models, intertemporally calculated and usually rather compact, as DIAM; DICE (Nordhaus[8]); or MERGE (Manne & Richels[9]).

Policy analysis simulation models raise problems when it comes to analysis with scarce resources (like fossil fuel resources) or environmental constraints (like a ceiling on CO_2 concentration). They tend to lead to « overshoot and collapse », and to have many ad hoc parameters. For these reasons, a lot of analysts prefer intertemporal models. This implies a modelling environment that is adequate to solve large optimisation programs. The *GAMS* language is often chosen, as it is a user friendly interface to powerful solvers. There are many different solvers, each adapted to different kind of optimisation problem: linear programming, non linear programming with *MINOS*, integer programming with *ZOOM*, and so on. But this choice is done at the expense of the commodity of a higher symbolic expression of the model, as possible in *Mathematica*, for example.

Many other researchers prefer to use more general programming languages such as FORTRAN or PASCAL. It allows them to do both optimisation and simulation for quite detailed models. The price for this freedom, of course, is that the complexity of code management increases considerably. This is a reason why pure simulation models only are often written is specialized languages like STELLA.

TABLE 2. Conceptual differences between Climate and Socioeconomic models

Climate	Socioeconomic
Prediction	Projection
Causality	Anticipations
Simulation	Optimisation
Recursive	Intertemporal

Those environments allows one to develop quickly and efficiently a processes-oriented model, in the same way that engineers modelize the production process of a factory.

In the long run, however, notwithstanding calculation limits of computers, the two types of models might converge as the equations are mainly the same, the main difference being more in the way the solution is sought.

2. IAMs in action: The WRE versus S controversy.

Our goal in that paragraph is just to show that IAMs are a useful tool for discussing climate change policy. We will do so by giving an very specific exemple that, we believe, is characteristic of the use of IAMs in our field.

Stabilisation of greenhouse gas concentrations implies switching from the present growing emissions trend to a decreasing emissions path. That raises many complex issues that can't be adressed all at the same time. In what follows, we neglected the multiplicity of countries and of greenhouse gases, to focus on the question of the timing of abatement: "How fast should we reform the energy systems to use less fossil fuels ?"

Wigley, Richels and Edmonds[2] (WRE) considered different emission pathways towards a given atmospheric CO_2 concentration, and explained that extensive early abatement would not be a cost-effective way to meet concentration targets of about 450 ppm or higher. Here we build upon their work by incorporating keys economic parameters into an explicit modelling framework which analyses least-cost pathways, and examine the sensitivity of pathways with respect to the target, inertia of the system, and discounting.

The IPCC's Second Assessment Report notes that « *The choice of abatement paths involves balancing the economic risks of rapid abatement now (that premature capital stock retirement will later be proven unnecessary), against the corresponding risks of delay (that more rapid reduction will then be required, necessitating premature retirement of future capital)*». WRE's paper examines different possible emission pathways towards stabilising atmospheric CO_2 concentrations, and presents reasons why pathways involving little early departure from reference emission trends would be cheaper than strong immediate departure due to the effects of discounting, existing capital stock and of technological progress. First, technical progress implies that in the future, reforming the energy system will be easier. Second, discounting makes the present value of any cost incurred less if it is deferred. Third and finally, the existing installed physical capital stock implies that any reform should start very slowly: there is inertia in energy systems.

We explore the balance between costs of too early abatement and economic risks of delay by incorporating some of the economic considerations highlighted by WRE directly into a stylised model that optimises the time-path of emissions under a stabilisation constraint. Our analysis highlights two particular features of the decision problem:

- The issue of capital stock utilisation and turnover reflects a broader question about the inertia of energy systems, that for consistency must be applied to the whole emissions pathway.
- The problem we face is to adopt a prudent strategy in the face of ignorance about the appropriate ultimate target. We are not likely to know soon at what concentration level dangerous interference with the climate system would occur, so the problem is stochastic in nature. We may for example adopt short term abatement targets consistent with a 550ppm concentration ceiling and be forced in a second step to switch toward a lower or higher ceiling, but lowering the ceiling would imply an accelerated abatement in order to outweight the inertia due to the timelife of greenhouse gases in the climate system.

Our numerical exploration of these issues shows that both inertia and uncertainty increase the potential cost of deferring emissions abatement, and the combination of these factors leads to much stronger abatement in early decades than either in isolation.

2.1. THE DIAM 2 MODEL

As a complete example of an IAM, we use a compact intertemporal optimization model, DIAM, which is developed from an initial exploration in[3]. Given assumptions concerning the discount rate, abatement costs, and the rate of autonomous technical progress in carbon saving technologies, DIAM determines the least-cost CO_2 emission pathway consistent with staying below a given or stochastic concentration ceiling.

TABLE 3 specifies the model.

TABLE 3. DIAM2 - a stochastic dynamic programming model on the Dynamics of Inertia and Adaptability for integrated assessment of climate change mitigation

t,u refers to time periods, s refers to states of the world
(s = "L"ow, "C"entral, "H"igh ceiling)

Exogenous parameters

$E^{ref}(t)$	Reference anthropogenic fossil carbon emissions
$M^{ref}(t)$	Reference CO_2 concentration path
$R(u)$	Atmospheric perturbation CO_2 response function
$p(s)$	Subjective probability distribution for concentration ceilings
$L(s)$	Levels of concentration ceilings
t_{info}	Date of resolution of uncertainty regarding the concentration ceiling
ρ	discount rate
r	rate exogenous decline in reduction costs (technical progress)
c_a, c_b	parameters of the additive cost function
c_k, D	parameters of the multiplicative cost function

Endogenous variables

$x(s,t)$	Abatement level (*Policy variable*)
$C(s, t)$	Abatement costs
$M(s, t)$	Atmospheric concentration of CO_2
$A(s, t)$	Acceleration of abatement
J	Total expected discounted abatement cost

Equations

$$\underset{\varepsilon}{\text{Min}}\, J = \sum_s p(s) \sum_t (1+\rho)^{t_0 - t}\, C(s,t) \tag{1}$$

subject to

$$\forall\, t \le t_{info},\ x(L, t) = x(C, t) = x(H, t) \tag{2}$$

$$M(s, t) = M^{ref}(t) - 0.471 \sum_{u = t_0}^{u = t-1} R(t-u)\, \varepsilon(s, u)\, E^{ref}(u) \tag{3}$$

$$M(s, t) \le L(s) \tag{4}$$

$$A(s, t) = x(s, t) - x(s, t-1) \tag{5}$$

$$C(s, t) = (1+r)^{t0 - t} \frac{E^{ref}(t)}{E^{ref}(t_0)} \left(c_a\, x(s, t)^2 + c_b\, A(s, t)^2 \right) \tag{6}$$

$$C(s, t) = (1+r)^{t0} - t\, \frac{E^{ref}(t)}{E^{ref}(t_0)}\, ck\, x(s, t)^2 \max\left(1, D\, A(s, t)\right) \tag{6 bis}$$

CO_2 emissions E(t) accumulate in the atmosphere according to a perturbation response function calibrated against Wigley's carbon cycle model. CO_2 emissions in the absence of any abatement increase by 2%/yr linearly up to 2100, approximating the IPCC central reference scenario IS92a. Deviations from the reference emissions are assumed to incur a positive cost. The model finds the pathway of abatement $x(t)$ that minimises the present value of total abatement costs, discounted at an annual rate ρ. Abatement costs decline due to assumed exogenous technological progress at a rate r, which we take as 1%/yr. We explore values of 3% and 5% per year for the societal discount rate.

The abatement cost at time t is a quadratic function of both the degree of abatement $x(t)$ and the rate of abatement $x'(t)$. By adding a term that depends upon the *rate* of abatement, the model captures inertia in the system. Costs are scaled according to the size of the reference system and assumed exogenous technical progress at rate r, to give equation 6.

Parameters c_a and c_b can be calibrated to various assumptions about abatement costs and this affects the economic acceptability of different stabilisation levels, but the optimal pathway under a specified stabilisation constraint does not depend on their absolute values, only upon the ratio c_b / c_a. Its square root $\tau = (c_b / c_a)^{½}$ is a duration that we can interpret as the 'characteristic time' for changes in the global energy system. If we interpret τ as the exponential half life time of equipment, then it is related to the annual depreciation rate of capital δ, by $\tau = (\ln 2) / \delta$.

Note that in reality there are also many other sources of inertia than capital stock lifetime in energy production. New energy sources have taken about 50 years to penetrate from 1% to only 50% of their ultimate potential. Time is also needed to remove market and institutional barriers to the diffusion of innovations and obstacles due to imperfect information and imperfect foresight. Moreover energy demand depends upon investments in buildings, transport and urban infrastructures whose effect may last more than 50 years. We take $\tau = 50$ years as our initial value.

The purely technical inertia of capital stock in the energy system corresponds to the lifetimes of household equipment, power stations, cars or refineries, typically 10-40 years. A value $\tau_{low} = 20$ years, which corresponds to capital depreciation at $\delta = 4\%$/yr, offers a lower value for inertia. Our results using this ratio are comparable to the results of economic / energy system models that do not embody institutional or larger infrastructural inertia. For example, as shown below, with $\tau = 20$ years we recover modest optimal reduction levels in 2020 when stabilising at 550ppm, as in the economic studies cited by WRE.

Note that even the lower value implies high transitional costs for extreme rates of change. For example, if we compact a 20% emissions reduction into the central year of a 20 year period, the cost over the period is about 13 times as high as when the reductions are spread evenly across the period; with $\tau=50$ years, the ratio is about 18. This could reflect the whole welfare impacts associated with sudden forced retirement of capital and associated shocks to labour and financial structures. In practice the model selects paths that never approach such extreme conditions, and transitional costs in the optimal paths are neither negligible nor predominant, as shown in TABLE 4.

The specification choosen in DIAM allows to describe a situation in which the permanent costs are low over the long run, and the transition costs high. An alternative specification of the adjustment costs would be to interpret x' as only related to the rate investment in the energy sector. As long as it is lower than the 'natural' rythm of replacement of capital $1/\tau$, there are no adjustment costs and the reduction cost is proportional to x^2. Above that rythm, costs would be multiplied by $\max(1,\ x'\ \tau)$. See equation 6bis. We also present in TABLE 4 the results with the 6bis cost function, to show that our core results are not very sensitive to the exact form of the cost function chosen.

2.1.2. Pathways under stabilisation constraints

We initially explore results under two concentration ceilings: 450ppm and 550ppm. The ceiling 450ppm leads to a total radiative forcing that may be roughly equivalent to a doubling of pre-industrial CO_2 concentrations, when other greenhouse gases are taken into account, which has been the benchmark for most climate model analyses of future equilibrium climate change. 550ppm represents the level given greatest attention in the WRE paper, which with other gases substantially exceeds a doubling of CO_2 equivalent.

FIGURE 2 illustrates the global emission pathways that minimise abatement costs under 450ppm and 550ppm for two cases: $\tau=50$ and $\rho=3\%$ (case A) and $\tau=20$ years and $\rho=5\%$ (case B), compared against the reference path. In 450A global CO_2 emissions increase little, returning to current levels shortly after 2020 (when they are 24% below reference levels); for 550A they peak at higher levels around 2050. 550B approximates the characteristics of the economic analyses cited by WRE, and their results. Our sensititives studies around 550ppm (TABLE 4) suggest that increased inertia results in stronger earlier abatement: with lower inertia (B), there is less initial abatement and emissions peak higher but then decline more steeply. Departure from the reference trajectory over the next 1-2 decades is modest for 550B, somewhat greater for 550A, and much greater for the 450 ppm cases especially 450A.

This is reflected in the cost of delay, which is positive in all cases. Delay until 2020 increases the total discounted costs by 8-10% for the 550ppm cases with $\tau=50$ years, but only by 2-3% if $\tau=20$ years; the increase is in the range 32-70% for the 450ppm cases (TABLE 4).

TABLE 4. Results of DIAM 2 for certainty scenarios and hedging strategies

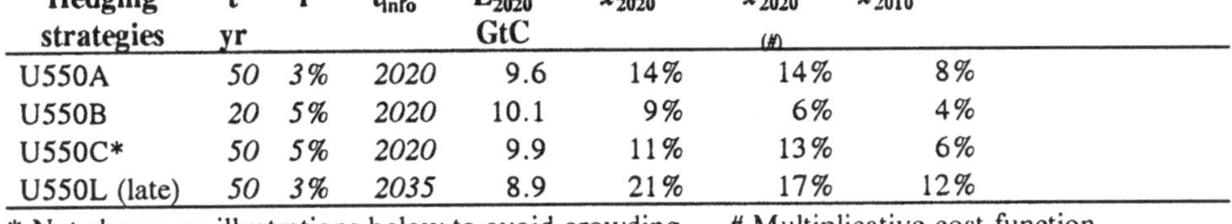

Certainty scenarios	t yr.	r	S ppmv	t_{stab}	x_{2020}	x_{2020} (#)	E_{max} GtC	t_{Emax}	Cost of delay
450A	50	3%	450	2060	24%	19%	8.7	2015	+70%
450B	20	5%	450	2050	19%	14%	9.2	2015	+32%
450C	50	5%	450	2050	20%		9.1	2015	+72%
450D	20	3%	450	2060	23%		8.8	2015	+25%
550A	50	3%	550	2100	7%	4%	11.5	2050	+10%
550B	20	5%	550	2080	3%	2%	12.6	2050	+2%
550C*	50	5%	550	2090	4%	2%	12.2	2050	+8%
550D*	20	3%	550	2090	5%	4%	11.9	2050	+3%
650A*	50	5%	650	2125	3%	0%	14.1	2070	+4%
Hedging strategies	**t yr**	**r**	**t_{info}**	**E_{2020} GtC**	**x_{2020}**	**x_{2020} (#)**	**x_{2010}**		
U550A	50	3%	2020	9.6	14%	14%	8%		
U550B	20	5%	2020	10.1	9%	6%	4%		
U550C*	50	5%	2020	9.9	11%	13%	6%		
U550L (late)	50	3%	2035	8.9	21%	17%	12%		

* Not shown on illustrations below to avoid crowding. # Multiplicative cost function.

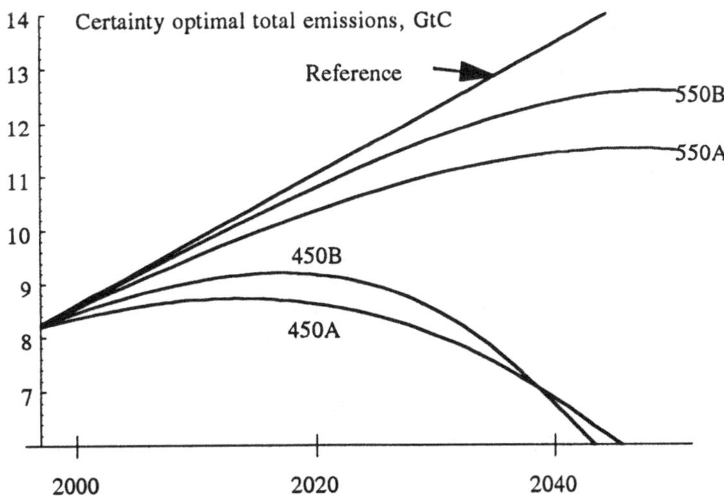

FIGURE 2. Optimal emissions pathways under stabilisation constraints

The FIGURES show pathways that minimise the total abatement costs for the given concentration target. The B cases have lower inertia and higher discount rate than A, both these factors leading to higher near-term optimal emissions: see TABLE 3.

FIGURE 3 provides insight into these results in terms of the abatement expenditure over time for $\tau=50$ trajectories, which for the reasons given we consider to be a more realistic representation of the full inertia in global energy systems (for $\tau=20$ results see TABLE 2). If abatement is not allowed to start until 2020 (but then follows an optimal trajectory), expenditure jumps sharply in the 450A case, to three times the peak level without such a delay. The increase for the 550A case is modest, and furthermore is heavily discounted as most of the expenditure arises after the middle of the next century. The benefits of early action, though positive, are not nearly as large when the time to stabilisation in the reference case is significantly greater than the characteristic time of the energy system (as in the 550ppm cases).

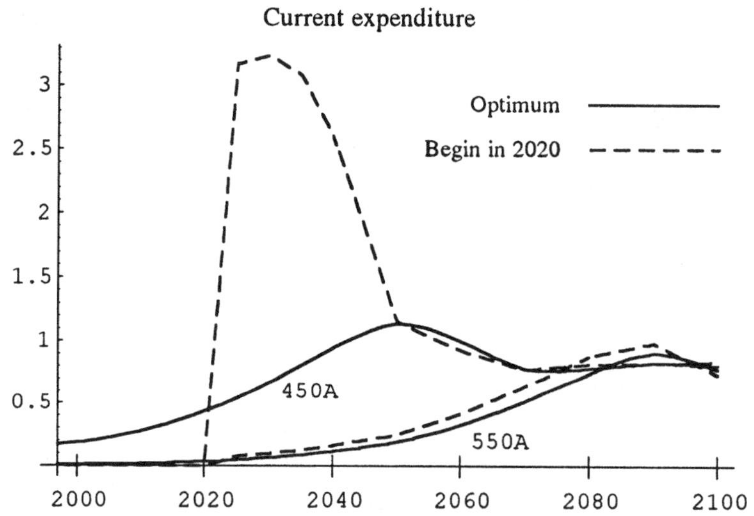

FIGURE 3. Optimal expenditure profiles under stabilisation constraints

Current expenditure profiles with (dashed lines) and without (solid lines) a 20 years delay, for 450 and 550 ppm stabilisation targets (case A). For 450 ppm, the cost peaks abruptly and much higher with defferal than without, due to higher adjustment costs in the 2020 - 2040 period. Results are not that sharp for 550 ppm, as the time available to stabilize at this level (or above) exceeds the characteristic time of the global energy system. Unit is % of 1990 world gross product, calibrated so that the total cost is comparable to the one in DICE [8], however the relative shape of emission and expenditure paths are independant of the scale of the abatement costs.

2.1.3 *Impacts of uncertainty in the stabilisation limit*

Huge uncertainty surrounds the climate problem; the Convention itself emphasises the need for policies to adjust in the light of accumulating knowledge. To analyse optimal trajectories when the stabilisation constraint is uncertain, we assume a probability distribution of a range of possible stabilisa tion contraints, and we assume that the uncertainty is resolved at time T_{info} . In this stochastic version of DIAM, before T_{info}, the model finds the pathway that minimises the expected value of abatement costs, that represents a 'hedging strategy' against the uncertainties.

The three cases U550A-C specified in TABLE 4 each represent cases in which the CO_2 stabilisation ceiling is distributed equally between 450, 550 and 650ppm (ie. with mean 550ppm) and the uncertainty is resolved in 2020. In each case, the optimal degree of abatement by 2020 is substantially greater than for the equivalent deterministic case for 500ppm: 9-14% against 3-7%. This is because the cost of switching to stay within 450ppm too late dominates the incremental cost of a path that follows a lower trajectory up to 2020, and then can relax abatement in the event of 'good news' about climate change (eg. allowing 650ppm or higher ceiling). This effect is all the more important if the resolution of uncertainty comes later: in U555L the uncertainty is not resolved until 2035, and abatement is greater again.

The qualitative conclusion is robust to the assumed degree of inertia and discounting: even with assumptions corresponding to most of the models cited by WRE, the abatement under uncertainty (U550B) at 2020 is 9% compared with 5% for the deterministic case.The difference falls below 2% only for absurdly low values of inertia (10-12 years).

These results reflect directly the inertia of large-scale industry and the cost of misallocating capital investment. In a situation of high uncertainty and inertia, policy conclusions cannot simply be referred to mean estimates of the constraint. If we were certain that atmospheric CO_2 concentrations could safely reach 550ppm or higher (which with other gases corresponds to over 650ppm in CO_2 equivalent), then deferring abatement by a couple of decades would be unlikely to increase total discounted abatement costs by more than 10%. However if 550ppm is a mean estimate of a much broader range of possible constraints, the costs of this sub-optimal trajectory may prove far greater than erring on the side of caution.

2.1.4. Discussion

These numerical experiments show the importance for policy of considering both inertia and uncertainty in systems that produce greenhouse gas emissions, so as to minimise future peaks in the rates of abatement and prevent possible draconian costs if tighter-than-expected constraints prove necessary. As with any such numerical projections, the qualitive conclusions are more important than the specific numerical results. If the atmosphere must eventually be stabilised, deferring all abatement is never optimal, but the costs of deferring abatement are higher under the conditions of high inertia and uncertainty that characterise the real-world climate problem. Inertia indeed has a 'Janus's' role, raising both the costs of premature abatement and the costs of further acceleration in the event of delayed action. It makes steady precautionary abatement more cost-effective: without any inertia, there would be no transition costs in switching from one path to another, action could be deferred until very close to the limit and uncertainty would matter far less.

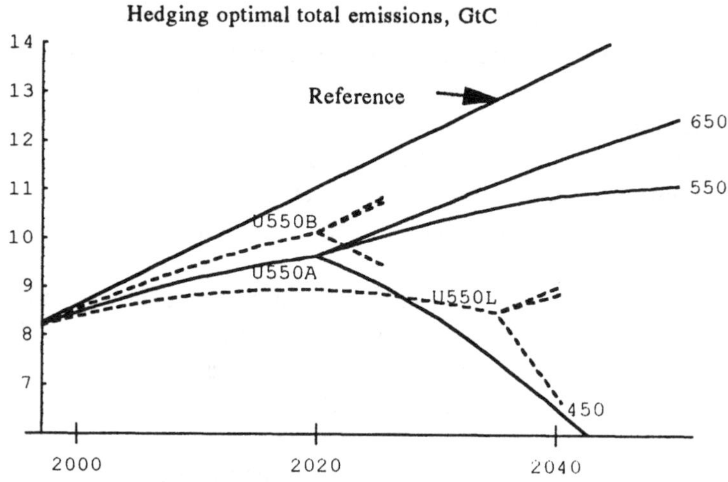

FIGURE 4: Optimal global emission strategies under stochastic constraints

Reference case : 2%/year linear increase (assumed least cost). For key see TABLE 3. The ultimate target is decided only in 2020 (2035 for U550L).

We also emphasise that our results reflect deviation from a reference case that is assumed to be optimal in the absence of constraints; the IPCC recognised that there are some 'free' emission reductions, and to the extent that this might make a least-cost path lower than our reference scenario, this would reduce absolute emission levels correspondingly. We also acknowledge that some precautionary abatement may be an effective way of learning how to make abatement less costly in the future, which is not captured in our analysis.

This analysis, like those of WRE and its references, treats the stabilisation constraint as inviolable and does not take any direct account of climate impacts. In reality the choice of stabilisation level reflects judgement about the balance of costs and risks. Deferring abatement also results in higher rates of climate change that may be significant. An analysis that seeks directly to trade off abatement costs against the full benefits of reduced climate change is economically more rigorous in theory, but in these circumstances, assumptions about the nature of the damage function and the degree of induced technical change, which are both very uncertain, become critical, and this would obscure the basic insights arising from WRE's analysis and our paper

We also note that precautionary abatement may be an effective way of learning how to make abatement less costly in the future. Our analysis has treated technical progress as exogenous; but experience demonstrates the role of learning-by-doing and economies to scale in the real world, which may also increase the costs of deferring abatement.

3. Conclusion: How to make these models more policy relevant ?

Using DIAM, we have been able to underline the economic risks of installing costly energy and transportation infrastructure that may have to be scrapped in future years, and the importance of understanding the balance of risks in systems of high inertia and great uncertainty. Early attention to the carbon content of new equipments reduces the exposure of both the environmental and the economic systems to the risks of unpleasant and costly surprises.

To conclude, I would like to summarize, first, aspects that economists think they understand clearly, to be distinguished, secondly, from aspects that we think important but not yet understood. Other aspects will be left to your personal reflections. With this division, we hope to make it obvious the distinction between economics as a science and a political tool. We also insist that if we do IAM, this is because we recognize that economic consideration alone is not sufficient for analysis.

3.1 THINGS WE THINK WE UNDERSTAND

Straightforward reasoning suggests that if we are to set a concentration target at or over 550ppmv, then waiting a decade or two is sensible (WRE remark). This arises from three economic facts rather well established from past experiences.

- Future generations are likely to be better off that we are, so what we have to do against climate change can be seen as the poor (us) giving their resources to increase the wellbeing of the riches (future generations).
- There has been a continuous technical progress and it does not seem to be slowing down on the contrary. In the information-oriented society that is to come soon, material goods and the carbon emission they imply will be less critical to the wellbeing of people.
- We have to account for the installed material capital: re-engineering industrial processes and rewamping plants are naturally done when they

become obsolete. Since before the end of the next century most capital will have been replaced at least two times, we should build upon that opportunity to avoid accelerating the obsolescence of capital. This also include the institutions, as rapid reallocation of work forces, for example, are often done at a considerably high social cost.

But then economics also make the following two points quite clear.

- Because new capital is always being installed, the issue of installed capital will be as serious in 20 years as it is today. So the converse of WRE remark is also true: if we wait, then we will be economically unable to achieve 450ppmv.
- Because of the huge uncertainties surrounding the climate change issues, looking at the 'best' way to achieve a given concentration target is not the best way to frame the issue. We should rather examine how to adopt a prudent strategy in the face of ignorance about the appropriate ultimate target.

In addition to those five points, one could regard with confidence the result on where and when flexibility:

- Allowing emission reductions to occur where they are least costly, and when they are least costly, can reduce the cost by a significant amount.

3.2 THINGS WE THINK ARE IMPORTANT BUT NOT SO CLEAR (YET?)

Although IAMs are needed to put some rationality in the decision making process, they are still far from being realistic. They still face challenges such as representation and valuation of climate change impacts, critical issues in developing countries, discount rates, technological change and uncertainty and irreversibility, and economic instruments to implement policies. These are only from an economist's point of view, and are explained in the 2 pages summary of Weyant's[1] paper distributed at the conference. Considerations of political sciences such as negociability, joint implemantation, joint equity and efficiency effects should be and actually are excluded from IAMs.

To tackle these challenges, scientific control of IAMs will be a primary issue, because by definition interdisciplinary research whereas people, as are organisations, are specialized. But the field is very young, less than fifteen years, and very active, so the progress rate is high. There will also be managerial, mathematical and computing obstacles. To overcome these will need lots of bright minds working together. This is a call to all the young scientists attending this conference. IAM are necessary and useful, they are also exciting to work with.

References

1. Weyant, P. J.: *Insights from Integrated Assessment*, Communication at the IPIECA Symposium on Critical Issues in the Economics of Climate Change, Paris, 8-11 October 1996.

2. Wigley, T.M.L., Richels, R. and Edmonds, J.: Economic and environmental choices in the stabilisation of atmospheric CO_2 concentrations, *Nature* Vol. 379, 18 January 1996.

3. Grubb, M.J., Chapuis, T. and Ha-Duong, M.: The economics of changing course: Implications of adaptability and inertia for optimal climate policy, *Energy Policy* **23** (1995), 417-432.

CLIMATE CHANGE AND INTEGRATED ASSESSMENT: THE TOLERABLE WINDOWS APPROACH

The ICLIPS Project: Integrated assessment of CLImate Protection Strategies

Dr. F. TÓTH, Dr. G. PETSCHEL-HELD, T. BRUCKNER
Potsdam Institute for Climate Impact Research
P.O. Box 601203, D-14412 Potsdam, Germany

Abstract. Increasing concerns about the potential risks involved in anthropogenic climate change over the past ten years have boosted the intensity of research efforts on all aspects of the problem and motivated an integrated analysis of their results. This paper is intended to provide a short overview of modeling approaches to integrated assessments of climate change. It also presents an innovative concept, the Tolerable Windows Approach (TWA), as it is adopted in a new research project called Integrated Assessment of Climate Protection Strategies (ICLIPS) at PIK. The approach is based on an inverse modeling concept that derives climate stabilization objectives from perceived unacceptable impacts and produces complete sets of permitted emission paths satisfying the corresponding climate change constraints. Initial results show that the approach is a promising new direction in integrated modeling.

The paper starts with a concise overview of the background to integrated assessment models and confirms the need for and policy relevance of the inverse approach. The concept of climate response functions, their sources and major types, and the difficulties involved in deriving them are addressed in Section 2. This is followed by a short presentation of the concept of TWA and the tolerable climate window used in this paper. Section 4 is devoted to the climate and atmospheric chemistry models as currently implemented. Results from applying the model to different specifications of two global climate windows are presented in Section 5. The paper ends with a short discussion of possible applications and related future development directions of the model system.

Key words. climate change, climate policy, integrated assessment, inverse modeling

1. Background and Objectives

Few other environmental problems have been characterized by such extremely differing views as global climate change. As far as risks are concerned, many assert that a continued emission of greenhouse gases in the atmosphere along the path observed over the past few decades would lead to profound changes in

J. Hacker and A. Pelchen (eds.), Goals and Economic Instruments for the Achievement of Global Warming Mitigation in Europe, 55–77.

the Earth's atmospheric system. This would involve a series of climate induced catastrophes ranging from disappearance of ecosystems to water resource disruptions, from unprecedented demise of people due to food shortage and coastal inundation to mass migration of environmental refugees. The other extreme maintains that changes will be slow, and human ingenuity and technological advances will make it perfectly possible to adapt to climate change or even benefit from it.

Partly motivated by the above perceptions but largely dominated by differing views on how modern economies and societies function, similarly extreme positions can be observed in the political process related to the management of the climate change problem. One position is that emissions of greenhouse gases should be reduced immediately and drastically in order to prevent climate change. The other camp cautions about the potentially high costs of draconian short-term measures and refers to the marginal role of annual emissions in the atmospheric budgets of greenhouse gases. This latter group emphasizes the insurance concept and the need for short-term strategies to prepare for long-term action. This debate revolves around the question how strong a signal economic and social actors need over the short-term in order to prepare themselves for drastic emission reductions over medium and longer term.

In the scientific analysis to support policy decisions, various components have proceeded largely in isolation for a long time. Atmospheric scientists and climatologists were working on their ocean and atmospheric models. Hydrologists, ecologists, and agriculturists attempted to assess impacts of a $2xCO_2$-equivalent climate on water supply, ecosystems, and food production. Economists and engineers were trying to work out the feasibility and costs of various emission reduction strategies. These research activities involved a diverse and often contradicting set of assumptions. Results were difficult to compare, and this has drastically reduced their policy relevance.

In the late 1980s, several research groups started to integrate various elements into a consistent framework. As the global warming problem has climbed ever higher on the social agenda, the need for policy-relevant results increased as well. Integrated assessments were conceived to be the appropriate tools to provide this service.

The traditional approach to integrated assessment modeling followed a forward direction (see FIGURE 1a). The analysis starts by summarizing assumptions about population growth, economic development, and other driving forces in the form of internally consistent scenarios of socioeconomic development. These scenarios are then used to project emission paths for various greenhouse gases into the future. Emissions, together with atmospheric removal and chemical transformation processes, lead to changes in concentrations of those gases. Simple climate models are then used to calculate changes in various climate attributes, typically mean temperature and, in some cases, precipitation values. Data describing changing climate patterns are feeding impact assessments in various impact sectors. Using observed market values, where appropriate, or imputed values, where goods and services related to the impact sector are not traded in the market, one can come to a rough damage assessment. Direct biophysical impacts or estimated monetary damages can then be related to the original scenario to assess to what extent would climate change impacts induce changes in the originally assumed patterns. There have been various attempts to package this information in the framework of a cost-benefit analysis (FIGURE 1b).

a) Customary sequence

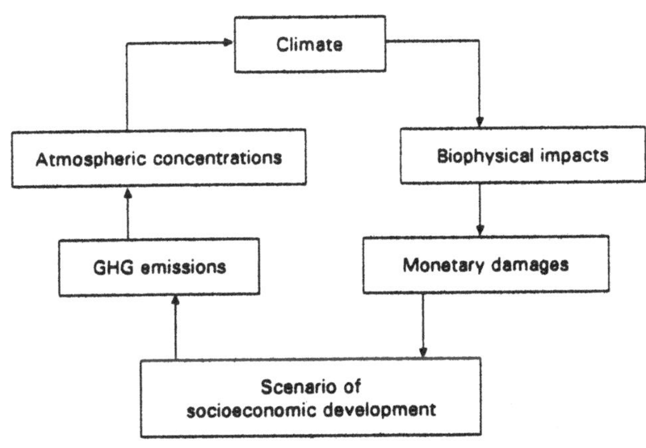

b) Cost-benefit framing

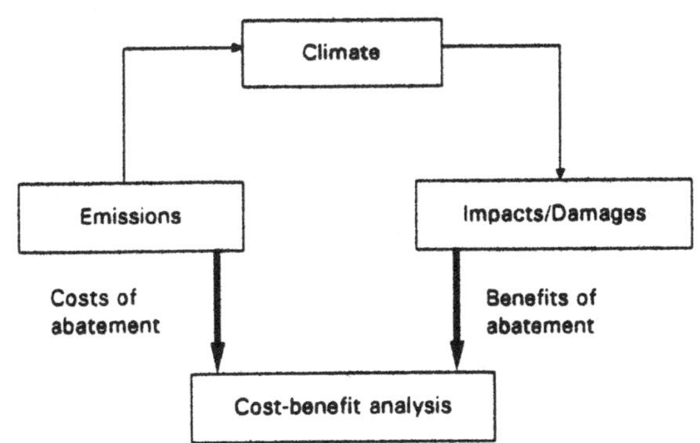

FIGURE 1. Framework for integrated assessments

Integrated assessments following the forward direction from cause to effect proved to be very useful for policy simulations. Models following this sequence are well-suited to assess environmental and socioeconomic consequences of various emission reductions strategies. The approach is less appropriate, however, if one is interested in searching for optimal policies. The objective in this case is to find the best strategy to reach an environmental or impact objective according to selected criteria. The two basic analytical frameworks here are cost-effectiveness (cost minimizing) and cost-benefit analyses.

The cost-benefit framing has major difficulties given our current state of knowledge about the climate system and impacts of climate change. First, despite a steadily increasing amount of research effort, little is known about potential impacts of climate change, especially outside the national accounts. Knowledge about the resilience and adaptive capacity of various ecosystems to climatic shifts and, especially, about the implications of the rate of climate change is still very vague. The second major difficulty involved in cost-benefit analysis is related to the monetary evaluation of non-market impacts, i.e., positive or negative changes in natural or social impact areas not directly involved in market transactions.

In addition to the above difficulties, questions raised by the policy process also point to the need of a novel approach. Policymakers would like to know the magnitude and kind of climate change entailing undesirable impacts in order to put appropriate policies in place. Article 2 of the Framework Convention on Climate Change, the most important political document on the issue so far, also invokes this requirement. It calls for the need "to achieve stabilization of greenhouse gas concentrations in the atmosphere at a level that would prevent dangerous anthropogenic interference with the climate system". Framing the central question for climate policy this way, two immediate practical questions follow. Where is the level of dangerous interference with the climate system, that is, at what level should greenhouse gas concentrations be stabilized? Given the stabilization target, what is the optimal emission trajectory to achieve it? These questions have given rise to a new family of analyses. They start from the damage/impact side in FIGURE 1a and proceed backwards. This technique has become known as the inverse approach.

Various attempts have been made over the past few years to explore these questions. Working Group I of the Second Assessment Report prepared by the Intergovernmental Panel on Climate Change (IPCC) has developed emission paths leading to stabilization of atmospheric concentrations of greenhouse gases at different levels. Wigley, Richels, and Edmonds (hence WRE) have defined a set of alternative emission paths (Wigley et al., 1996a) leading to the same stabilization levels in concentrations but involving slightly different total emission budgets and, more importantly, different time schedules of actual emission reductions (see FIGURE 2).

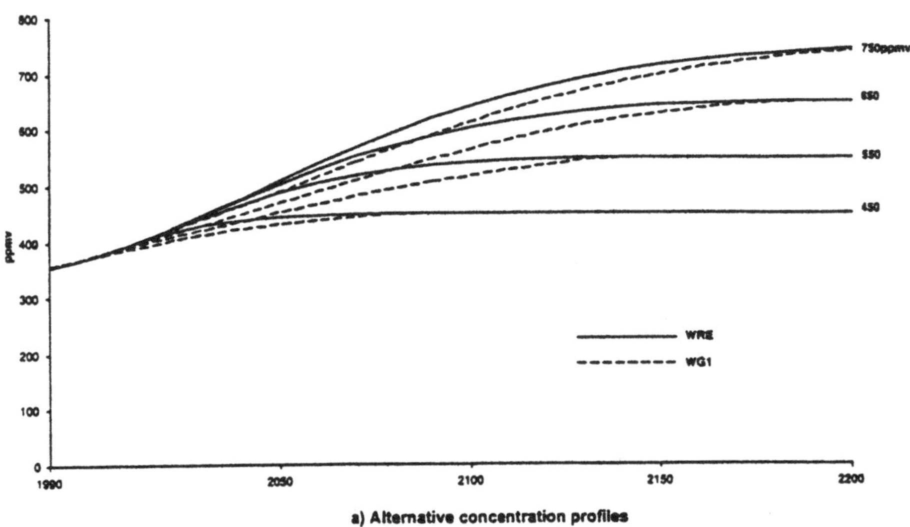

a) Alternative concentration profiles

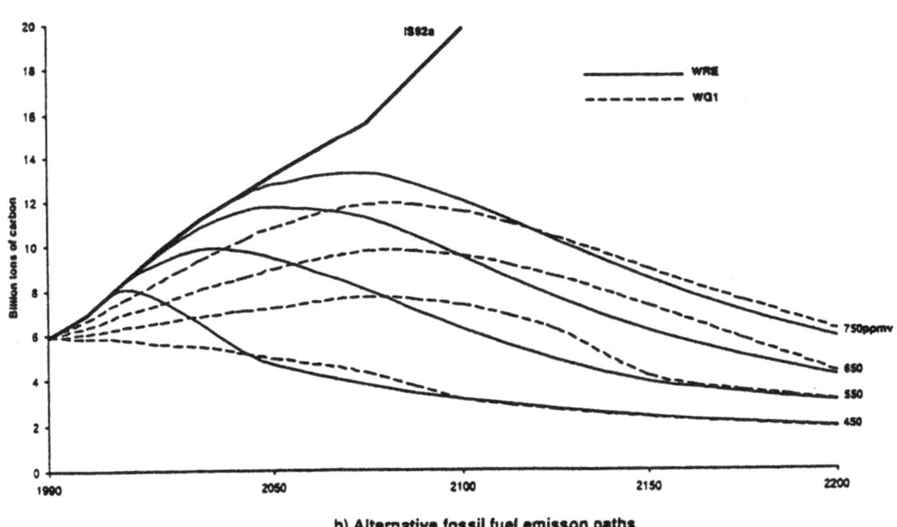

b) Alternative fossil fuel emisson paths

FIGURE 2. Alternative routes to CO_2 stabilization
Source: Manne and Richels (1997).

Wigley et al. have shown that these alternative paths achieve the same long-term environmental objective as those of IPCC WGI but at a much lower cost. It is widely debated what are the interim implications of the alternative paths and what are the short-term policy measures required to guarantee the implementability of the tail end of the WRE paths.

While searching for an optimal emission strategy towards a stabilization target is a useful exercise, its policy-relevance is limited by the fact that concentration targets are not real decision variables.

Taking the inverse analysis a step further, the Scientific Advisory Council on Global Change to the Federal Government of Germany (WBGU) has defined the climate protection target in climatic terms (WBGU, 1995). The Council argued that most ecosystems as we know them today and other life-supporting environmental systems relied upon by humanity have evolved in the geological period of the late quaternary. According to this argument, dangerous anthropogenic interference with the climate system cannot be excluded beyond a level where human emissions force climate out of the range that characterized the late quaternary. The last interglacial maximum of the global mean temperature is estimated to be 16.1° C. The Council added half a degree tolerance to that and defined the upper limit for global mean temperature at 16.6° C. The second constraint on anthropogenic climate change, as defined by WBGU, is derived from expert judgment. To allow natural and socioeconomic systems to adapt to changes in climate, the Council assumes that the rate of change should not exceed 0.2° C per decade.

As the above thresholds involve normative judgments, they should be open to scientific and social discussion. The relevance of climate history at geological time scales for today's societies and ecosystems, for example, is heavily debated. Nevertheless, an additional interesting feature of the WBGU analysis is the so-called Tolerable Windows Approach, an inverse computation technique aimed at defining complete sets of possible emission paths which would keep the global climate within a pre-defined window. This approach became the conceptual cornerstone of a new integrated assessment project at PIK with the objective to specify climate protection goals on the basis of socially defined intolerable impacts. It is called Integrated Assessment of Climate Protection Strategies (ICLIPS) and it serves as the basic source of information for the rest of this paper.

2. Climate Response Functions

One innovative feature of the ICLIPS project is that it makes an explicit attempt to derive climate protection objectives from possible impacts of climate change which societies may want to prevent or avoid. Best available information from sectoral and regional climate impact assessments are synthesized in the form of climate response functions. Impact sectors include socioeconomic areas and components of the natural environment highly valued by humans which are assumed to be sensitive to changes in climate. Response functions indicate how these systems react to climate change. They contain simple representations of highly aggregated information and are usually derived from detailed impact assessment studies. They can be formally presented as $I = I(C,S)$ where I indicates the impact, C includes relevant climatic variables (e.g., annual mean temperatures, seasonal temperatures) and S represents significant socioeconomic variables (e.g., GDP/capita, inequality indicators, poverty).

A closer look at the various impact areas reveals two major types of systemic responses to climate change. Some systems are characterized by a smooth response function within plausible intervals of relevant climatic variables. The tolerable level of climate change forcing in the particular system depends on the perception and judgment of social actors, especially those directly affected. In this case, it is a social decision problem to demarcate beyond which level implications of climate change become perceived as dangerous. This implies a desirable level of maximum permitted anthropogenic forcing of the global climate system.

The second type of climate change impacts can be characterized by discontinuous response functions. These systems are assumed to undergo a phase change beyond a given forcing. This level of forcing implies a threshold at which the system would switch from one qualitative behavior to another. Here the social decision problem is whether human societies would want to refrain from crossing such thresholds.

Turning to systems characterized by smooth response functions, it is important to distinguish between impact sectors dominated by human management (such as agriculture, agroforestry) and those dominated by natural biogeochemical processes (natural ecosystems, hydrology). The profound difference between the two cases is related to the fact that the vulnerability of sectors under the influence of human management is changing over time as it is closely associated with the development level, institutional capacities, and technological capabilities of human society. Natural systems, in contrast, have a more or less constant climate sensitivity.

The special difficulty involved in the assessment of climate impacts and hence in the formulation of response functions for managed systems is the need to distinguish between the biophysical sensitivity and socioeconomic vulnerability of these systems. Taking agriculture as an example, biophysical sensitivity is primarily climate-driven. Phases of crop growth, the amount of yield, or the risk of crop failure heavily depend on vagaries of weather. A persistent shift in weather patterns as manifestation of global climate change will, ceteris paribus, inevitably affect yields and risks of crop failure. Orthogonal to this axis, socioeconomic vulnerability of the society behind this agricultural system is largely development-dependent. The vulnerability concept combines two components: to what extent is society affected by implications of climate change in the first place, and what is the adaptive capacity of the society to respond to and mitigate impacts of climate change.

Underlying arguments about changing socioeconomic vulnerability are simple. One is historical: the agricultural sector in developed countries of North America and Western Europe are much less sensitive to climate fluctuations today than they were 50-100 years ago. The other argument is cross-sectional: in today's world economy, agriculturists, particularly small and subsistence farmers, in poor countries are much more heavily affected by weather fluctuations than their counterparts in rich countries. Vulnerability is reduced as institutional capacity and financial resources (e.g., alternative sources of income, commercial insurance, state support) are more readily available in rich societies. In spite of this plausible relationship, it remains a major challenge to define and assess vulnerability of various impact sectors of future societies with respect to climate change.

The issue is further complicated by the capacity of managed systems to adapt to shifting weather patterns. In this respect, it is useful to distinguish between autonomous adaptation and deliberate adaptation. The first one takes place automatically and is largely cost-less. Switching to a different cultivar of the same crop which is better suited to higher July temperatures, for example, is a case of autonomous adaptation. Even changing the capital stock instigated by the need to switch to a different crop could easily be accommodated within the regular investment-renewal cycle of agricultural capital. Many adaptation measures, nevertheless, would require deliberate action and the associated price could be high. Securing water supply by keeping the guaranteed minimum flow at the same level under changing temperature and precipitation regimes would require additional water storage facilities. Financial and environmental costs of such measures could be substantial even for rich societies. Costs of these deliberate adaptation measures would nonetheless significantly depend on whether they are undertaken as preemptive measures (before negative impacts become apparent) or in a reactive mode (after much damage and degradation has already occurred).

Ongoing research in the ICLIPS project attempts to formulate climate response functions for agriculture, human health, water resources, sea level rise, and natural ecosystems.

While some systems are assumed to be characterized by smooth response functions within plausible ranges of climatic attributes, they can also involve discontinuous changes. In some regions, for example, boreal forest may become unviable beyond a certain level of warming. This does not imply, however, that boreal forests as ecosystem will disappear everywhere.

Many components of the Earth's geophysical system have shown major qualitative changes and these are well detectable in the Earth's geological history. New results indicate that some of these changes occurred in a split of a second by geological measures. While the exact mechanisms of these changes are still poorly understood, there is increasing concern that anthropogenic greenhouse forcing could trigger such changes in the future. Examples of these changes include the North Atlantic deep-water formation, the so-called conveyor belt phenomenon delivering huge amounts of heat to the Northeast Atlantic and keeping Western and Northern Europe much warmer than they otherwise would be. Additional examples include the South Asian monsoon patterns, sea-ice albedo, and breakdown of ice shields.

Information synthesized in both types of climate response functions will be used by policymakers and various social actors. These functions should help them make their judgments about tolerable impact levels in various sectors and various regions with respect to climate change. This information is then used by an innovative model to define permitted climate windows. Tolerable climate windows define changes permitted in climatic attributes such as temperature, rate of temperature change, precipitation, rate of precipitation change, sea level rise, rate of sea level rise. This model and the concept behind it is discussed in the next sections.

3. A tentative tolerable window for climate and socioeconomic evolution

The tolerable windows approach seeks to investigate implications of and trade-offs among several constraints related to different domains in the climate-society system. Tolerable windows are formulated through normative choices by policymakers on the basis of scientific insights. In order to facilitate such a normative decision, it is necessary to translate climate change from a relatively academic and aggregated level of description (e.g., global mean temperature change) to one which is relevant for social decision problems and thus for policymakers (e.g., regional loss of wetlands). Unfortunately, a quantitative description of possible climate impacts, satisfying this requirement is not yet available for pertinent impact categories (cf. IPCC, 1996b).

Instead of impact windows, we therefore selected a family of climate windows as a starting point for our exemplary application of the tolerable windows approach. This kind of tolerable window is located at an intermediate point in the chain of causes and effects, that starts with socioeconomic evolution and ends with climate impacts. One advantage of this provisional choice is that the necessary input variables are provided easily by the climate model of the ICLIPS model system. Moreover, the climate windows to be analyzed here are derived from that defined by the German Advisory Council on Global Change (WBGU) in its statements to the First and Third Conferences of the Parties to the Framework Convention on Climate Change (WBGU, 1995; WBGU, 1997). Adopting a precautionary viewpoint and taking into account our ignorance about possible impacts of climate change, the Council declares a global mean temperature change of more than 2°C (relative to the pre-industrial value) and a rate of temperature change of more than 0.2°C per decade to be intolerable. The Council also assumes that adaptive capacities of ecosystems and social communities are declining as we are approaching the upper temperature limit (see FIGURE 3a). The selected climate window reflects the goal to preserve nature in its present form and to avoid intolerable climate adaptation costs.

It has to be stressed that it is not a scientific task to define such a tolerable window. We therefore refrain from evaluating the usefulness of this climate window in this paper. The tolerable windows approach is meant as a decision support tool rather than a decision making system. Thus, we take the window as an input to our analysis and elaborate solely on its consequences. The definition of the window itself has to be stipulated by political or social judgements on the basis of scientific insights. In the case of the chosen climate window, the necessary expert judgement was made by WBGU. Members of this Council are scientists. They are appointed by the German government and have the mandate to "consider ethical aspects of global environmental change" as well (WBGU, 1997). In our view, a body, such as the WBGU, located at the interface between science and policy and jointly controlled by scientists and policymakers, seems to be exactly the type of forum where defining of tolerable windows could be successful.

In this paper, we adopt the Council's approach to some extent. Yet instead of restricting ourselves to a single window, we want to carry out a sensitivity analysis using five different windows at all. These windows are formulated in terms of global mean temperature (T), the rate of temperature change (DT), global mean sea-level rise (S), and the rate of sea-level rise (DS). We generalize the functional form of the Council's window and write:

$$
\begin{aligned}
& T \leq T_{\max} \\
& DT \leq \begin{cases} DT_{\max} & \text{if } T \leq T_{trans} \\ DT_{\max}\sqrt{\dfrac{T_{\max} - T}{T_{\max} - T_{trans}}} & else \end{cases} \\
& S \leq S_{\max} \\
& DS \leq DS_{\max}
\end{aligned}
\tag{1}
$$

with parameters DT_{max} , DS_{max} for the maximal rate, T_{max} , S_{max} for the absolute value, and T_{trans} for the transition value from linear to non-linear behavior of the underlying generalized response function. In this study, we will make five different assumptions concerning the tolerable climate window with parameters given in TABLE 1.

TABLE 1. Parameter values for the climate window defined by Equation (1). The corresponding windows are shown in FIGURES 3a and b. Dashes indicate that the corresponding constraints are not effective.

Name	DT_{max} *(°C/dec)*	T_{max} *(°C)*	T_{trans} *(°C)*	S_{max} *(cm) rel. 1990*	DS_{max} *(cm/dec)*
WBGU	0.2	16.6	15.6	–	–
WBGU SLR	0.2	16.6	15.6	30.0	3.0
LARGE	0.3	17.6	16.6	–	–
LARGE SL	0.3	17.6	16.6	30.0	–
LARGE SLR	0.3	17.6	16.6	30.0	3.0

The tolerable windows approach is used to provide an integrated assessment of climate protection strategies. It is therefore not sufficient to take into account the possible impact of climate change only. A balanced assessment should consider the maximum tolerable burden imposed on mankind by climate change mitigation measures as well. In order to include this requirement in our analysis, we initially specify a maximum rate of emission reduction of 10%/yr (see Toth et al., 1997). Once again, this restriction is arbitrary and only meant as an example as no one seriously asserts that emissions can be halved in seven-year periods. Therefore no numerical result discussed in this paper should be considered as policy advice.

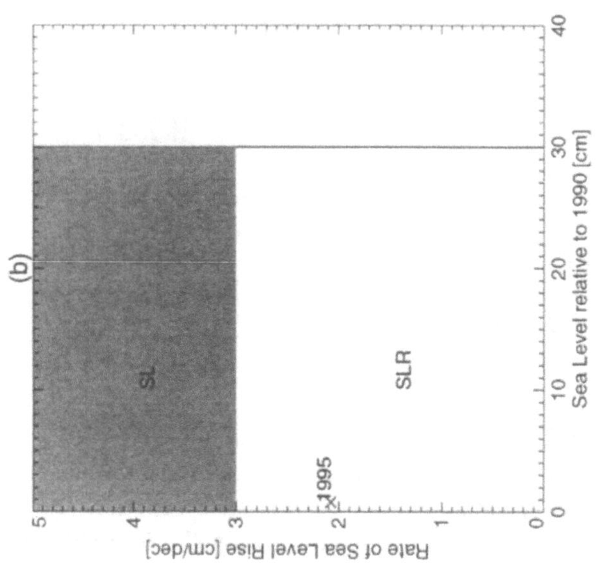

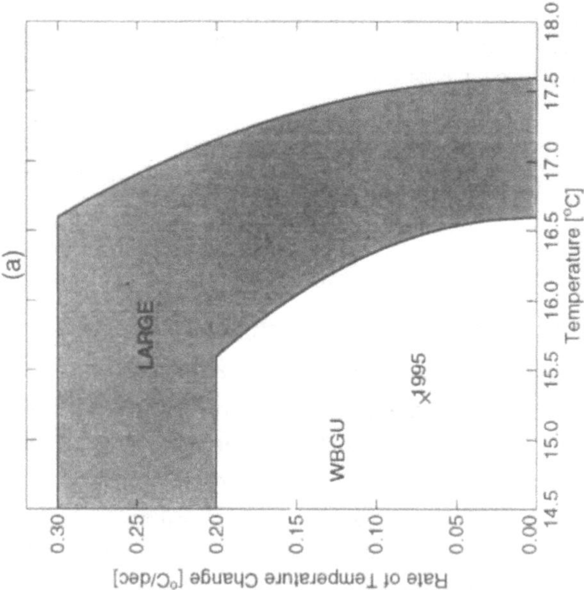

FIGURE 3. Tolerable windows of climate evolution used in this paper. Five combinations between temperature and sea-level rise restrictions are considered: WBGU, WBGU SLR, LARGE, LARGE SL, and LARGE SLR.

Strictly speaking, it would be necessary to explore such constraints in socially relevant indicators like welfare or private consumption losses by using an adequate socio-economic-energy model. Such models are being developed as part of the ICLIPS project but they are not yet available.

4. The coupled climate and greenhouse-gas cycle model

The objective of the ICLIPS climate and greenhouse-gas cycle model is to describe the relationship between emissions of various direct and indirect greenhouse gases and climate change. Appropriate tools to achieve this goal are complex and rather detailed atmospheric-chemistry and coupled atmosphere-ocean general circulation models, respectively. However, computational (and financial) costs of running these models are very high relative to their precision compared to adequately simplified (so-called "reduced-form") models. In order to find an appropriate level of complexity for these reduced-form models, we have to take into account the following peculiarity of the tolerable windows approach: it is possible to show that a typical result of a tolerable windows approach exercise is gained by solving a large number (e.g., several hundreds or even thousands) of independent optimization problems. The related high computer time requirements must be kept in mind when deciding about the complexity of individual models used in the framework of the tolerable windows approach. Fortunately, using the advanced scaleable parallel supercomputer at the Potsdam Institute for Climate Impact Research, we are able to describe the most relevant aspects of anthropogenic-climate interaction.

For climate system modeling, a family of coupled sub-modules is employed in the current version of the ICLIPS climate model (ICM). It includes all major greenhouse gases (CO_2, CH_4, N_2O, halocarbons, tropospheric and stratospheric O_3, and stratospheric water vapour). In addition, it takes into account the radiative effects of aerosols originating from SO_2 emissions and from biomass burning. The input to ICM consists of emission paths for CO_2, CH_4, N_2O, CFC11, CFC12, HCFC22, HFC134a, and SO_2 and the output is given by transient global-mean temperature change, rate of temperature change, global-mean sea-level rise, and rate of sea-level rise. A forthcoming version of the model will include regionalized climate variables according to their role in regional vulnerability to climate change (IPCC, 1996b).

ICM consists of **biogeochemical sub-modules** for turning emissions into concentrations (whereby carbon dioxide, well-mixed gases with well-defined lifetimes, aerosols, and gases not directly emitted are treated differently); **radiative transfer sub-modules** for calculating radiative forcing from concentrations; a **climate sub-module** (in the strict sense) for translating radiative forcing into global-mean temperature change; and **sea-level rise sub-modules** for calculating sea-level change from thermal expansion of oceans and ice melting. The entire scheme of ICM is depicted in FIGURE 4.

The carbon cycle is described according to a model developed by Svirezhev and Brovkin (Svirezhev, 1996; Svirezhev et al., 1997). In this model, annually averaged carbon cycle is composed of four large carbon pools: atmosphere, ocean, terrestrial vegetation and pedosphere, connected to each other by carbon fluxes. Special features of this carbon cycle model include the following: CO_2 uptake by vegetation is assumed to be a function of changes in temperature and CO_2 concentration (CO_2 fertilisation effect), and soil decomposition rate, describing the release of CO_2 from soil to the atmosphere, is temperature dependent. Land-use change is modeled by using an averaged time-dependent residence time of carbon in the vegetation. The model is calibrated to historical concentration and temperature records.

We adopted parts of the MAGICC climate model (Wigley, 1988; Wigley and Raper, 1992; Wigley, 1994; Osborn and Wigley, 1994; Wigley et al., 1996) to simulate the circulation of all greenhouse gases (except CO_2, see above) and to describe the radiative forcing (Shine et al., 1990). The main reason is that these submodules are identical or very similar to the "simple models" (IPCC, 1997) used by the Intergovernmental Panel on Climate Change for scenario analyses reported in its Second Assessment Report (IPCC, 1996a). Concentration changes of well-mixed gases (N_2O and CH_4) are calculated by simple box models. In the case of methane, the effective lifetime depends on the methane concentration itself; this effect is taken into account by the model (IPCC, 1997). Halocarbons are divided into four groups as proposed by Wigley (1994). These groups are defined on the basis of similarities in chemical compositions, and in projected future emissions according to IPCC scenarios. Each group is represented by a key member, for which the concentration change is calculated explicitly. The total radiative forcing for a group is estimated by scaling up the forcing of the respective key member. According to Wigley (1994), it is useful to consider the following groups: CFC11 group (strong ozone depleters), CFC12 group (moderate ozone depleters; long lifetime), HCFC22 group (moderate ozone depleters; short lifetime), and HFC134a group (non-ozone depleting gases). Changes in the concentration of tropospheric ozone (indirectly produced by chemical reactions involving CH_4 and other gases like CO, NO_x, and volatile organic compounds), stratospheric ozone (depleted by chemical reactions involving halocarbons), and stratospheric water vapour (increased by emissions of CH_4) are not treated explicitly (IPCC, 1997). Instead, the radiative forcing due to changes in these concentrations are calculated directly by using other gases (e.g., halocarbons

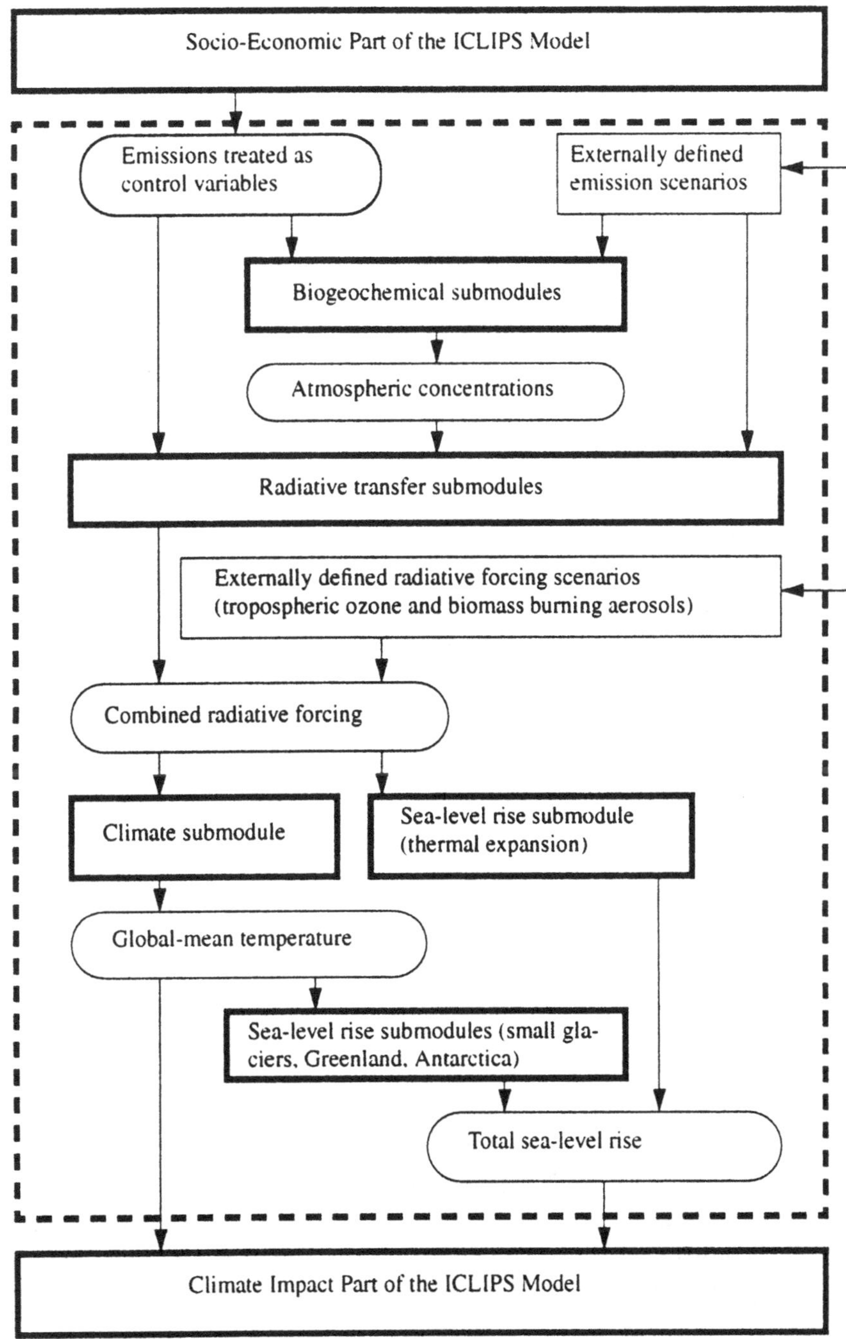

FIGURE 4. General structure of the ICLIPS Climate Model (ICM) and its embedding into the general ICLIPS modeling frame.

and methane) as proxy variables for the full complex chemistry. Changes in aerosol concentrations mainly stem from anthropogenic sulphate (SO_2) emissions and biomass burning. Due to their short lifetime, aerosol concentrations respond instantaneously to changes in emissions (IPCC, 1997). Therefore, aerosol concentrations are not computed explicitly. Global (sulphate) emissions are linked directly to radiative forcing.

5. Results and Discussion

Formally the tolerable windows approach is implemented by applying so-called *differential inclusions* (Deimling, 1992, Aubin and Cellina, 1984, Kurzhanski and Vályi, 1997). They are defined as

$$\dot{\mathbf{x}} \in \mathbf{F}(\mathbf{x},t), \quad \mathbf{x} \in \mathbf{D}(\mathbf{x},t) \tag{2}$$

where t denotes time. Here **x** is the state vector containing the concentrations of different greenhouse gases, the global mean temperature, four variables for sea-level rise, and the carbon content of dead and living biomass as well as of soils and the ocean. The left hand side **F**(**x**,t) is a set-valued function given by a set of "policy options" prescribed by socio-economic constraints as they are discussed above. The tolerable climate window is given by **D**(**x**,t) where restrictions on rates of changes are expressed in terms of state variables. We have developed an effective numerical algorithm (Toth et al., 1997, Petschel-Held et al., 1998) to obtain a *necessary* representation of the entire set of admissible functions **x**(t) fulfilling Eq. (2). The algorithm computes the envelope of all admissible functions and the interior of the envelope is called the *funnel* of tolerable climate evolution. The funnel is a multi-dimensional object and the simplest way to visualize it is to project it onto the most relevant state variables as a function of time. These projections are called *necessary corridors.*

FIGURES 5 and 6 depict such projections for the two different temperature constraints, WBGU and LARGE, and for four variables: CO_2-equivalent greenhouse gas concentrations, global mean temperature, sea level relative to 1990, and global energy and cement-related CO_2 emissions. In case of temperature constraints alone, the white area is forbidden. Additional prohibitions due to the constraints in sea-level rise are encoded by the dark (magnitude only) and by dark and medium gray shaded areas (magnitude and rate of rise), respectively. The projection of the funnel onto the emissions subspace might be called the *CO_2 emission corridor.* As the results have been obtained by simultaneously reducing CO_2, SO_2, CH_4, and N_2O, the corridors for other trace gases can be obtained by a simple scaling procedure. In the initial phase, i.e. until reduction is started, greenhouse gas emissions are assumed to follow the IS92a scenario of IPCC (IPCC1992). Land use change are extrapolated from historical records and halocarbon emissions are treated according to the updated IPCC scenarios (IPCC, 1996a). The depicted corridors and projections represent *necessary* conditions for the time evolution of the respective variable to fulfill the specified socio-economic and climatic constraints: each admissible evolution lies within the corridor, but not any arbitrary emission path lying within the limits is admissible. On the other hand, every *point* within the domain can be reached by at least one admissible control path. See Toth et al. (1997) and Bruckner and Petschel-Held (1998) for a detailed discussion on necessary and sufficient conditions.

The main results for the WBGU domain (FIGURE 5) are:

1. Considering only emission related to energy and cement production, effective emission reduction has to commence in 2017 at the latest due to the limitation in the rate of temperature change. The longer the reduction is delayed, the sharper the turn and the longer it has to last with a higher rate.

2. The restriction on the magnitude becomes effective only after 2115. The rate of sea-level rise chosen here has no influence on the emissions corridor.

3. If there is no restriction on sea-level rise, a simple analysis of the climate model yields the maximal admissible equilibrium CO_2-equivalent concentration at 442 ppm. Yet within the time horizon investigated in this paper, the concentration can be slightly higher, i.e. about 450 ppm. With restrictions on the sea level, however, admissible concentrations are much lower: after reaching its absolute maximum of 441 ppm in 2110, it definitely has to decrease continuously to 405 ppm in 2200. Due to the long-term melting effects of the West Antarctic and Greenland ice shields there is no climatic equilibrium if general sea-level rise is restricted. Similarly global mean temperature declines from 16.4°C in 2115 to ca. 16.2°C in 2200.

The LARGE window of tolerable temperatures implies the following main differences in the results:

1. The latest time to start reductions shifts to 2053 without sea level constraints, 2047 with limit on the absolute amount of sea-level rise, and to 2024 with restrictions on the rate of sea-level rise. Note, however, that in the case of sea-level constraints, long-term emission constraints are equal to those of the WBGU temperature domain, i.e. emissions have to be less than 2.8 GtC/yr in 2200.

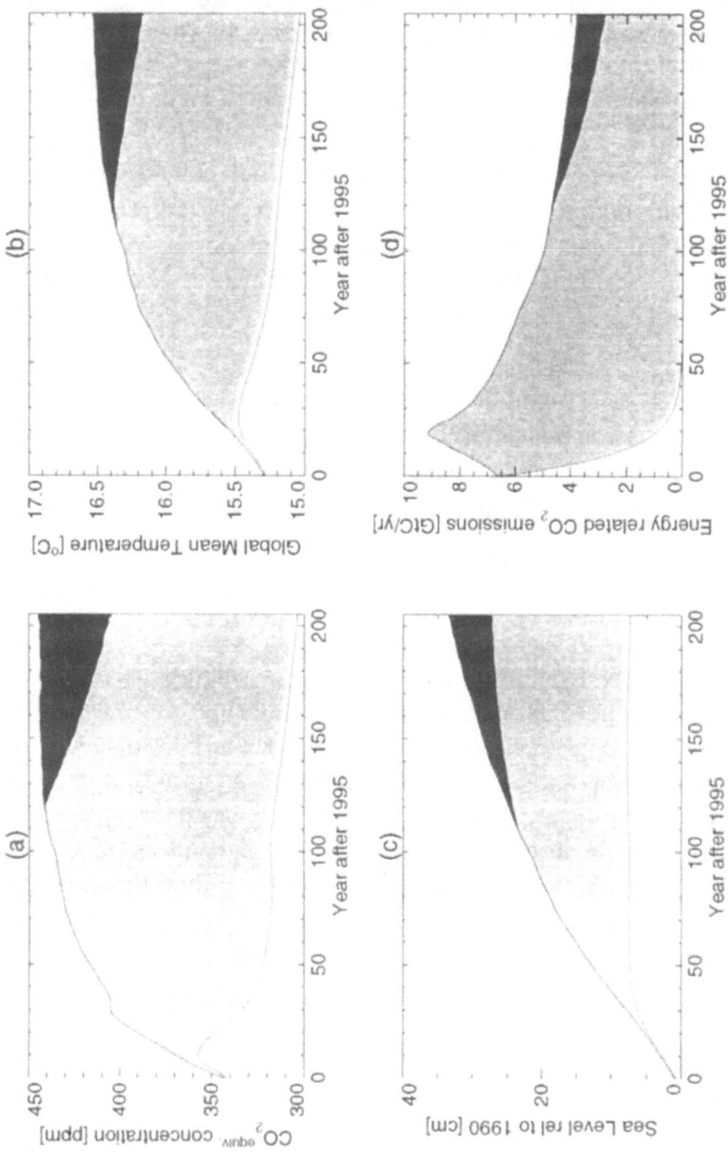

FIGURE 5. Necessary Corridors of climate evolution for the WBGU constraint. Without restrictions in sea-level rise the corridors are constituted by the union of all shaded areas. With restriction in sea-level rise only the light gray corridor is allowed.

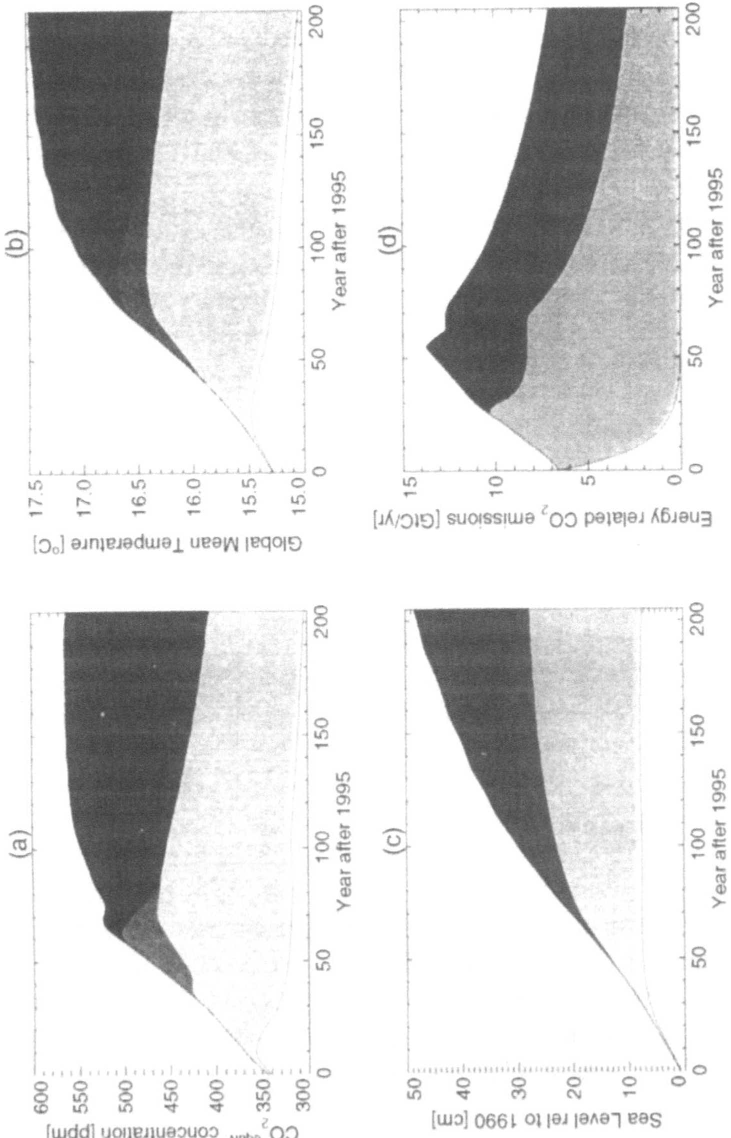

FIGURE 6. Corridors for the LARGE temperature constraints. In case of restriction of the magnitude of sea-level rise the dark area is forbidden. Additional restrictions due to limits in the rate of sea-level rise are depicted in medium gray.

2. The equilibrium concentration without sea level restrictions now amounts to 556 ppm which will not be exceeded within the next 200 years.
3. If emission profiles are chosen appropriately and emissions are not restricted due to sea-level rise, it is possible to keep emissions rather high (approx. 12.5 GtC/a) far into the next century. This requires emissions to be stabilized not later than in 2035. Similarly, in case of constraints in the rate of rise, emissions might be stabilized at about 8.3 GtC/a if they are frozen beyond 2007.

TABLE 2 summarizes our main results in terms of the maximum energy and cement-related CO_2 emissions for the different assumptions on tolerable windows. As one can see, emission reductions before 2045 is needed only if either the rate of temperature change or the rate of sea-level rise have to be restricted. Limitations on the absolute values are important only in the long term.

TABLE 2. Maximal achievable energy and cement-related CO_2 for the five different climate windows as defined in Equation (1) and TABLE 1.

CO_2 emissions in GtC/yr	*2030*	*2050*	*2100*	*2200*
WBGU	7.4	6.4	4.9	3.8
WBGU SLR	7.4	6.4	4.9	2.8
LARGE	11.6	13.7	10.2	7.0
LARGE SL	11.6	11.5	5.3	2.8
LARGE SLR	8.8	8.3	5.3	2.8

6. Applications and future plans

The funnel of all permitted emission paths itself provides useful insights for climate policy. On the impact side, different funnels derived from various response functions and perceived critical levels/rates of climatic change can be compared. Trade-offs between avoided impacts and the size of the room to maneuver with emission controls can be analyzed. Correspondingly, on the emission side, long-term implications of short-term emission reduction targets can be assessed.

The mathematical formulation of the model allows extensions of the TWA in various directions. The first possibility is to define and formalize a set of relevant decision criteria related to emission control. This would lead us to a multi-objective decision problem in which political concerns over various repercussions of climate protection can be formulated. At a general level, economic efficiency of abatement strategies, equity in sharing the associated burden, political feasibility of implementation, and similar criteria could be formulated and integrated into the ICLIPS model. Defining these criteria and their relative weight is again a social decision problem. The model, however, would be able to determine whether there exists a feasible solution at all and, if so, what are the trade-off relationships between various decision criteria and different climate windows.

Conventional policy optimization models usually reduce the number of decision criteria to one and seek to find the best possible strategy in terms of cost-effectiveness. This implies searching for a single path within the emission corridors depicted in FIGURES 5 and 6 that can be implemented at the lowest possible cost. In order to undertake this analysis in a meaningful way, an economic model is needed that keeps track of economic activity and emissions in the absence of control measures. The economic model should also contain information about the costs of abatement measures across regions, time scales, and severity of reduction. One important component of the ICLIPS project is to develop such an economic model and integrate it with the climate model presented in this paper.

The Tolerable Windows Approach presented in this paper is a new concept in integrated assessment of climate protection strategies. The approach responds to requirements of the policy process to help define climate policy on the basis of perceived critical level of climate change. Deriving climate windows from response functions is a critical input to the analysis. The core of TWA is an inverse modeling technique that provides complete sets of all permitted emission paths that keep the climate system within the predefined windows. Additional model components, currently under development, will permit extensions of the analysis to include more detailed assessments of socioeconomic issues involved in controlling GHG emissions.

A general consensus in the integrated assessment community holds that, with a view to the complexity of issues involved in the climate change issue and our ignorance about its many aspects, the best integrated models can offer to policymakers is insights about the nature, basic relationships, and key dynamic features of the problem. This paper has shown that, even in its early phase of development, the TWA approach can make a useful contribution to the debate and provide valuable insights for the policy process in its search for appropriate management strategies.

References

Aubin, J.and Cellina, A. (1984) *Differential Inclusions,* Springer Verlag, Berlin.

Bruckner, Th. and Petschel-Held, G. (1998) Application of differential inclusions in environmental sciences, To be submitted.

Deimling, K. (1992) *Multivalued Differential Equations,* No.1 in De Gruyter Series on nonlinear analysis and applications, De Gruyter, Berlin.

Hasselmann, K., Sausen R., Maier-Reimer E., Voss R. (1993) On the cold start problem in transient simulations with coupled atmosphere-ocean models, *Climate Dynamics* **9**, 53-61.

IPCC (1992) Intergovernmental Panel on Climate Change (Ed.) *Climate Change 1992.* The supplementary to the IPCC Scientific Assessment", Cambridge University Press, Cambridge.

IPCC (1996a) Intergovernmental Panel on Climate Change (Ed.) *Climate Change 1995,* Contribution of Working Group I to the Second Assessment Report of the IPCC, Cambridge University Press, Cambridge.

IPCC (1996b) Intergovernmental Panel on Climate Change (Ed.) *Climate Change 1995,* Contribution of Working. Group II to the Second Assessment Report of the IPCC, Cambridge University Press, Cambridge.

IPCC (1997) Intergovernmental Panel on Climate Change (Ed.) An Introduction to Simple Climate Models used in the IPCC Second Assessment Report, IPCC Technical Paper II, IPCC.

Kurzhanski, A. and Vályi I. (1997) *Ellipsoidal Calculus for Estimation and Control,* Birkhäuser-Verlag, Boston.

Manne, A. and Richels, R. (1997) On Stabilizing CO_2 concentrations - cost-effective emission reduction strategies. Paper presented at the IPCC Asia-Pacific Workshop on Integrated Assessment Models, United Nations University, Tokyo, Japan, 10-12 March, 1997.

Osborn, T. J., Wigley, T.M.L. (1994) A simple model for estimating methane concentration and lifetime variations, *Climate Dynamics* **9**, 181-193.

Petschel-Held, G., Schellnhuber, H.J., Bruckner, Th., Hasselmann, K. (1998) The Tolerable Windows Approach: an inverse integrated assessment of climate change, submitted to Climatic Change.

Shine, K.P., Derwent, R.G., Wuebbles, D.J., Morcrette, J.-J. (1990) Radiative forcing of climate, in IPCC, *Climate Change. The IPCC Scientific Assessment,* Cambridge University Press, Cambridge.

Svirezhev, Y. (1996) Model description and basic equations, in Y. Svirezhev (Ed.), Toys: Materials to the Brandenburg Biosphere Model /Gaia, PIK Report 14, Potsdam-Institute for Climate Impact Research, pp. 9-15.

Svirezhev, Y., Brovkin, V., von Bloh, W., Schellnhuber, H.-J., and Petschel-Held, G. (1997) Optimization of reduction of global CO_2 emission based on a simple model of the carbon cycle, submitted to Environmental Modeling and Assessment.

Toth, F.L., Bruckner, Th., Füssel, H.-M., Leimbach, M., Petschel-Held, G., Schellnhuber, H.-J. (1997) The tolerable windows approach to integrated assessments, Proceedings of the IPCC Asia-Pacific Workshop on Integrated Assessment Models, Tokyo, Japan, 10-12th March 1997.

Warrick, R. A. and Oerlemans, J. (1990) Sea level rise, in IPCC, *Climate Change. The IPCC Scientific Assessment,* Cambridge University Press, Cambridge, pp. 257-282.

WBGU (1995) Scenario for the derivation of global CO_2 reduction targets and implementation strategies, Statement on the occasion of the First Conference of the Parties to the Framework Convention on Climate Change in Berlin, German Advisory Council on Global Change (WBGU), Bremerhaven.

WBGU (1997) Targets for Climate Protection, 1997, Study for the Third Conference of the Parties to the Framework Convention on Climate Change in Kyoto, German Advisory Council on Global Change (WBGU), Bremerhaven.

Wigley, T.M.L. (1988) Future CFC concentrations under the Montreal Protocol and their greenhouse-effect implications, *Nature* **335**, 333-335.

Wigley, T.M.L. and Raper, S.C.B. (1992) Implications for climate and sea level rise of revised IPCC emission scenarios, *Nature* **357**, 293-357.

Wigley, T.M.L. and Raper, S.C.B. (1993) Future changes in global mean temperature and sea level, in Warrick, R.A., Barrow, E.M., Wigley, T.M.L., *Climate and Sea Level Change: Observations, Projections and Implications*, Cambridge University Press, Cambridge, New York, Victoria, p. 111-133.

Wigley, T.M.L. (1994) *MAGICC (Model for the Assessment of Greenhouse-gas Induced Climate Change): User's Guide and Scientific Reference Manual,* National Centre for Atmospheric Research, Boulder, Colo.

Wigley, T.M.L., Raper, S.C.B. and Salmon, M. (1996) Source code of the MAGICC Model (as used in MiniCam, Version 2.0), Climate Research Unit, University of East Anglia, Norwich.

Wigley, T.M.L., Richels, R., and Edmonds, J. (1996a) Economic and Environmental Choices in the Stabilization of Atmospheric CO_2 Concentrations, *Nature* **379**, 240-243.

CLIMATE CHANGE AND GLOBAL CHANGE: THE SYNDROME CONCEPT

The QUESTIONS Project: QUalitativE Dynamics of Syndromes and TransiTION to Sustainibility

Dr. G. PETSCHEL-HELD AND F. REUSSWIG
Potsdam Institute for Climate Impact Research
P.O. Box 601203, D-14412 Potsdam, Germany

Abstract. The problematique of global climate change is usually investigated in a ceteris paribus framework, i.e., impacts and driving forces are embedded in the world as it presents itself nowadays. In parallel, however, quite a number of other phenomena of global environmental change take place: soil degradation, shortage of freshwater resources, or the massive loss of biodiversity are just a few examples. Furthermore drastic socio-economic developments take place: the globalization of economy and society, the still increasing international indebtedness, the hefty progress in information technologies and many others more. All these complexly interconnected *symptoms* of Global Change call for an integrated research strategy which can serve as the information and evaluation base for rough but robust global environmental management strategies. In this paper we want to present a research approach which from our point of view is capable to include most of the relevant trends and symptoms of Global Change. The approach is based on the assumption that Global Change can be decomposed into archetypical patterns of civilization-nature interactions, called *syndromes of Global Change* (Schellnhuber et al., 1997). Besides a general discussion of the syndrome concept we sketch some qualitative ideas how this approach can be applied to evaluate climate change mitigation strategies with respect to synergies and side-effects.

1. Introduction

Climate Change can be regarded as a research field on its own, including relevance, complexity, hierarchies of disciplinary knowledge, their related methodologies, and so forth. We clearly need this sort of disciplinary or sectoral research, not only for a better understanding of the underlying processes but also for political action in order to prevent possible damages. The results of the last climate summit in Kyoto in December 1997, disappointing in the eyes of many scientists and environmentalists, underline the persistent necessity of this type of climate research.

J. Hacker and A. Pelchen (eds.), Goals and Economic Instruments for the Achievement of Global Warming Mitigation in Europe, 79–95.

Typically the investigations and analyses of global climate change are carried out in a ceteris paribus framework. The driving forces, i.e. the energy-economy system, and the impacts of climate change are always considered within the world as it is now (IPCC, 1996a). Nevertheless major impacts are expected to occur beyond 2030 and it can therefore be anticipated that the impacts of climate change affect an altered and changing world. In particular it is probable that global environmental change in general will give rise to a number of synergies which might be important. For example, if the geobiophysical growth conditions change with the climate, the impact on the agricultural sector might be significantly increased by an ever increasing soil degradation (WBGU, 1993). Similar examples can be given for most of the other climate impact sectors (see Toth et al., 1998). It is clear that there is no way to construct an "analogue world model" which would allow a sufficient forecast of all relevant aspects of development. Nevertheless the argument shows the necessity for integrated research strategies, focusing on the globally important interlinkages of the whole Earth System, including natural systems and civilization processes. Examples for the phenomena to be taken into account are:

- Industrialization and globalization of the economy.
- Soil degradation of all types.
- Reduction of natural ecosystems by area and quality, implying significant loss of biodiversity.
- Pollution of freshwater resources and coastal zones.
- Global dissemination of allochthonous species, pests and disease vectors.
- Population growth.
- Amplification of world-wide disparities regarding affluence, sanitation and education - not to speak of „imponderables" like human dignity.

Accepting the necessity of an integrated analysis of all the pertinent interdependencies, the question arises, how this can be done. So far, Global Change (GC) research can be characterized by two kinds of strategies (Schellnhuber et al., 1997): On the one hand there are the rather traditional disciplinary-sectoral approaches, which endeavor to determine single *facets of Global Change* in their global manifestation. Examples of these are the high-tech measurement campaigns to determine fluoride and bromide-containing components in the stratosphere, or the large-scale mapping of the effects of wind erosion on soil resources (Pimentel et al., 1995). The results of this research are impressive and are an essential basis for any sort of systematic view of the overall spectrum of problems. The simple accumulation of such results per se, however, cannot reflect the complex character of the system under investigation as it does not sufficiently take into account the synergetic interdependencies.

In contrast to the - in the best sense - "reductionist" approach just mentioned, the so-called "world models" do not push the individual (dimensional, sectoral, disciplinary, etc.) determinants of the system to the center of analysis, but rather the *"wiring" of the segments*. This approach to Earth System analysis makes use of the simulation of more or less sophisticated simulations of the planet in the laboratory of virtual reality and owes its existence largely to the advent of electronic computers. Prominent representatives of the adolescent school of "integrated modeling" are WORLD3 or IMAGE2.0 (Meadows et al., 1992; Alcamo, 1994). The protagonists of this school hope that with progressing geographical explicitness and process connectivity the digital copies will be ever better able to mimic the dynamic character of the original. This hope may possibly prove deceptive, since the chosen approach of *analogous modeling* by reproduction of the quantitative structure of the system may gain forecasting and hindcasting power only when the degree of sophistication becomes excessive. In this case the simulation model completely loses its character as a heuristic instrument, as its dynamics are no easier to understand than those of its original.

We thus think that a *combination of both approaches* can help us on the way to Earth System analysis. In most cases, this will require sacrificing quantitative rigor significantly - but not drastically - in favor of *qualitative, intuitive* and *typifying* aspects. The basic idea here is that the overall phenomenon "Global Change" should not be divided into regions, sectors or processes, but be understood as a *co-evolution of dynamic partial patterns*. These patterns are bundles of interactive processes which appear repeatedly and widely spread in typical combinations and are called "*syndromes of Global Change*".

These syndromes are not only useful in interconnecting Global Change phenomena with regard to human usage of natural systems, but might also serve as a tool for integrating Climate Change research into the wider field of Global Change. The next sections will present some key features of the syndrome concept and then try to give an idea of how this integration of the climate component might look. Work on the syndrome concept and its actual data-based realization is currently subject of the PIK core project QUESTIONS. Thus the discussion we present here is still preliminary - this is, as will be shown in the spirit of the approach.

2. Key Features of the Syndrome Concept

2.1. SYMPTOMS AND SYNDROMES

In medicine the term syndrome refers to a typical co-occurrence of different symptoms, like a cough or sneeze in the case of a cold. Each of these symptoms could be observed as a single phenomenon or in combination with others. But the medical expert can identify syndromes or diseases by typical patterns of symptoms. We have adopted this idea for the analysis of Global Change, i.e. the philosophy of the syndrome concept rests on the assumption that GC phenomena cannot be resolved into isolated changes occurring in single Earth System spheres, such as the hydro-, atmo- or anthroposphere. Rather, the interactions of processes in all spheres, especially the social driving forces, their direct or indirect effects across sectoral borders and the feedback loops that "re-import" anthropogenic changes have to be taken into account. Here it might be remarked that humans use natural systems and functions throughout the globe in very different ways, in such variety that any attempt to describe these interactions globally has to fail. Nevertheless one can find quite similar "failures" or problematic co-evolution patterns of man-environment interaction. The type of land use in sub-Saharian Africa is quite similar to that in the Middle East or in parts of South Asia. Conversely the different pattern of agriculture in Western Europe is similar in nature to that in North America or parts of Russia. We therefore structure the variety of interlinkages into a limited number of *archetypical patterns of civilization-nature-interactions*. Due to their relation to 'human failures' within the management of nature and its resources, syndromes of Global Change represent typical patterns of non-sustainability on a global scale.

There is a wide range of definitions of *Sustainable Development*, a notion quite well known since the Rio conference in 1992 and currently propelled by many economists and ecologists - and even some politicians (Quarrie, 1992). Many of these definitions are conflicting; some include even contradictory demands. It seems that much of this controversy is due to the fact that our knowledge base is still too narrow for making strong positive statements about all relevant aspects of *the* sustainable Earth System pathway. Our approach tries to avoid this by focusing on *non-sustainability* rather than sustainability and thus to a *class of sustainable future developments*. In this sense, the syndromes are intended to characterize the multidimensional domain that humankind should avoid to touch during future development processes, i.e. most of the environmentally induced woes are included in the catalogue of syndromes (s. next section). In doing so, modern society and its correlated policy process gains a sort of a development corridor left by the syndromatic borderlines. Thus the syndrome approach is not only designed for scientific purposes, but for scientifically based policy advice (see also the Tolerable Windows Approach in Toth et al., 1998).

To achieve this a basic vocabulary of syndrome analysis has to be developed. To find a common language is a key problem in transdisciplinary research. Simply using terms of the contributing sciences is very time consuming and leads in most cases to misunderstanding and/or redefiniton necessities throughout the research process. Our research project, including such different disciplines as physics, biology, climatology, economy, sociology and others, decided to formulate a common meta-language for Global Change description and analysis. Logically, this language is situated above the fine-tuned terminology of the disciplines involved, thus referring to Global Change phenomena on a highly aggregated manner, but is still rooted in disciplinary research and able to adapt new results. This language includes two basic elements: first we identify so called symptoms (or trends) of Global Change, serving as a semantical basis. Second we focus upon the interlinkages and causal connections between these symptoms, which represent the syntax of our language.

We are currently operating with about 80 symptoms, including the following:

- urban sprawl
- increasing significance of NGOs
- terrestrial run-off changes
- deposition and accumulation of waste
- increasing mobility
- tropospheric pollution
- increasing consumption of energy and resources.
- ...

The names of the symptoms have to be understood more as guiding headlines than as definitions. These symptoms, taken from different spheres (e.g. atmosphere, biosphere, anthroposphere, etc.), focus on qualitative and quantitative changes of the Earth System and usually include the states and the rates of change of the quality or quantity concerned. The time-scales of the symptoms are typical for the specific sectors and usually range from months (economic sector) to decades (climate, soils). Similarly the spatial scales range from small patches (ecosystems) to countries (economy, society) or even continents (hydrology) or the globe (climate).

The starting point of any syndrome analysis is the intuitive formulation of the patterns by inspection, i.e. within a group of interdisciplinary scientists the principle elements of the patterns, their differences and pecularities are formulated (Reusswig and Schellnhuber, 1997). This is corrobarated by intensive literature studies and expert interviews. In the next step either a survey of corresponding case studies, an examination of respective theories or general analyses of sub-elements of the syndrome are used to formulate the global network of interrelations which is a graphical representation of the relevant symptoms and their interactions (Schellnhuber et al., 1997) and serves as a mental map of the entire field of GC problems. These networks of interrelation qualitatively define the syndromes as 'clinical pictures' of the Earth System. Yet in the course of the detailed analyses based upon actual data, to be sketched in the next section, these patterns might be modified. Thus the syndrome concept is a rather dynamic research programme, open for modifications and extensions.

The following list gives an overview of the 16 syndromes currently under research. These syndromes have been formulated in a two year discussion process involving scientists at PIK, the German Advisory Council on Global Change and other partners. The names refer to functional patterns found worldwide, not to specific places or events. The functional structure of each syndrome, including human driving forces and natural systems reactions are indicated by the short descriptions. Syndromes can be further grouped by regarding the way in which humans and social systems use and misuse nature. This leads to three major types: *utilization syndromes, development syndromes* and *sink syndromes*.

TABLE 1. Syndromes of Global Change and their short definitions. These patterns of non-sustainable development can be grouped according to basic human usage of nature: as a source for production, as a medium for socio-economic development, as a sink for civilizational outputs. For more details see Schellnhuber et al. (1997).

Utilization Syndromes	
SAHEL SYNDROME	Overuse of marginal land
OVEREXPLOITATION SYNDROME	Overexploitation of natural ecosystems
RURAL EXODUS SYNDROME	Degradation through abandonment of traditional agricultural practices
DUST BOWL SYNDROME	Non-sustainable agro-industrial use of soils and bodies of water
KATANGA SYNDROME	Degradation through depletion of non-renewable resources
MASS TOURISM SYNDROME	Development and destruction of nature for recreational ends
SCORCHED EARTH SYNDROME	Environmental destruction through war and military action

TABLE 1. Cont'd

Development Syndromes	
ARAL SEA SYNDROME	Damage of landscapes as a result of large-scale projects
GREEN REVOLUTION SYNDROME	Degradation through the transfer and introduction of inappropriate farming methods
ASIAN TIGER SYNDROME	Disregard for environmental standards in the course of rapid economic growth
FAVELA SYNDROME	Socio-ecological degradation through uncontrolled urban growth
URBAN SPRAWL SYNDROME	Destruction of landscapes through planned expansion of urban infrastructures
DISASTER SYNDROME	Singular anthropogenic environmental disasters with long-term impacts

Sink Syndromes	
SMOKESTACK SYNDROME	Environmental degradation through large-scale diffusion of long-lived substances
WASTE DUMPING SYNDROME	Environmental degradation through controlled and uncontrolled disposal of waste
CONTAMINATED LAND SYNDROME	Local contamination of environmental assets at industrial locations

2.2. METHODS AND MEASUREMENT

As a new and relatively unprecedented piece of transdisciplinary research Earth System analysis by syndromes has no proven set of well defined methodologies. The "soft identity" of the syndromes of Global Change and their transdisciplinary composition demands specific and innovative methods of investigation such as

- decomposition of complex functional networks,
- qualitative reasoning concepts,
- modeling of fuzziness and uncertainty,
- knowledge acquisition strategies, and
- set-value analysis.

Furthermore Geographical Information Systems (GIS) are used, as syndromes are not only functional patterns but also geographical ones.

If the Global Change phenomenology can be decomposed into 16 syndromes, and if many (not necessarily all) problems of GC have clearly quantitative - take only the increase of greenhouse gas (GHG) concentrations in the atmosphere or the number of people below the poverty line - aspects, then syndromes have quantitative and measurable aspects as well. The methodology used (e.g. Fuzzy Logic) enables us to cope with qualitative knowledge as well. Nevertheless worldwide available data sets - e.g., as provided by the World Bank, the World Resource Institute or other research groups - are widely used in order to fill the conceptual framework of trends and interactions with quantitative or qualitative data and to represent the thus obtained model in world maps. This enables us - and regional and/or functional experts we are consulting - to compare our results with any kind of "ground truth".

For measurement purposes we distinguish between

- the *disposition* of a given region for a specific syndrome, determined by structural peculiarities of that region, persisting over medium and long time-scales, including natural and social factors (like soil quality, climate, political system, cultural values). These factors are selected according to the relevant syndrome and are combined in a consistent logical argument in favor or against a structural precondition for the syndrome-specific mechanisms. For example, consider the SMOKESTACK SYNDROME: The diffusion of airborne pollutants leads to different levels of acidification in the soils, according to the regional circulation patterns and the buffer capacity of soils - the latter two represent factors for the disposition towards the syndrome. In most cases, these arguments are complex logical clauses and are evaluated in fuzzy-logic based evaluation trees and are mapped with GIS techniques. High disposition values indicate a high proneness of the region for the outbreak of the syndrome, logically they offer a probability assessment.
- the *exposition* factors for a syndrome include endogenous or exogenous triggers (like natural catastrophes, civil strife or sudden political events) that actuate the outbreak of the syndrome in a specific region.
- the *intensity* of a syndrome finally measures its presence in a region already affected, based upon the regional expression of the dynamic behavior as it is induced by the characteristic network of symptoms and interactions.

It is clear that syndromes, although being separate objects of GC research, can reinforce each other, as, for example, the SAHEL SYNDROME might lead to emigration of (poor) people to cities, thus reinforcing the FAVELA SYNDROME. How this concept might work in order to integrate the specific phenomenon of Climate Change into GC research in general will be discussed in the next section.

3. Effects of Syndromes on Climate

As discussed in the previous section, syndromes of Global Change represent entire patterns of interdependencies of civilization-nature interactions. This not only includes effects like soil-degradation, shortage and/or pollution of water resources, or local air pollution, but also anthropogenic climatic change. Yet instead of analyzing the causes of climate change alone, our approach is capable of embedding the enhanced greenhouse effect into the entire phenomenon of global environmental change. In the rest of this section we want to highlight some of the elements of this interconnection, though only preliminary results can be presented as detailed research is still in its launching phase. In particular we focus on qualitatively discussing the effects on climate of various syndromes. FIGURE 1 gives an overview, not only on the climate effects of the syndromes, but also on their sensitivities to climate changes as they will be discussed in the next section.

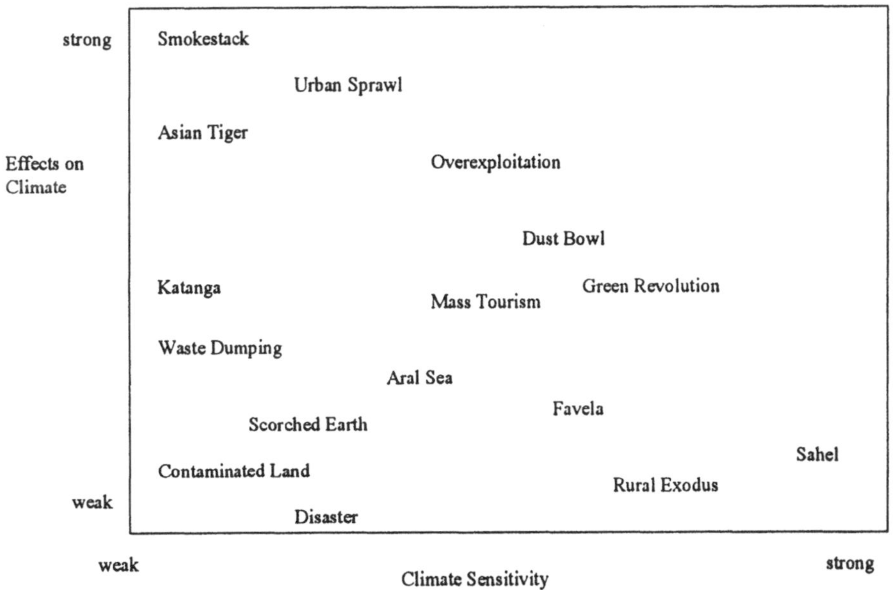

FIGURE 1. Assessment of the relative relevance of syndromes with respect to their effects on climate and their sensitivity to climate change. For the definition of the syndromes see TABLE 1.

First it has to be noted that the effects of syndromes on climate can happen on various scales: globally, regionally, and/or locally. Most prominent in the current discussion is the - globally effective - anthropogenically enhanced greenhouse effect, mainly caused by the increasing combustion of fossil fuels and net emissions of trace gases by land use changes (IPCC, 1996). Examples of human-influenced regional climate change range from the continued desertification, e.g., in the Sahel region of West Africa, to the regional effects due to emissions of aerosols - an effect often integrated within the global greenhouse considerations. Local effects are again dominated by land use changes, e.g., by the construction of large dams and the reservoir filling connected with it (McCully, 1996), or a direct anthropogenic heating, sometimes known as urban heat islands (Oke, 1973).

According to this typology of global, regional, and local climate changes we can identify the following syndromes to be primarily responsible (compare FIGURE 1):

a) Global

In its generality, the SMOKESTACK SYNDROME describes the remote effects of substance emissions following disposal in the environmental media air, water, and soil (WBGU, 1996). The basic human aspect of this syndrome is the fatal strategy to dispose undesired substances into air or water. Within this strategy only weak and thus tolerable environmental impacts are expected due to the high resolutions of the pollutants in the media. Nevertheless this strategy fails, either due to major effects of small concentration changes of the pollutants (greenhouse gases, heavy metals, etc.) and/or due to the long-lasting life-cycles and corresponding accumulation effects (acidification of soils, Waldsterben, etc.) with a possible exceeding of critical loads. Therefore, with respect to climate change, the SMOKESTACK SYNDROME not only covers the industrial emissions of greenhouse gases effective globally, but also of aerosols as the major regional active trace gases and furthermore responsible for local and regional air, water and soil pollution.

Other important syndromes with respect to global climate change are the ASIAN TIGER, URBAN SPRAWL, the OVEREXPLOITATION, the GREEN REVOLUTION, and the DUST BOWL SYNDROME. All these syndromes contribute to an increasing concentration of greenhouse gases in the atmosphere (CO_2, CH_4, N_20, CFCs, etc.), either directly or by land-use changes. It is most important to keep in mind that beside the effects on the global climate these syndromes involve other major environmental effects of which the most relevant are listed in TABLE 2 (together with the syndromes responsible for local and/or regional influences).

b) **Regional**

As already discussed above, humans affect regional climate mainly by land use changes. Yet in contrast to the global effects the climate sequels are due to rather large scale changes in the morphology, e.g., roughness, changes of albedo (desertification), and disruptions of the regional water balance (Nisbet, 1991). These effects play a dominant role, e.g., in the OVER-EXPLOITATION SYNDROME which describes the excessive utilization of natural resources, in particular fish, forest, and rangelands. Major socio-economic elements of this syndrome, e.g., for the 'forest-variety', are the international indebtedness of the respective country, a forest-industry with a rather high traditional relevance, e.g., in terms of economic revenues and/or labor, and a high demand for wood, e.g., wooden houses, construction techniques, etc.. Forests are very important for the regional water-cycle, especially in the tropics, which implies another cause-effect chain for the impact of this syndrome on the regional climate. Other syndromes connected with major land-use changes are the DUST BOWL and the GREEN REVOLUTION SYNDROME which are also active globally, e.g., by emission of laughing-gas N_20. Yet also emissions of climatically relevant short lived gases and particles, i.e., aerosols, are effective on a regional level only. This implies contributions by the SMOKESTACK, ASIAN TIGER, and URBAN SPRAWL SYNDROMES.

c) **Local**

The principle mechanisms of human influences on local climate are similar in nature to the ones for the regional scale. Yet in many cases the change of morphology is effective for the local climate only, e.g., in the case of reservoir filling (ARAL SEA SYNDROME), only the climate in the immediate vicinity of the new water-body is effected (BMZ, 1984). It has to be stressed that the borderlines between regional and local effects are often rather fluid and thus only useful for analytic purposes. Most of the syndromes influencing the regional climate are also active on a local scale. Further mechanisms might yet be important: it has been shown that there is a significant correlation between the size of a city and its direct heating capacity (heat island effect; Oke, 1973). The URBAN SPRAWL SYNDROME is therefore not only active due to the emissions of GHGs, but also by its direct heating capacity. As is indicated in the name of the syndrome, its dynamics are mainly due to the increased movement of people, industry, and services into planned suburbs and areas for industrial and commercial use, often located far out of the center.

TABLE 2. Overview on the climate effects of the most relevant syndromes and their environmental side effects. Further the scale (*G*lobal, *R*egional, *L*ocal) and the basic mechanisms of climate effects are listed.

Syndrome and scale of climate effected	Mechanism of Climate Effects	Primary Environmental Effects aside from Climate Change	Major Regions affected (according to various sources)
SMOKESTACK (G,R)	Emission of greenhouse gases including aerosols	Acidification of soils and water and other impairments of natural metabolisms, ecotoxicological effects on organisms including humans etc.	Western Europe, Northern America, Japan, some parts of China and Russia
ASIAN TIGER (G,R)	Emission of greenhouse gases including aerosols	Local Air, Water and Soil Pollution, direct impacts on human health, etc.	South and South-East Asia
URBAN SPRAWL (G,R,L)	Emission of greenhouse gases, land use changes (albedo, roughness), heat islands	Loss of soil functionality (e.g., water filter) due to surface sealing, pollution of ground- and surface water, increase of local air pollution and tropospheric ozone, etc.	Most cities in the developed world and in countries in transition
OVEREXPLOITATION (G,R,L)	Land use changes, direct inference in the water cycle, change of albedo	Loss of bio- and genetic diversity, conversion and/or degradation of ecosystems, in particular forest and fish resources, etc.	Boreal and temperate forest regions in Canada and Siberia, wide areas of tropical forests in Brazil and South-East Asia
DUST BOWL (G)	Land use changes, fertilization with emission of N_2O, emission of methane from rice paddies, livestock, etc.	Major soil degradation (erosion, chemical degradation, compression) and critical nutrient and toxic load of water bodies	Europe, including the capital intensive agriculture in Eastern Europe, North America, Parts of China, Australia and Southern Africa

TABLE 2. Cont'd

Syndrome and scale of climate effected	Mechanism of Climate Effects	Primary Environmental Effects aside from Climate Change	Major Regions affected (according to various sources)
GREEN REVOLUTION (G)	Land use changes, fertilization with emission of N_2O, emission of methane from rice paddies, etc.	Major soil degradation (erosion and chemical degradation) and critical nutrient and toxic load of water bodies	India, China, Pakistan, Parts of North Africa and Central and Southern America
ARAL SEA (R,L)	Land use changes, intervention into local water cycles	Degradation of ecosystems with loss of biodiversity due to reservoir filling or changes in flood regimes, impact on water resources (temperature, salinity, etc.)	Large dams and reservoirs around the world (e.g., Aswan). Hot spots in South-East China, Northern India, United States and South Brazil

The discussion has shown that there is a close interrelation between climate change and global change in general. This close relationship implies - spoken from a system-theoretical point of view - the necessity to embed the evaluation of any mitigation measure into a systemic framework. Any maneuver intended for the mitigation of one or the other major trend or symptom of environmental change might have impacts on other trends, either attenuating or reinforcing. Consider, for example, an extension of the hydropower usage which, in the first place, has a mitigation effect on anthropogenic climate change due to the substitution of carbon-intensive primary energy sources. Yet it might be accompanied by an essential local or regional climate change and/or other major environmental and social effects known as the ARAL SEA SYNDROME (WBGU, 1997). This relation is inverted in case of a reduction of aerosol emissions: positive effects with respect to pollutant depositions have to be contrasted with a decreased cooling effect within the global greenhouse. As an example for a 'positive synergism' between climate change mitigation measures and a cure for global environmental change in general, consider a moderate, adequately compensated, increase in energy prices. On the one hand this leads to a reduction of GHG emissions. On the other hand it might reduce individual traffic in general and thus the need for new roads, i.e., no further sealing of soils (see above). In a nutshell, the syndrome concept calls for *synergetic measures* able to soothe syndromes in their complex interrelation.

The discussion presented here is rather tentative and cannot yet serve as a baseline for actual policy advice. Yet the syndrome concept in general seems to be rather adequate to investigate the general efficacy of environmental policy measures, especially with respect to any unintended environmental side-effects. A respective research project has been launched at the Potsdam Institute.

4. Sensitivities of Syndromes to Climate

In the previous section we have qualitatively discussed the major driving sources of climate change as they can be systematized within the syndrome concept. It is obvious that this is only part of the story: the feedback of climate on human and natural systems is equally important. As this feedback is directly positioned at the civilization-nature interface, the syndrome-concept is capable of structuring this interaction in a similar way as it does for the climate effects. Yet we have to distinguish between two temporal scale of climate impacts: a shorter one where climate change might act as a *exposition factor* for a syndrome and a longer one which might influence the disposition of a syndrome. A prominent example of the first case is the frequent occurrence of droughts in the Sahel region in the late sixties and early seventies which probably has triggered the vicious cycle between poverty and soil degradation, in our context known as the SAHEL SYNDROME. As the investigation of exposition factors and elements, however, is still more preliminary than for syndromes in general, we henceforth concentrate on the analysis of possible shifts in *syndrome disposition*, i.e., on the sensitivities of the syndrome specific mechanisms, symptoms, and interactions to climate.

If we consider the sensitivity to climate we do not have to distinguish between climate changes on different scales: even a global climate change influences the local or regional systems by its local distinction only, e.g., a global mean temperature change of say 2°C might has a variety of locally different impacts, depending on the actual temperature and/or precipitation change in the region under consideration. We have to remark that the analysis of the sensitivity of syndromes to climate change is part of the general field of climate impact research. But instead of analyzing single sectors it focuses on the functionality of the important patterns of civilization-nature interaction and can thus include the synergetic effects induced by climate change as they shortly have been discussed in the introduction. This highlights some new and possibly rather massive aspects of anthropogenic climate change. We have to stress, however, that, as we did for the effects on climate, we will discuss the subject rather qualitatively. A prototypical quantitative analysis for the sensitivity of the disposition of the SAHEL SYNDROME to climate change see Lüdeke and Moldenhauer (1998).

As can be seen from FIGURE 1 and TABLE 1 most of the syndromes sensitive to climate change are related to the usage of natural resources, especially agricultural goods (crops or pasturing). Most sensitive is probably the SAHEL SYNDROME, which describes environmental degradation due to the overuse of marginal land. One major factor of marginality, however, is the climate, in particular the risk of droughts and/or inter-annual variability in general (Cassel-Gintz et al., 1997). The disposition of the SAHEL SYNDROME has been assessed by the construction of a logical decision tree which has been evaluated using fuzzy-logic (Schellnhuber et al., 1997). Within this assessment it turns out that besides climate, a poor quality of soils, a steep or hilly landscape, and the surface water availability are the major natural determinants for marginality. This means that in case of climate change, soil erosion might be a major synergetic factor for an increase in the disposition towards the SAHEL SYNDROME. Furthermore socio-economic aspects are important: It is the poor and alternative missing subsistence farmer who is most vulnerable to a loss of his natural resources. Therefore a wrong local policy concentrating on the western development paradigm alone is often a major factor behind the syndrome.

TABLE 3 gives a first preliminary overview on possible synergies within syndrome disposition in the course of climate change. Note that these factors also include social and economic trends which might be considered as much more suited for adaptation measures. Further it is important to remark that climate sensitivity does not necessarily imply an increase of the disposition, it might equally lead to its decrease, e.g., in case of a rise in water availability or a general improvement of agricultural growth conditions.

As in the case of the syndromatic effects on climate, the analysis of the interlinkages in terms of climate vulnerability allows to give hints on 'no regret' adaptation options, i.e., with *no* negative side effects. For example, the support for an autonomous economic development in the countries and regions disposed to the SAHEL SYNDROME is not only a good development policy, but also an effective and robust climate change adaptation policy. Contrarily, a highly mechanized progress of agricultural techniques might be good in terms of the DUST BOWL SYNDROME, but might be inappropriate for the poor and marginalized peasants involved in the SAHEL SYNDROME, unable to afford any of the techniques. Vice versa, the provision of irrigation systems might be countermined by climate change and is therefore a risky strategy which has to be supervised carefully.

TABLE 3 Possible synergies in the syndrome dispositions between climate change and other, syndrome relevant, symptoms of Global Change. Note that synergies might be different in sign, i.e., an improvement of one factor might set off the negative impact of the other.

Syndrome	Character of Climate Sensitivity	Possible environmental and socio-economic synergies
Sahel, Green Revolution, Dust Bowl	Geobiophysical growth conditions including surface water availability	Soil degradation, water pollution, further increase of socio-economic disparities, general saturation of agricultural productivity, world prices of crops, etc.
Favela	Climate sensitivity of water and vector-borne diseases, increased frequency of extreme events, etc.	Water and air pollution, missing improvement of the public health system and infrastructure, etc. (Kropp et al., 1997)
Mass Tourism	Climatic 'attraction' of the region, availability of resources, etc.	Water and air pollution, increased socio-economic disparities, globalization etc.

5. Outlook

Future syndrome research with respect to climate change has to fulfill the following tasks:

1. Completion of the set of syndromes under detailed investigation (disposition, exposition, intensities).
2. Comparison with climate models in order to improve expert input of the climate sector processes.
3. Semi-quantitative modeling of core syndromes, including calibration and hindcasting.
4. Closer attention towards syndrome coupling in order to, e.g., prevent unintended consequences of mitigation strategies.

In general it will be necessary to develop the strategic and the political dimensions of the syndrome approach. Many climate policy recommendations up to now suffer from too narrow angles, just focusing upon interlinkages with only few further sectoral aspects. In order to come to a more integrated GC policy it will be necessary to include climate issues in other GC aspects (e.g. soil degradation, water scarcity etc.), The syndrome concept seems to be a very promising novel transdisciplinary approach, able to formulate - at least in the middle range and in qualitative terms - some robust policy options for climate change mitigation.

References

Alcamo, J. (Ed.) (1994) IMAGE 2.0, Kluwer Academic Publishers, Dordrecht.

BMZ (1984) Bundesministerium für wirtschaftliche Zusammenarbeit: *Ökologische Auswirkungen von Staudammvorhaben*, Weltforum-Verlag, Köln.

Cassel-Gintz, M.A., Lüdeke, M.K.B., Petschel-Held, G., Reusswig, F., Plöchl, M., Lammel, G. and Schellnhuber, H.J. (1997) Fuzzy logic based global assessment of the marginality of agricultural land use, *Clim. Res.* **8**, 135-150.

IPCC (1996) Intergovernmental Panel on Climate Change (Ed.) *Climate Change 1995*, Contribution of the Working Group I to the Second Assessment Report of the IPCC, Cambridge University Press, Cambridge.

IPCC (1996a) Intergovernmental Panel on Climate Change (Ed.) *Climate Change 1995*, Contribution of the Working Group II to the Second Assessment Report of the IPCC, Cambridge University Press, Cambridge.

Kropp, J., Lüdeke, M.K.B. and Reusswig, F. (1997) Global Analysis and Distribution of Unbalanced Urbanization Processes by Usage of the Fuzzy Logic Concept: The Favela Syndrome, submitted to International Journal for Systems Research and Information Science.

Lüdeke, M.K.B., and Moldenhauer, O. (1998) The climate sensitivity of the Sahel Syndrome, Manuscript Potsdam Institute for Climate Impact Research.

McCully, P. (1996) *Silenced Rivers*, Zed Books, London.

Meadows, D.H., Meadows, D.L. and Randers, J. (1992) *Beyond the Limits: Confronting Global Collapse, Envisioning a Sustainable Future*, Chelsea Green Press, London.

Nisbet, E.G. (1991) *Leaving Eden. To protect and manage the Earth*, Cambridge University Press, Cambridge.

Oke, T.R. (1973) City Size and the Urban Heat Island, *Atmospheric Environment*, **7**, 769-779.

Pimentel, D., Harvey, C., Resosudarmo, P., Sinclair, K., Kurz, D., McNair, M., Crist, S., Shpritz, L., Fitton, L., Saffouri, R. and Blair R (1995) Environmental and economic costs of soil erosion and conservation benefits, *Science*, **267**, 1117-1123.

Quarrie, J. (Ed.) (1992) *Earth Summit 1992*, Regency Press, London.

Reusswig, F. and Schellnhuber, H.J. (1997) Die globale Umwelt als Wille und Vorstellung. Zur transdisziplinären Erforschung des Globalen Wandels, in A. Daschkeit und B. Schröder (Eds.): *Umweltforschung quergedacht. Perspektiven integrativer Umweltforschung und -lehre*, to appear.

Schellnhuber, H.J., Block, A., Cassel-Gintz. M., Kropp, J., Lammel, G., Lass, W., Lienenkamp, R., Loose, C., Lüdeke, M.K.B., Moldenhauer, O., Petschel-Held, G., Plöchl, M. and Reusswig, F. (1997) Syndromes of Global Change, *GAIA* **6**, 19-34.

Toth, F., Petschel-Held, G., and Bruckner, Th. (1998) Climate Change and Integrated Assessment: The Tolerable Windows Approach, this volume.

WBGU (1993), German Advisory Council on Global Change: *World in Transition. Basic Structure of the Man-Environment Interaction*, Economica, Bonn.

WBGU (1996), German Advisory Council on Global Change: *World in Transition. The Research Challenge, Springer*, Berlin.

WBGU (1997), German Advisory Council on Global Change: *World in Transition. Ways Towards Sustainable Management of Freshwater Resources*, Springer, Berlin.

SECTION II

FROM SCIENTIFIC RESULTS TO THE POLITICAL AGENDA

A COMPARISON OF GHG INVENTORIES AND REDUCTION GOALS FOR DIFFERENT COUNTRIES IN EUROPE*

GORDON McINNES
European Environment Agency
Kongens Nytorv 6, DK-1050 Copenhagen K, Denmark

* The opinions expressed in this paper are of a personal nature and do not necessarily reflect the views of the EEA, the European Commission or any other Community Institute.

1. Introduction

This paper will present an overview of the role of the European Environment Agency (EEA) in environmental information at the European level for policy making and assessment as well as present and analyse some of the latest information on national, EU15 and European emission estimates, projections and reduction goals for the main greenhouse gases (GHGs).

2. European Environment Agency

The European Environment Agency was established in Copenhagen in 1993 under Council Regulation 1210/90 to provide the European Community and Member States with (Article 1(1)): "objective, reliable and comparable information at the European level enabling them to take the requisite measures to protect the environment, to assess the results of such measures and to ensure that the public is properly informed about the state of the environment".

The Agency's work programme is adopted by its management board, which consists of representatives of its 18 member countries (EU15, Norway, Iceland and Liechtenstein), the European Commission (DGXI and JRC) and two nominees appointed by the European Parliament. The work programme is designed to provide information in support of policy making and assessment which describes the present and foreseeable state of the environment in terms of the pressures on the environment and the quality and sensitivity of the environment. Priority is given to a number of areas of work specified in the Agency's regulation, including air quality and atmospheric emissions.

J. Hacker and A. Pelchen (eds.), Goals and Economic Instruments for the Achievement of Global Warming Mitigation in Europe, 99–112.

The Agency has adopted the DPSIR framework as the basis of its work to provide reports, databases and online (Internet-based) services in relation to the main environmental problems including Climate Change. The DPSIR framework connects together the chain of **D**riving forces (or underlying causes of the problems, for example industry and transport), **P**ressures (or the pollutants released by these driving forces), the resulting **S**tate (or quality) of the environment, the **I**mpact these pollutants have on for example human health, ecosystems and buildings to the **R**esponses from society - such as environmental policies and targets (see FIGURE 1).

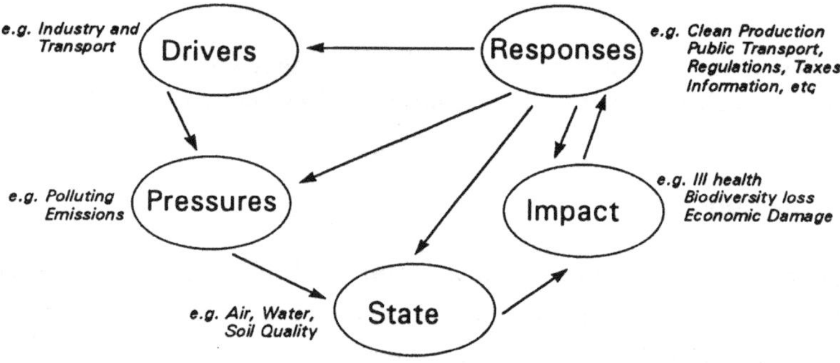

FIGURE 1. The DPSIR framework
Source: EEA

Using the DPSIR framework, the Agency works with experts in the member countries, the European Commission and various international programmes/ activities to connect national monitoring through assessment to reporting. Thus nationally funded monitoring provides the basis for the information and services the Agency provides to the policy agents and the public and, in turn, the Agency's reporting feeds back into improving the quality of national monitoring by highlighting gaps and inconsistencies (see FIGURE 2).

Monitoring	Monitoring, measurement and observation carried out within countries, normally by public authorities
Data	Data and databases resulting from such Monitoring
Information	Summaries, presentation and (statistical) analysis of relevant data from the monitoring networks
Assessment	Analysis, interpretation and presentation of information to explain (parts of) the causal DPSIR chain as the basis of EEA Reporting
Reporting	Publication of information and assessments from the EEA Work Programme

FIGURE 2. From monitoring to reporting
Source: EEA

3. CORINAIR

The Agency has inherited the CORINE databases including Corinair, the Emission Inventory Methodology for Europe, used by over 30 countries in Europe as the basis for complete, consistent, transparent and timely emission estimation and reporting. The Corinair project is being continued through the Agency's Topic Centre on Air Emissions (ETC/AE) which works with national experts as well as experts from DGXI and various international programmes including IPCC and EMEP to help prepare annual emission estimates as required for reporting under various international commitments. These include the UNFCCC and EU Monitoring Mechanism, as specified in the IPCC Guidelines and the UN LRTAP Convention, as specified in EMEP Guidelines and the joint EMEP/Corinair Atmospheric Inventory Guidebook.

Corinair is based on a detailed technology based nomenclature - SNAP, Selected Nomenclature for Air Pollution - which has been developed as SNAP97 to be fully consistent with IPCC and EMEP 1997 Guidelines. Full use of SNAP, the EMEP/Corinair Guidebook and the IPCC Guidelines will provide consistent annual emission inventories (FIGURE 3).

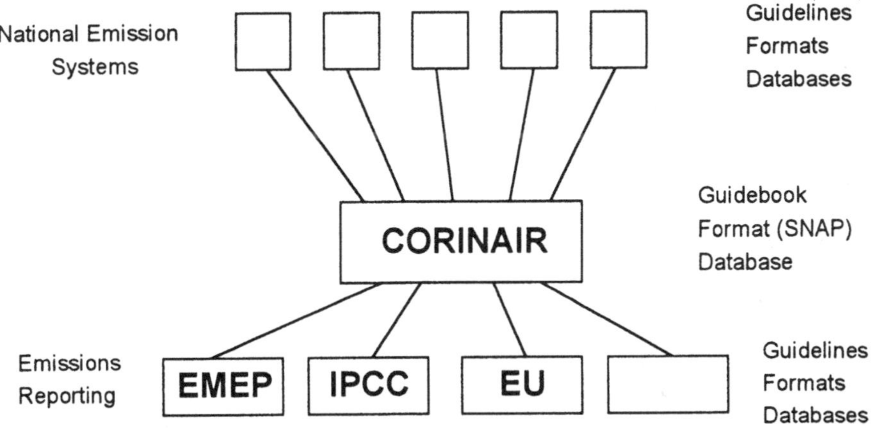

FIGURE 3. Corinair methodology

This should be achieved in most countries by the end of 1997 after several years of cooperation and development between the various international programmes and the member countries. For the moment some inconsistencies remain and these are being assessed by ETC/AE and discussed with the national experts.

4. Current state of Greenhouse Gas Inventories

GHG inventories are compiled by national authorities for reporting to the Climate Change Secretariat under UNFCCC, DGXI under the Monitoring Mechanism and ETC/AE under the Agency work programme. In addition, Eurostat provides estimates of CO_2 emissions from fossil fuels from energy dated supplied by EU member states. To date, there have been differences between these different inventories as understanding and reporting requirements have developed. These differences have been used as a basis for verifying or cross-checking the emission estimates and help to improve the understanding and the reporting guidelines. As indicated above, consistency should be achieved by the end of 1997. In the meantime, extracts from the currently available inventories will be used in this paper to highlight various features of national emissions and reduction goals. The main differences and inconsistencies between inventories will be mentioned where appropriate.

5. National and European GHG emissions

The emission estimates for the three main GHG are presented in TABLE 1. The CO_2 emissions provided in this table can be regarded as the emission targets for stabilisation in 2000, although countries may revise annual emission estimates for 1990 and subsequent years as new information becomes available.

TABLE 1. EU15 greenhouse gas emissions 1990 (Gg) (as required under UNFCCC)
Source: EEA-ETC/AE (1997)

EU 15 member			CO_2 Emission Estimates				
States - 1990 -	CO_2	Removals	CH_4	N_2O	NO_x	CO	NMVOC
Austria	61.876	13.300	587	12	197	1.333	491
Belgium	113.405	2.057	634	31	339	1.127	331
Denmark	52.002	940	426	11	277	772	172
Finland	53.800	-	246	18	295	487	213
France	378.379	33.218	3.017	182	1.909	11.355	3.156
Germany	1.014.155	30.000	5.682	226	2.640.	10.743	3.155
Greece	84.217	NE	353	14	356	1.465	228
Ireland	30.719	5.160	811	29	115	429	180
Italy	436.300	-	3.889	116	2.034	9.258	2.401
Luxembourg	13.300	-	24	>1	23	171	19
Netherlands	152.750	1.500	1.104	51	574	1.072	444
Portugal	44.485	-	660	7	282	976	212
Spain	226.420	23.170	2.181	94	1.164	4.734	1.123
Sweden	55.445	34.368	324	9	336	1.204	526
United Kingdom	610.854	10.377	4.402	113	2.710	6.419	2.419
EU-15	3.328.107	154.090	24.341	913	13.251	51.545	15.071

In TABLE 2 the emission estimates for 1994 and 1995 are presented as they have been submitted within the framework of the Council Decision for a Monitoring Mechanism of Community CO_2 and other Greenhouse Gas Emissions.

TABLE 2. Emissions of CO_2 in 1994 and 1995 (as required under UNFCCC and EU Monitoring Mechanism) and changes since 1990
Source: EEA-ETC/AE (1997)

CO_2	1994 Gg	Change from 1990 (%)	1995 Gg	Change from 1990 (%)
Austria	59 467	-3,9%	62 019	0,2%
Belgium	120 392	6,2%	112 194	-1,1%
Denmark	63 123	21,4%	59 549	14,5%
Finland	59 253	10,1%	56 050	4,2%
France	373 086	-1,4%	385 346	1,8%
Germany	904 500	-10,8%	885 200	-12,7%
Greece	88 298	4,8%	89 929	6,8%
Ireland	33 324	8,5%	33 931	10,5%
Italy	421 052	-3,5%	436 497	0,0%
Luxembourg	11 994	-9,8%	10 001	-24,8%
Netherlands	160 900	5,3%	169 400	10,9%
Portugal	46 871	5,4%	-	
Spain	231 370	2,2%	-	
Sweden	58 500	5,5%	58 108	4,8%
United Kingdom	581 017	-4,9%	547 817	-10,3%
EU-15	**3 213 147**	**-3,5%**	-	-

From 1990 to 1994 CO_2 emissions decreased in a few countries (Germany, United Kingdom, Italy and Luxembourg) to provide an emission reduction of approximately 3 percent for EU15, mainly due to short-term factors like the temporary decrease of industrial and economic growth rates, the restructuring of industry in Germany, the closing of coal mines in the UK and the conversion of power plants to natural gas. However CO_2 emissions increased in most EU15 countries between 1990 and 1994 despite the introduction of some measures including carbon/energy taxes (Denmark, Sweden, Finland, Netherlands). Reporting is incomplete for 1995 but from the results available both further increases and decreases relative to 1990 can be seen.

FIGURE 4 shows that, on the basis of the Commission's business-as-usual scenario, stabilisation of EU15 CO_2 emissions by 2000 relative to 1990 may not be achieved. It will be close but it will not be clear until 2000-2001 whether or not stabilisation has been achieved. FIGURE 4 also indicates that CO_2 emissions may increase by 8% in 2010 relative to 1990 under the business-as-usual scenario.

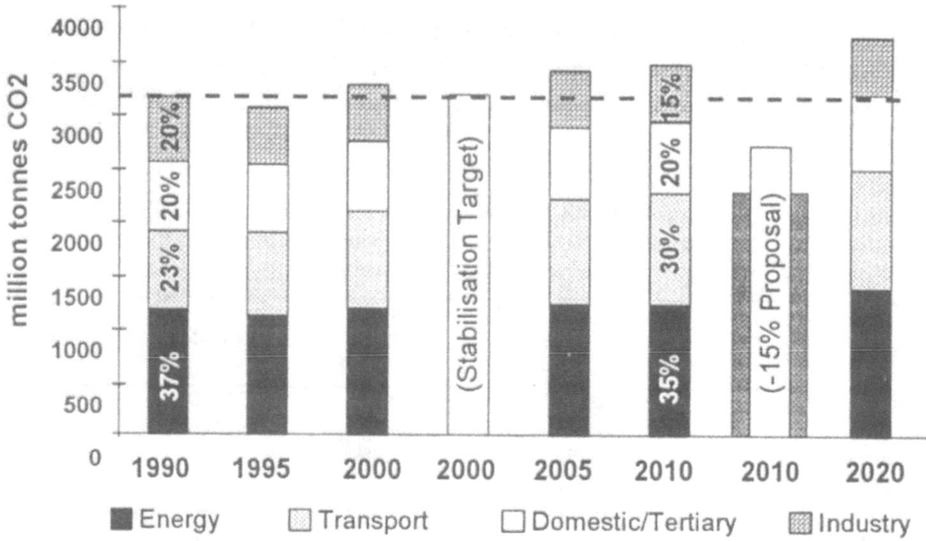

FIGURE 4. CO_2 from fossil fuels EU15 trends (business-as-usual scenario) and targets

FIGURE 5 presents the contribution of the EU Fifth Environmental Action Programme target sectors to total European greenhouse gas emissions in 1994 by pollutant. Greenhouse gas emissions can be expressed in terms of their global warming potential (GWP) using the term CO_2 equivalents (IPCC, 1995).

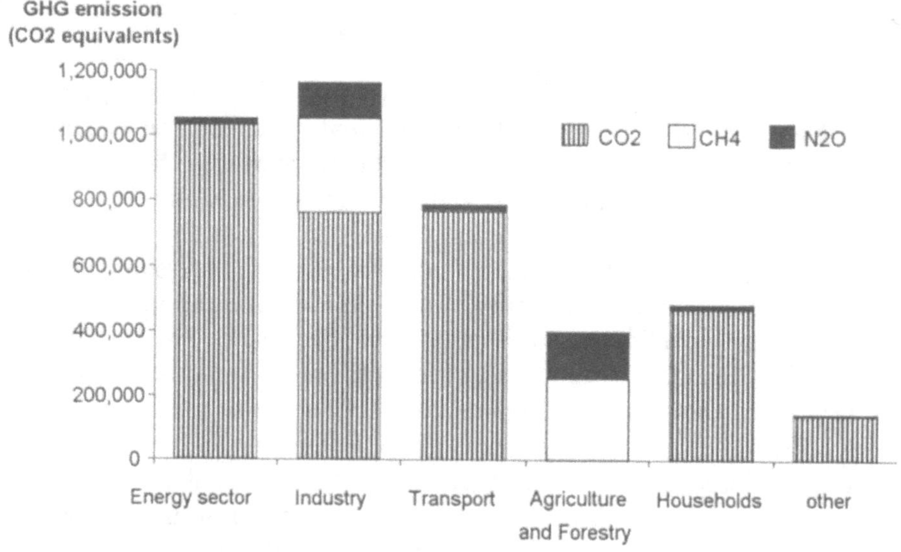

FIGURE 5. Contribution by different pollutants and target sectors (industry includes waste treatment) to greenhouse gas emissions in Europe (EU15, Corinair 94) in 1994

Based on a 100 year time horizon the global warming potentials used here are: CO_2=1, CH_4=21 and N_2O=310 (CO_2 equiv.). Over 80% of the total greenhouse gas emissions (expressed in CO_2 equiv.) is due to CO_2, whereas 12% is in the form of methane and 7% in the form of N_2O. The contribution of the 5 EAP target sectors to the total emission of greenhouse gases is as follows:

- energy sector : 26%
- industry : 29%
- transport : 20%
- households : 12%
- others : 13%

Combustion sources, both stationary and mobile, dominate the greenhouse gas emissions, mostly as CO_2. Significant contributions are from methane emissions from waste treatment and disposal (incorporated in the target sector industry) and methane and nitrous oxide emissions from agriculture and forestry.

Comparison of Eastern and Western Europe in FIGURE 6 shows a few specific differences in the contribution of the source sectors, although the total per capita emissions in both parts of Europe are comparable. Emissions due to the sectors energy and industry in Western Europe tend to be lower, whereas the emissions due to transport in Western Europe tend to be higher as compared to Eastern Europe. This is partly due to the higher energy efficiency and the higher road traffic volumes in Western Europe. The average total per capita greenhouse gas emission in Europe in 1990 was about 12 Mg CO_2 equiv./capita/year. Comparison between Northern and Southern countries indicates that emissions are lower in southern Europe due to higher temperatures (and hence less space heating), lower use of cars or lower industrial activity.

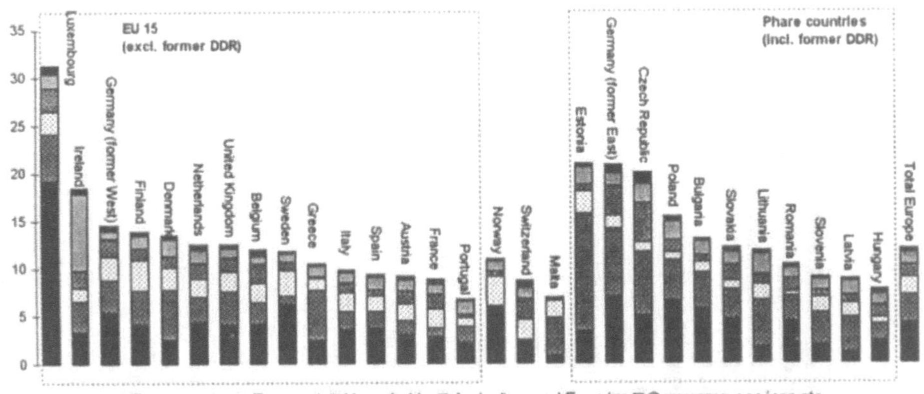

FIGURE 6. Per capita total greenhouse gas emissions in Europe in 1990 by target sector (Mg CO_2 Equiv./capita/year)

Figure 6 is reproduced additionally as Figure J in the colour section of the book.

FIGURE 7 shows the CO_2 emissions from combustion of fossil fuel for EU15. The global average annual per capita emissions of CO_2 due to the combustion of fossil fuels is at present about 4 tonnes, in EU15 about 9 tonnes and in the USA about 20 tonnes. Within EU15, only Portugal, Spain and Sweden have CO_2 per capita emission rates similar to the global average.

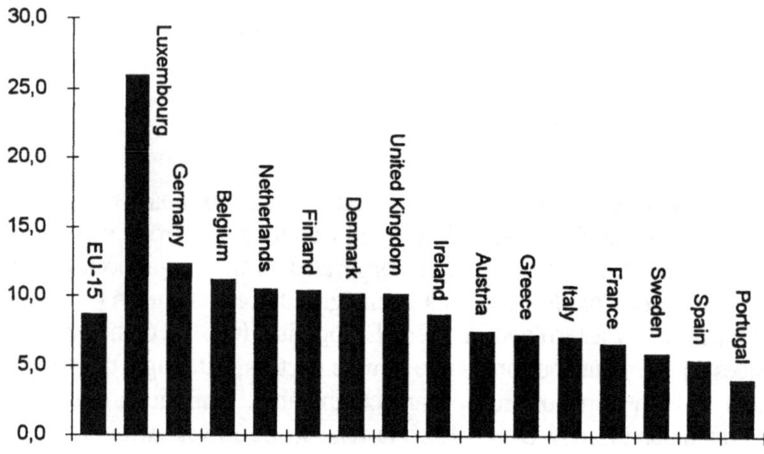

FIGURE 7. Per capita emissions of CO_2 emissions from fossil fuel in EU15 in 1990 (Mg CO_2/capita/year), Source EU15: Eurostat

FIGURE 8 presents CO_2 emissions per unit GDP (kg/$/year) for a range of European countries and shows that in general the emission rates are higher in eastern Europe than in western Europe since, as indicated above, total emissions are similar but GDP is much lower in eastern countries.

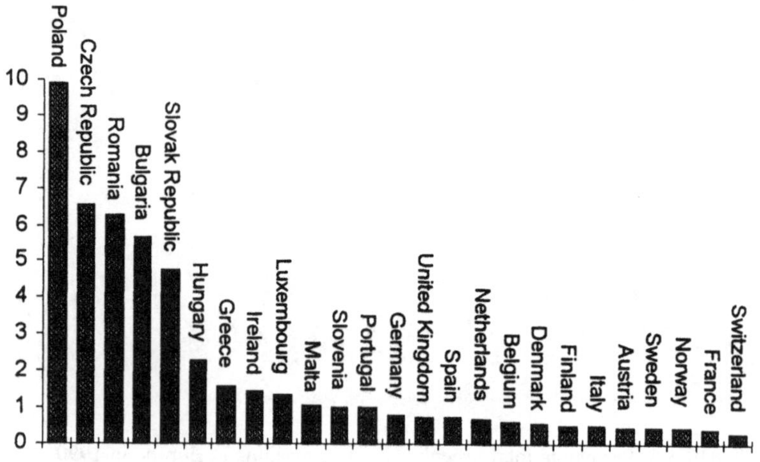

FIGURE 8. CO_2 emissions per unit GDP in Europe in 1990 (kg CO_2/$/year)

6. GHG emission reduction targets

The EU Council of Ministers in June 1996 concluded that a protocol or another legal instrument should be agreed under the UNFCCC including commitments for Annex-I Parties regarding policies and measures and quantified emission limitation and reduction objectives within specified time frames as called for in the Berlin Mandate. Other conclusions were:

- global average temperatures should not exceed 2 degrees Celsius above pre-industrial level and concentration levels should not exceed 550 ppm CO_2. This means that the concentrations of all GHGs should also be stabilised;
- it is feasible for the Community as a whole to reach reduction of CO_2 emissions by the year 2010 compared to 1990 level, through the implementation of policies and measures identified by Member States and the Commission, at national and Community level.

In March 1997 the Council of Environment Ministers proposed that developed countries should reduce greenhouse gas emissions to 15% below 1990 levels by 2010. The target is based on the combined reduction of the main GHGs (CO_2, CH_4, N_2O), taking into account their global warming potential. Current commitments (see TABLE 3) from Member States would enable the EU as a whole to reduce its emissions by 10% by 2010.

TABLE 3. EU Member States GHG emission targets for 2010

Member State	GHG emission target 2010 (% change from 1990 level) for CO_2, CH_4, N_2O together
Austria	- 25
Belgium	-10
Denmark	-25
Finland	0
France	0
Germany	-25
Greece	+30
Italy	-7
Ireland	+15
Luxembourg	-30
Netherlands	-10
Portugal	+40
Spain	+17
Sweden	+5
United Kingdom	-10

Some EU Member States would be allowed to increase their emissions because this would be off set by decreases by other Member States. Further policies and measures will be identified to enable EU countries to deliver an overall reduction of 15% by 2010, should developed countries agree to that target at Kyoto.

The result of these targets is summarised in FIGURE 9 which compares the 1990 and 2010 emissions in CO_2 equivalents per year for each of the member states. Emission rates become more consistent across EU15 but remain significantly above the current global average.

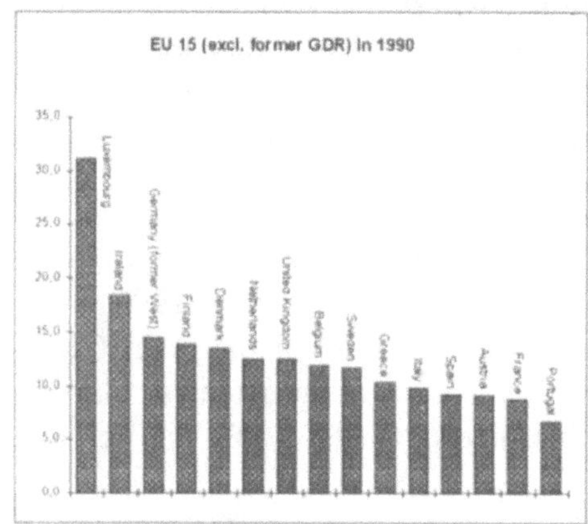

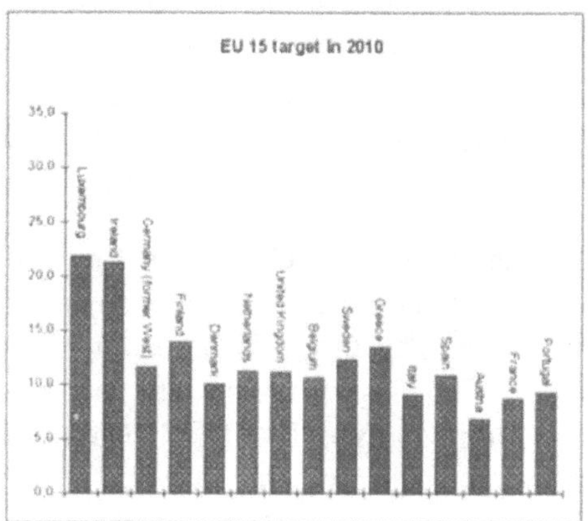

FIGURE 9. Per capita total greenhouse gas emissions in EU15 in 1990 and 2010 (Mg CO_2 Equiv./capita/year)

The current national targets deliver only a 10 percent reduction by 2010. A further 5 percent reduction will need to be found from additional national and common and coordinated Community policies and measures including the following eight types of measure (type 9 on sinks has been included to link to TABLE 4):

1. renewable energy programmes
2. more use of combined heat and power
3. energy efficiency improvements
4. limitations of emissions of other GHGs (including HFCs, PFCs and SF6)
5. reduction of CH_4 and N_2O emissions
6. progressive reduction/removal of fossil fuel and other subsidies, tax schemes and regulations which counteract an efficient use of energy
7. increases of minimum tax levels as a way to reduce GHG emissions and improve energy efficiency and
8. modal switch in transportation and
9. enhance sinks (mainly through afforestation)

TABLE 4 summarises which of these nine types of measures have been adopted or proposed in each of the Member States to reduce or restrain growth in GHG emissions. In some but not all cases, the effects of the adopted and proposed measures have been quantified and made available in national communications. It can be seen that almost no measures to address other GHGs or to reduce/remove barriers to improving energy efficiency have been considered (although it is possible that this latter type of measure may be subsumed under other measures, for example improving energy efficiency).

TABLE 4. Main types of measure adopted or proposed in each EU Member State

Member State	1	2	3	4	5	6	7	8	9
Austria	X						X		X
Belgium	X	X	X			X	X	X	
Denmark		X	X	X	X		X		
Finland	X	X	X		X		X	X	X
France			X				X	X	X
Germany	X	X	X	X	X		X	X	X
Greece	X	X	X					X	X
Ireland		X	X		X			X	X
Italy		X	X				X		X
Luxembourg[A)]		X	X				X	X	
Netherlands	X	X	X	X	X		X	X	X
Portugal		X	X				X	X	X
Spain	X	X	X				X	X	X
Sweden	X	X	X		X		X		X
United Kingdom		X	X	X	X		X		

NOTE: [A)] Luxembourg will also restructure its steel industry which is its major source of CO_2

However, even a 15 percent reduction in EU15 GHG emissions may not be a sufficient contribution to maintain climate change within sustainable limits.

Assuming that the EU maximum temperature objective refers to the year 2100, allowable emissions can be computed for the year 2010 (see TABLE 5). The results depend on the allowable temperature increase per decade. Scientists have proposed to use a provisional limit for sustainability of 0.1°C temperature rise per decade and a provisional limit of a 2 cm rise in sea level per decade to prevent damage to coastal zones, wetlands and coral reefs caused by too rapid climatological changes. Increases above this limit will cause major risks for ecosystems, food production and sensitive coastal areas.

TABLE 5 shows that if the sustainable limit of 0.1 °C temperature increase per decade up to 2100 is added as a climate target, emissions in industrialised countries (Annex-1 countries, including EU) should be reduced by *at least* 30-

55% in 2010 compared to 1990 levels, depending on the trend in emissions in the non-industrialised countries.

This would for example mean for CO_2 emissions from fossil fuel a reduction of the annual average per capita emission (EU15) of 8.8 tonnes in 1990 to 5.8 and 3.7 tonnes respectively in 2010 (accounting for some increase in population).

The ranges of allowable global emissions in 2010 (the so-called emission corridors) include the minimum and maximum reductions according to the timing of policy measures. Following the upper limit of the emission corridor maximum actions are required after 2010 (i.e. *global* reductions of emissions of 2% per year). Only the maximal allowable emissions in 2010, according to the upper limit of the corridor, are presented in TABLE 5. If the lower limit of the corridor is followed, alternative strategies (including lower emission reductions) are still open after 2010. The table also shows the emissions corridors for the considerably higher rate of temperature increase per decade (0.15 °C/decade).

TABLE 5. Maximum allowable CO_2 equivalent emissions scenarios and associated minimum emission reductions for Annex-1 Parties in 2010
(based on EU objective of 1.5°C temperature increase between 1990 and 2100)

Required rate of temperature increase 1990-2100[a]) (°C/decade)	Global emission corridor in 2010 (Gt C CO2 equiv.)	Max. allowable emissions Annex-1 Parties [b]) in 2010 (index 1990=100)	Min. emission reductions Annex-1 Parties[b]) in 2010 (index 1990=100)
0.1	7.6 - 9.5	45 - 70	30 - 55
0.15	7.6 - 12.3	90 - 120	[c]) - 10

a) Including (unavoidable) violations of the temperature increase concerned in two decades. A temperature increase of 0.1 degrees Celcius per decade could be regarded as a limit for sustainability. An increase of 0.15 degrees Celcius per decade is substantially above this level

b) Range presents non-Annex-1 Parties baseline emissions of 5.3 - 7 Gt C equivalent CO2 in 2010 and only includes the upper limit of the emission corridor

c) The lowest baseline emission value for non-Annex-1 Parties means that in this case (upper limit of emission corridor) that no emission reduction by Annex-1 Parties is required

Source: IMAGE computations by RIVM using the so-called safe landing method (Alcamo and Kreileman, 1996)

7. Conclusions

The European Environment Agency and its Topic Centre on Air Emissions is working with national experts in over 30 European countries to improve the quality and consistency of GHG emission estimation and reporting. Some uncertainties and inconsistencies remain but are being gradually removed and consistent reporting should be achieved by most countries by the end of 1997.

The 1990 baseline for national emissions of the main GHGs has been established although minor adjustments may continue to occur as new information becomes available and until full consistency is achieved across European countries. The use of fossil fuels is the main source of CO_2, a major source of CH_4 and a minor contributor to N_2O. Industry and agriculture are major contributors to CH_4 and N_2O.

CO_2 emissions in Southern European countries are lower than in the north due to higher temperatures (and hence less space heating), lower use of cars or lower industrial activity. CO_2 emissions are similar in Western and Eastern European countries since emissions from lower transport activity are offset by poorer energy efficiencies in eastern countries.

EU15 contributes significantly above global average per capita emission rates to the main GHGs. Current per capita CO_2 emission rates in Portugal, Spain and Sweden are similar to the global average per capita rate but increases are being proposed in these countries through to 2010. CO_2 emission rates per unit GDP tend to be higher in Eastern Europe than in Western Europe.

The European Community is preparing its negotiating position for Kyoto to reduce GHG emissions by 15% in 2010 relative to 1990. This will reduce EU15 emissions from 4122 million tonnes CO_2 equiv. in 1990 (of which 3328 million tonnes or 81% is CO_2, 12% is CH_4 and 7% is N_2O) to 3504 million tonnes CO_2 equiv. in 2010. Per capita emission rates in 2010 should be more consistent across EU15 since countries with relatively high rates will reduce their emissions while those with relatively low rates will increase their emissions. However, the EU15 per capita emission rate will remain above the current global average emission rate.

EU15 member states have adopted or are proposing to adopt a range of recognised measures to reduce or restrain their GHG emissions after 2000. These are mainly focused on use of alternative fuels, energy efficiency improvements and modal shifts in transportation. Measures to reduce non-CO_2 GHGs and to remove/reduce barriers subsidies and other schemes which counteract an efficient use of energy are not so well developed or not explicitly reported in national programmes.

Since CO_2 (and other GHG) emissions will rise under the business-as-usual scenario, significant additional measures will required to reverse this rise and help achieve a 15% reduction in GHG emissions in EU15. Further reductions, if they prove to be necessary to limit climate change to sustainable levels, will be even more difficult.

The European Environment Agency will be monitoring and reporting on progress.

ALLOCATING EMISSION REDUCTION TARGETS IN THE EUROPEAN UNION - ISSUES AND PROPOSALS

Dr. UTE COLLIER
Friends of the Earth Trust
26-28 Underwood Street, London N1 7JT, United Kingdom

Acknowledgements: This paper is based on the results of a study carried out by the author and Dr. Michael Grubb (Royal Institute of International Affairs, London) for DG XI of the European Commission. The study was commissioned when the author was working as a Research Fellow at the European University Institute in Florence, and the opinions expressed in this paper do not necessarily represent the views of Friends of the Earth.

1. Introduction

There are large variations in per capita greenhouse gas emissions between the signatory countries of the UN Framework Convention for Climate Change (FCCC). Even amongst industrialised nations (Annex 1 countries), emissions, as well as projected abatement costs, vary considerably. As the negotiations move towards the adoption of legally binding emission targets at the Kyoto Conference of the Parties in December1997, one of the main dividing issues between parties continues to be the choice between 'flat rate' and differentiated targets.

The issue of differentiation involves a complicated set of issues related to questions of effectiveness, equity and efficiency. Even the parties who support differentiation have differing views on which criteria are appropriate for allocating differentiated emission targets. The European Union (EU) has always argued for differentiation among its 15 member states but its current negotiating position is a result of political bargaining rather than a formula based on equity and efficiency principles.

The aim of this paper is to examine the most important issues associated with allocating emission reduction targets and to discuss the scope for developing indicators for differentiation. Its focus is on the EU and its member states but international developments are also highlighted. The paper starts with an analysis of the rationale for differentiation before discussing different approaches to defining indicators.

J. Hacker and A. Pelchen (eds.), Goals and Economic Instruments for the Achievement of Global Warming Mitigation in Europe, 113–127.

Proposals for a sectoral indicator approach are examined. The results of the analysis and the conclusions highlight the problems associated with the definition of fair and transparent indicators. The focus of the paper is on energy-related CO_2 emissions, as data is more extensive and reliable for these emissions than for other gases.

2. The rationale for differentiation

Flat rate emission targets, while attractive because of their simplicity, are not necessarily the most economically efficient means of achieving emission reductions. As numerous studies have shown, different countries face different marginal costs of CO_2 abatement (see e.g. Kram and Hill, 1996). Flat rate targets do not minimise overall costs. There thus have been suggestions that mitigation needs to aim at overall minimum economic cost and equalise marginal costs of abatements in all countries. According to Greene (1996), such allocative efficiency aspects have not overly influenced target setting in other environmental agreements (although within the EU, the Large Combustion Plant Directive provides a precedent of some sort). However, as climate change abatement threatens to affect economies much more than in the case of other environmental problems (e.g. acid rain or ozone depletion), the cost allocation issue is even more pertinent in the case of climate change and needs to be addressed. As Lunde (1995) points out, the higher the political and economic stakes in international negotiations, the more likely are nations to think and act in terms of relative gains and losses.

Flat rate targets are also not equitable, as the abatement costs faced by countries are not necessarily related the overall burden they impose through past and present emissions. In some countries, the achievement of a certain target may be politically infeasible, as past efforts might make further emission reductions particularly difficult and politically sensitive (e.g. if they would have to come mainly from the transport sector which shows continuous fast growth). Finally, flat rate targets are not necessarily the best outcome in environmental terms. There is a danger that agreement can only be reached on a 'lowest common denominator target' which will require very little from some countries and is unlikely to lead to the kind of target needed to reduce emissions by the levels recommended by the IPCC. Alternatively, countries may agree on a target knowing that they would be unlikely to achieve it.

Differentiation is thus in principle preferable for reasons of:

- allocative efficiency
- equity
- effectiveness

While differentiation might be preferable in principle, in practice it poses considerable problems. Indicators have to be chosen as a basis for differentiation, a process which can be politically difficult. There is a growing literature on the idea of burden sharing and definition of potential indicators (see e.g. chapter 3 in IPCC, 1996, Grubb et al, 1992, Hayes and Smith, 1993, Ridgley, 1996), although not all of this discussion is relevant to the EU context. Much of the global discussion has centred around the concept of equity, considering issues such as responsibility for past and current emissions, ability to pay for abatement, future allowable emissions and intra-generational equity. There are a myriad of ways for measuring equity and, as Kawashima (1996) has pointed out, even if equity principles could be agreed on, theoretical ideals of equity can only serve as an allocation mechanism in the long term, as in the short-term, emissions have to be reduced, at least to some extent, within existing infrastructures, as otherwise immense economic costs could arise.

3. Proposals for differentiation for the Kyoto negotiations

The climate convention provides a basis for differentiation arguments. Article 3.1 states:

> *The Parties should protect the climate system for the benefit of present and future generations of humankind, on the basis of equity and in accordance with their common but differentiated and respective capabilities.*

The main purpose of the inclusion of this article was to ensure that developed countries would recognise their historic responsibility and take a lead in greenhouse gas abatement. However, the artcile has been increasingly referred to as a justification for differentiation among Annex 1 countries.

A number of Parties have submitted proposals for target differentiation in the run-up to Kyoto. Not surprisingly, differentiation has been proposed by those parties who would fare badly under a flat rate approach. Most of the Parties (France, Iceland, Norway, Japan, Korea and Switzerland) have made proposals which benefit countries with low per capita emissions, while others (Australia, Iran) are looking for exemptions for exceptionally high per capita emissions. In general, it can be said that the suggested differentiation formulae tend to be particularly advantageous to the Parties who have proposed them.

TABLE 1 summarises the various differentiation criteria under the current proposals. As is shown, Australia and Iran are insisting on the largest number of

TABLE 1. Differentiation proposals under UNFCCC

Country/ Criteria	CO_2 per capita	CO_2 per unit of GDP	GDP per capita	share of/ potential for renewables	cumulative emissions of GHG	emission intensity of exports	fossil fuel intensity of export	defence budget	population growth	special circum-stances
Australia	✔	✔	✔(growth)			✔	✔		✔	
France	✔									
Iceland & Norway	✔	✔	✔	✔						
Iran			✔(growth)	✔	✔		✔	✔	✔	✔
Japan	✔									
Korea	✔(equal right solution)	✔(inverse GDP elasticity of GHG)	✔(equal capability solution)		✔					
Poland	✔	✔	✔		✔					
Switzer-land	✔									

criteria, with the Iranian 'special circumstances' leaving room for exemptions based on any criteria that suits a specific Party. Norway would do quite well under two of its proposed indicators (CO_2 per capita and CO_2 per unit of GDP) but not under the GDP per capita indicator. Per capita CO_2 emissions is the indicator favoured by the largest number of Parties. Three Parties (France, Japan and Switzerland) favour per capita CO_2 as the only indicator which would exempt some of the richest countries in the world from action. Countries with high per capita emissions (especially the US) are unlikely to agree to per capita differentiation.

It seems unlikely that the parties will be able to agree on a differentiation formula in the short negotiation time left until Kyoto. The Convention on Long Range Transpoundary Air Pollution (LRTAP) set some kind of precedent, although the issues involved were in many ways less complex. Under LRTAP, the Parties initially accepted flat rate reduction targets for SO_2 and NO_x. Subsequently, it took 12 years to agree differentiated commitments for further reductions. At the time of writing (September 1997), a flat rate target with trading and joint implementation thus appeared a more likely outcome of Kyoto than differentation. However, the differentiation issue will continue to have currency, especially as future reduction targets will need to become increasingly stringent. Also, at EU level, the differentiation issue is far from solved.

4. The EU and differentiation

The EU is no newcomer to the differentiation debate. Differences in emission characteristics and economic circumstances are not as broad in the EU as at they are at the global level, but they are nevertheless substantial enough to make it more difficult and/or expensive for some member states to achieve emission reductions than for others. It might improve the performance of the EU as a whole if efforts were concentrated in those countries and those sectors in which emission reductions are most easily and cheaply achieved. Furthermore, issues of fairness and equity also play a role in the EU context and the less developed member states have long argued for less stringent obligations and/or compensation within an EU 'burden-sharing' arrangement.

Differentiation and target sharing were briefly considered as an option in 1990 during the discussions about a common target for stabilisation. However, in the end the target was the outcome of an uneasy agreement based on existing national targets, as attempts to set up a 'burden-sharing' arrangement failed. The current inter national negotiations for legally-binding post-2000 reduction targets raise the issue of target-sharing again and more forcefully, placing the EU under considerable pressure to illustrate how reductions would be shared among member states.

In March 1997, the Council adopted a negotiating position a reduction of 15% from 1990 levels by 2010, and reached agreement upon a distribution of emission reductions between member states, with indicative targets ranging from minus 30% (Luxembourg) to plus 40% (Portugal).

The full Council agreement, whilst of great significance, does not fully resolve the issue of target-sharing. The individual targets only amount to an EU reduction of 10% from 1990 levels by 2010 for a basket of three gases (CO_2, CH_4 and N_20); and they are only indicative, not having any legal force or redress against governments that are applying softer and perhaps unrealistic criteria. The remainder of this paper presents a new approach to EU differentiation based on a project carried out for DG XI of the European Commission.

5. Criteria for allocation indicators

Analysis of indicators is greatly facilitated by a clear definition of the *criteria* against which indicator proposals might be assessed. If initial allocations are made tradeable, then efficiency can be achieved through trading of the initial obligations; which is one mechanism under discussion in the Kyoto Protocol negotiations. Allo-cation then however determines the distributive aspects of the agreement, making it of central importance to the acceptability of such a system. For indicators to be of use in the allocation negotiating process, we suggest they need to meet four main criteria:

1. They must be *defensibly fair and politically feasible*. Indicators are unlikely to be completely fair according to several different possible interpretations of 'fairness' simultaneously. What matters is that they be defensibly fair, so that negotiators feel they have obtained a reasonable allocation that can be accepted as fair enough in subsequent domestic ratification debates. For this reason, defensibly fair is very closely related to the question of political feasibility; governments will not agree to something they feel is not sufficiently fair.

   A specific aspect of this may be related to 'North-south' divisions in the Union. Indicators that result in allocations exceeding 'business-as-usual' emissions in richer, northern countries and being lower than 'business-as-usual' emissions in the poorer southern countries (which we here call 'converse') might be considered unfair and cause political difficulty, even if they correctly reflect differing efficiencies.

2. Indicators should be *transparent*. It is vital that negotiators and governments can understand the basis upon which the allocation is decided and can explain that to others.

3. Indicators should seek to *minimise the number of arbitrary assumptions*. The more arbitrary assumptions involved, the more protracted is likely to be the process of negotiation, and the more difficult it may be to justify the results. The negotiation process may introduce some ad-doc adjustments to meet political imperatives, but this should be minimised in the basic indicators.

4. They must seek to *reflect agreed trends*. Indicators for the year 2010, for example, should not be based wholly on what is visible in 1990 or 1995, but nor can they rest heavily on controversial economic or other projections. They should to the extent possible reflect *agreed trends*. The main examples in the EU would appear to be:

   - Differential population growth (higher in some southern European countries)
   - Differential economic growth (higher in the Cohesion countries)
   - Single European Market (market harmonisation)
   - Energy systems liberalisation (Electricity Liberalisation Directive, June 1996 and ongoing negotiations on gas liberalisation).

6. Reflecting national differences

Looking at the different national circumstances in EU countries, we can draw up an approximate list of issues that need to be taken into account when attempting to find a fair and efficient solutions:

- *past effort:* e.g France's CO_2 emissions have declined by 25% from 1973 to 1990, Swedish ones by 40% from 1970 to 1990, due to the expansion of nuclear power and energy efficiency efforts;
- *scope for further reductions*: this is to some extent related to past efforts, but also to other variables such as renewable energy potentials, where e.g. the UK is much better placed than Germany;
- *proportionality*: responsibility for largest shares of past and present emissions, both in total terms (Germany and the UK are in the lead here) and on a per capita level (Luxembourg, Germany, Denmark, Belgium);.

- *current level of economic development*: the cohesion countries - Spain, Greece, Ireland and Portugal - are lagging behind, economic growth to catch up with the rest of the EU will invariably mean some emission increases, furthermore restricted ability to pay for expensive abatement options
- *role of energy intensive industries in the national economy:* they are especially important in Belgium, Germany, Luxembourg and the Netherlands, severe emission limitations raise issues of national competitiveness;
- *reliance on domestic coal resources:* e.g. Spain, Germany, implications in social (employment) terms as well as security of supply, balance of payments;
- *social and institutional dimensions*: e.g. varying distributive effects of policy measures, different public perceptions of the climate change problem, Southern countries are usually said to have weak environmental institutions;
- *other factors:* e.g. climatic differences and heating demand (North-South divide), size of country and population densities (affecting transport demand);
- *the likely economic and social burden imposed by climate change impacts:* e.g. the Netherlands would suffer most from sea level rises, Spain from temperature increases, but there is a high uncertainty factor.

Quite obviously, a rather complex formula would be needed to take into account all these differences and to achieve efficient and equitable differentiation. As the next section discusses, most proposed differentiation approaches fall far short from such an ideal.

7. Approaches to defining indicators in the EU

Equal per capita entitlements are in the EU most strongly supported by France and Sweden. France has suggested stabilisation on a level of 7.5 tonnes of CO_2 per capita (tCO_2/cap, equivalent to 2.05 tC/cap). In the EU, under such an approach, the largest percentage emission reductions would have to take place (in descending order) in Luxembourg, Germany, Denmark, Belgium, Finland, the Netherlands and the UK. However, two of the largest emitters, France and Italy would not be required to reduce their emissions and could even increase them, which would put a disproportionate burden on the other member states if an overall target for the EU of 5 or 10% reduction was to be achieved.

A CO_2 emissions per unit of GDP approach would hit some of the cohesion countries (Greece and to some extent Ireland and Portugal) particularly hard, which would not be acceptable on equity grounds. CO_2/GDP figures cannot reveal whether high CO_2 emissions are due to e.g. inefficiencies in production processes or simply reflect a pre-dominance of energy intensive industries. In the cases of countries with energy-intensive industries, national competitiveness could be threatened if the required reductions were too severe. Again, France and Italy would fare well with the CO_2/GDP approach. While it is true that in France, emission reduction opportunities are more limited (at least in the electricity sector), in Italy there is a clear reduction potential in the electricity sector which is almost entirely fossil fuel based.

The per capita or the GDP approaches are thus rather crude measures of national performance and fail to adequately reflect issues such as reduction potentials or meet criteria of cost-effectiveness. Nor are they necessarily 'fair', considering they would exempt two of the richest countries in the EU from action. There is thus a need for an approach which provides a clearer reflection of national situations.

Indicators could be based on criteria of *cost* distribution. However, costs would have to be evaluated on the basis of projected trends and abatement cost estimates using the result of economic models. This further complicates the assessment of relative burdens, as different models, based on different assumptions, can yield very different results. There is most unlikely to be agreement on the correct model or assumptions to use. This approach also clearly violates the criteria of transparency.

The recent EU negotiations on burden-sharing were in part stimulated and supported by the *Triptique* analysis carried out by a team of reasearchers at the University of Utrecht on behalf of the Dutch government (Blok, Phylipsen and Bode, 1997). However, this included too many arbitrary assumptions and member states made it plain that they did not accept this as a formal basis for target-sharing and some of the national targets were only weakly related to those arising from the Triptique analysis. At the same time, the ad-hoc basis of country-by-country target-sharing is certainly not one that could be repeated across the full Annex 1 group.

8. The case for sectoral differentiation

This paper argues for a disaggregation of national emission characteristics through the use of sectoral indicators. While this was also done in the triptique analysis, we use different approaches for the industry and electricity sectors. Different sectors of activity exhibit different growth patterns in CO_2 emissions, ease of control and institutional characteristics. In some cases, sectors exhibit large similarities between countries, while in other they differ substantially. Disaggregating among them may make reaching agreement easier by breaking the negotiations down into more manageable chunks. Sectoral analysis concerning the electricity and manufacturing sectors could also open possibilities for negotiated agreements with relevant industries. Furthermore, sectoral indicators could provide the basis for the allocation of tradeable emission permits at an industry level.

Sectoral analysis can then either be used as a basis for estimating differentiated total national obligations, or to define national obligations for several sectors separately. However, the latter approach is likely to have some disadvantages in that it reduces the flexibility countries have, in particular in the light of unforeseen developments. However, this issue is clearly important for the electricity sector in which cross-border trade is becoming increasingly important.

While such an approach is a distinct improvement on using simple economic indicators, there are nevertheless drawbacks. First, the available data are sometimes unsatisfactory. Secondly, it is unlikely that any simple numerical index, even if consisting of a number of sectoral indicators, will capture all the factors that affect the distribution of emissions and comparative costs of abatement, still less the political differences which affect national perspectives and willingness to negotiate emission commitments. Thirdly, there are many ways to construct indices (i.e. which data to use, how to weight individual indicators), which could result in lengthy discussions and further delays to finding an EU agreement on emission reductions. It could also tempt countries to advance their own preferred solution through another, apparently more 'rational' means. Notwithstanding these limitations, sectoral indicators could serve as an aid to negotiations by providing what is accepted as an objective indication of real differences, and thus a focal point on which differentiation negotiations can rest.

The following section provides a summary of the sectoral approach developed by the author and Michael Grubb (Grubb and Collier, 1997).

9. Sectoral indicators - a possible approach

9.1. DOMESTIC AND COMMERCIAL SECTOR

An underlying principle of long-run convergence of per-capita emissions is proposed, leading to a total EU 30% reduction in these sectors by 2030 (15% by 2010). Currently, there are considerable differences in domestic sector per capita emissions (less so for the commercial sector) due to different heating requirements and differences in appliance ownership and use. However, it is assumed that many of these differences will disappear in the long-run. Heating requirements in Northern Europe will converge as energy efficiency programmes greatly reduce energy needs in countries which are currently inefficient. Meanwhile, in Southern Europe the growth of air conditioning is expected to cancel out lower heating requirements on climatic grounds.

9.2. TRANSPORT EMISSIONS

Transport energy demand depends upon population, wealth, car efficiency, trip distance, town planning, and the provision and quality of public transport. Conestion and constraints on time, and environment suggest that transport demand is unlikely to continue unconstrained expansion with rising affluence, making an explicit link with GNP inappropriate. Two main indicators could be considered:

- Emissions per passenger-kilometre (private and passenger transport) and tonne-kilometre (freight). However, data, both in projections and retrospectively, may be a serious problem.

- The alternative is a simple criterion of per-capita convergence. However, it may be considered as penalising countries with large travel needs arising from large land area (as noted for Sweden). Corrections could be considered, most probably for land area or more specifically, population density.

We chose the later option because of unavailability of data. A convergence of per capita emissions to the 1990 level was assumed, reflecting the huge growth rates assumed in business-as-usual scenarios (+49% by 2020). Real reductions might be difficult to achieve.

9.3. INDUSTRIAL EMISSIONS

The defining context for industry is the Single European Market. Free trade across Europe should lead to continued convergence of emissions intensity in each specific sub-sector, for which intensity may be defined as emissions per unit of value added or per unit of physical output (eg. tonnes of steel). Differences in aggregate energy intensity reflect very different industrial structures, which as noted have persisted and are likely to continue to do so for reasons of physical geography (eg. transport links) and economic and cultural differences.

We therefore propose that indicators should be based on *convergence of energy intensities*. The simplest measure would be an aggregate index of carbon emissions per unit of manufacturing GNP. Probably however this is too simple because of its failure to reflect structural differences. Emissions per ton of produced energy intensive material is more disaggregated and needs much more data, but it could just focus on the five most energy-intensive sub-sectors, ie. paper & pulp, chemicals, stone, clay & glass, ferrous metals, and non-ferrous metals. Thus we propose that these industrial sectors be separated (according to standard SEC categories) and that allocation is based upon convergence of subsectoral physical energy intensities:

Allocation = SUM ($I_n W_n$) + $I_r E_r$

where I_n is the physical emissions intensity of major energy using subsector *n*
W_n is the physical output of that subsector
I_r is the emissions intensity of all remaining subsectors relative to economic value-added
E_r is the value added from those remaining subsectors

Numerical application of this proved complex. Comparison of two sources - Eurostat and UN data - revealed differences in energy intensities, in some cases very large ones, and neither set appears fully consistent; nor are data on sub-sectoral carbon intensities available. Also, the available data suggest surprising wide variations in energy and carbon intensities even at the subsector level in 1990, with very high intensities in Greece. This means that moving too quickly to equalise sub-sectoral intensities would be converse (ie. requires poorer countries to do substantially more). For these reason, we propose again linear convergence on 2030, this time of sub-sectoral carbon intensities.

9.4. ELECTRICITY SECTOR

Here the defining context is electricity market liberalisation. It is argued that there is likely to be substantial convergence of per-capita emissions from power generation. The lower per-capita emissions from Cohesion countries are likely to grow with increases in demand; those from Sweden in particular may rise with decline in nuclear power; whilst the very high per-capita emissions in Denmark, Germany and the UK are likely to reduce. A convergence to 30% of 1990 CO_2 emissions is assumed by 2030.

This proposal is clearly controversial. However, if emission reductions are to be achieved, the electricity sector has to play a major role and it is in this sector where there is the greatest scope for fast emission reductions (as has already happened in the case of the UK).

10. Some results

In order to make our proposals comparable with the March Council conclusions, we have modelled a 10% reduction in total EU CO_2 emissions for 2010, which equals a 14% per capita emission reduction over all four sectors. According to our analysis, to achieve a 10% reduction in CO_2 emissions by 2010, the largest emission reductions are required by Luxembourg (-31%) and Belgium and Denmark (both -20%), followed by Germany (-19%) and the Netherlands (-16%). Emission increases would be allowed in a number of member states, including Portugal (+15%), Sweden (+9%) and France (+7%). As TABLE 2 illustrates, these results differ considerably from the indicative targets to which member states agreed under the negotiating position of the March 1997 Environment Council (although it must be kept in mind that these are not exactly comparable as the latter are based on a basket approach, rather than CO_2 only).

TABLE 2: 2010 emission reductions in comparison

Country	Sectoral approach (CO_2)	EU Council (basket)
Austria	+4%	-25%
Belgium	-20%	-10%
Denmark	-20%	-25%
Finland	-15%	0%
France	+7%	0%
Germany	-19%	-25%
Greece	-10%	+30%
Ireland	-9%	+15%
Italy	-3%	-7%
Luxembourg	-31%	-30%
Netherlands	-16%	-10%
Portugal	+15%	+40%
Spain	+5%	+17%
Sweden	+9%	+5%
UK	-15%	-10%

Austria has accepted reductions much higher than those calculated by our analysis.Conversely, Belgium, Finland, Greece, Ireland, Portugal and Spain are committed to much less than would be expected from our figures. The emission reductions demanded by our analysis from Greece and Ireland (arising from the high industrial sector carbon intensities found) are hard to justify considering the state of economic development in these two countries and a case can be made for adjustments, taking into account differing GDP rates.

Compared with EU business-as-usual (BAU) figures (Commission of the EC, 1997), our results require considerable emission reductions in the electricity and domestic sectors. In these sectors, emissions are expected to increase by 2% and 4% respectively, while we assume reductions of 17% in both sectors. Meanwhile, for the industrial sector our reductions (10%) are actually below the BAU figure of 15%. We believe that the BAU figure is rather unrealistic, making overoptimistic assumptions about changes in industrial structure and/or industrial energy efficiency improvements. For the transport sector, our analysis returns per capita emissions to 1990 levels by 2010, which means a 4% increase in total emissions. In the EU BAU scenario, a 39% increase is assumed. Quite clearly, measures to stop the emissions growth in this sector are crucial to achieving emission reductions post-2000.

Some of the sectoral emission reductions in a few countries reveal how very specific country circumstances can yield results which should not necessarily be translated directly into emission targets:

- the very low electricity sector emissions for France and Sweden mean that emission increases from this sector of 78% and 139% respectively are allowed within the overall target. Clearly, such increases would be environmentally undesirable and the establishment of a tradeable permit system could be considered as a means of encouraging the maintenance of low carbon emissions in this sector;
- the high carbon intensity of the Greek and Portuguese industrial sector indicate that industrial emissions in these two countries should be cut by 29 and 16% respectively. However, this might be hard to justify politically, as these countries are much less developed than Northern European countries;
- it seems unlikely that the Portuguese domestic sector will need to grow by the 108% allocation, considering the limited heating requirements in this country.

There is a danger that because of such skewed results, which can also occur with a different choice of indicators, there could be endless negotiations about the allocation of differentiated targets. These problems must be kept in mind when deciding on indicators for target allocation.

11. Conclusions

This paper has argued that in order to be acceptable, indicators have to be fair, politically feasible, transparent and reflect agreed trends, as well as minimising the number of arbitrary assumptions. However, in reality, it is difficult to chose a set of indicators which fulfills all these criteria, in particular in view of data availability problems.

Undoubtedly, some member states will find it politically difficult to agree to the emission reductions suggested by the analysis presented. Even if indicators can be proven to be fair and transparent, different political priorities might prevent countries from accepting their 'fair' share in emission reductions. Allocation indicators cannot be a panacea for solving differentiation and burden-sharing problems in the EU. However, they can provide a starting point for political negotiations. The difficulty in allocating emission reduction target must not be allowed to get in the way of reaching an agreement at Kyoto, as to avoid dangerous climate change, the growth in emissions has to be checked as soon as possible.

References

Blok, K., Phylipsen, G.J.M., Bode, J.W. (1997) The triptique approach: burden differentiation of CO_2 emission reduction amongst EU member states, Discussion paper for the EU Ad Hoc Group on Climate, January 1997.

Commission of the European Communities (1997) *Communication from the Commission: the Energy Dimension of Climate Change*, COM (97) 196 final.

Greene, O. (1996) 'Lessons from other international environmental agreements, in Paterson, M. and Grubb, M. (eds) *op. cit*

Grubb, M. and Collier, U. (1997) *Developing Indices for Differentiating CO_2 Emissions in the European Union: Issues and Proposals*, Report to DG XI, June 1997.

Grubb, M., Sebenious, J., Magalhaes, A. and Subak, S. (1992) 'Sharing the burden' in Mintzer, I. (ed) *Confronting Climate Change*, Cambridge: Cambridge University Press, pp. 305-22.

Hayes, P. and Smith, K. (1993) *The Global Greenhouse Regime: Who Pays?*, London: Earthscan.

IPCC (1996) *Climate Change 1995. Economic and Social Dimensions of Climate Change*, Cambridge: Cambridge University Press.

Kawashima (1996) 'The possibility of differentiating targets: indices and indexing proposals for equity', in Paterson, M. and Grubb, M. (eds) *op. cit.*

Kram, T. and Hill, D. (1996) 'A multinational model for CO_2 reduction: defining boundaries for future CO_2 emissions in nine countries', *Energy Policy* 24 (1), pp. 39-51.

Lunde, L. (1995) 'Greenhouse burden-sharing after Berlin: economic ideals and political realities', in Grubb, M. and Anderson, D. (1995) *The Emerging International Regime for Climate Change: Structures and Options after Berlin*, London: RIIA.

Paterson, M. and Grubb, M. (1996) *Sharing the Effort: Options for Differentiating Commitments on Climate Change*, London: RIIA.

Ridgley, M. (1996) 'Fair sharing of greenhouse gas burden', *Energy Policy* 24 (6), pp.517-530.

THE EU POSITION IN THE NEGOTIATION PROCESS TOWARDS A PROTOCOL ON THE LIMITATION AND REDUCTION OF GREENHOUSE GAS EMISSIONS*

Elements for intervention

CHRISTOPH BAIL
Global Environment, European Commission
Rue de la Loi 200, B-1049 Bruxelles, Belgium

* Reflects the position end of July 1997. The views expressed are personal and do not commit the European Commission.

1. Introduction

History international climate negotiations (very brief):

- 1992: UNFCCC
- 1995: Berlin Mandate
- 1997: Kyoto Protocol

Negotiating in Ad hoc Group of the Berlin Mandate (AGBM):

- Main task AGBM sessions: Bridge the gap towards a credible Protocol with legally binding targets and some agreements on policies and measures (PAMs).
- In order to have a stronger negotiating position and to influence the process, groups of like-minded countries are working together. The main players are the EU, JUSCANZ, G77 and China, AOSIS, OPEC. Groups are mainly formed on the basis of shared interests. JUSCANZ and G77 are rather divergent groups. US is the most important state actor.
- EU speak with a single voice and has leadership role.
- State of the art: various proposals from various countries or groups of countries were tabled last year with negotiating positions on targets, PAMs and flexibility. Since there are only two session left before the meeting in Japan, the focus will be on issues for possible convergence - agreement.

J. Hacker and A. Pelchen (eds.), Goals and Economic Instruments for the Achievement of Global Warming Mitigation in Europe, 129–135.

It will be very difficult to reach agreement and decide upon a strong protocol in Kyoto, considering the big differences in negotiating position that some Parties have submitted. Nevertheless put all effort in it to make it a success.

2. Targets

COP-2 (Geneva) agreed that targets Annex I countries in specific time frames will be set in Kyoto (2005, 2010, 2020). (The most important) negotiating positions so far:

2005

EU: at least - 7,5% for Annex I countries (1990). Agreed upon last June Council.
AOSIS: - 20% for Annex I countries (1990 levels)

2010

EU: - 15% in a basket of 3 gases (CO_2, CH_4, N_2O) on the basis of 1990.
Internal burden sharing. Agreed upon last March Council.

Russia: Maintain GHG emissions in the period 2000-2010 at 1990 levels. For period after 2010: make decision on reductions before 2007.

Apart from the EU and Russia, no other major Party to the Convention has tabled a quantified proposal for 2010. Waiting for other Parties to come forward with their proposal. Most Parties favour a target for all Annex I countries, but did not yet agree on specified figures. However, it is expected that none of them is willing to go as far as the EU.

2.1. FLAT RATE APPROACH?

Various proposals suggest that 'flat rates', as proposed by the EU and US, are not equitable and that a differentiated approach should be developed in the Protocol. Some alternatives:

Australia: ensure equity by equalising percentage changes in per capita economic welfare losses resulting from mitigation action, and to rely therefore on indicators such as population growth, GDP per capital growth, emission intensity of GDB etc.

Japan: Each Party should select one or two of the following options, either a flat rate with respect to 1990 emissions or a per capita emission cap.

Switzerland: set according to GHG emissions per capital levels.

2.2. EU VIEW ON THIS

The EU experience on burden sharing confirms that from a practical point of view differentiated targets are not easy to negotiate. The EU called upon all Annex I parties - for the sake of simplicity - to accept individually and jointly the same target for 2005 and 2010. The EU is prepared to accept any arrangement between non EU and Annex I Parties which preserves jointly the 15% target for 2010.

It will be difficult if not impossible to agree any formula before Kyoto. Therefore our objective would be that all Annex I Parties put their cards on the table (overall reduction and individual commitments) and that we start negotiating in the light of these proposals.

Targets for *CEECs and Russia*: These countries are often arguing that reduction targets will put restrictions to their economic growth. They assume that if GDP is increasing again, CO_2 emissions will automatically follow. However, analysis showed that this is not the case. In a lot of CEECs GDP is increasing, while CO_2 emissions are still decreasing. This means that CEECs can accept a firm reduction target as well as the rest of the Annex I countries. The great potential for energy efficiency makes it relatively easy for them to achieve the targets.

Targets for *non Annex I countries*: very sensitive issue. Annex I Parties should take the lead in accepting binding commitments (Berlin Mandate). Right to develop. After 2005, restrictions for all Parties to the Protocol.

3. Policies and Measures

Necessary to achieve reduction goals!

3.1. VARIOUS OPTIONS WITH SOME EXAMPLES

- *Regulation*. Will be very difficult, if not impossible to agree upon on the international level. In certain sectors, such as heavy industry, subject to global competition, necessary (energy saving standards).
- *Financial instruments*. 1) Incentives. Will be very difficult to agree upon on the international level. Possibly initiative on aviation tax. 2) Abolishing of subsidies, seems more acceptable to most Parties, but faces strong resistance from vastest interests, e.g. coal miners.
- *Communicative instruments*. Awareness raising campaigns etc. Almost no resistance, but low effect on reducing CO_2 emissions.
- *Voluntary agreements*. No legal status. Popular with industry. Essentially limited to no regret measures.

3.2. POSITIONS OF PARTIES

The EU tabled a proposal containing of the following elements:

A - PAMs common to all Parties Annex X
B - PAMs high priority for co-ordination with other Annex X Parties
C - National PAMs to be given high priority for inclusion in national programmes this proposal is the most elaborated one, which include the most concrete-binding PAMS (Annex A and/or B).

1. renewable energy programmes
2. programmes regarding CHP
3. energy efficiency standards and improvements in appliances
4. energy efficiency in buildings
5. fuel efficiency improvement and CO_2 emission reduction from passenger and freight) vehicles
6. energy efficiency improvement and greenhouse gas emission reduction in heavy industry and large combustion plants
7. limitation of emissions from HFCs, PFCs and SF6s
8. reduction in methane emissions from landfills and other sources
9. reduction of N_2O emissions in the chemical industry
10. progressive reduction/removal of fossil fuel and other subsidies, tax schemes and regulation which counteract energy efficiency
11. increase of minimum tax levels as a way to reduce green-house gas emissions and to improve energy efficiency

Switzerland proposed a list of internationally co-ordinated measures.

Canada suggests that Parties commit themselves to certain PAMs (which are extremely soft, like info sharing and reporting and voluntarily agreements). In addition to co-ordinated actions, a menu of PAMs could be included in the Protocol.

Japan tables a proposal which contains five themes, under which Parties can choose from the given options. The themes are very much related to efficiency and technology.

Russia proposes PAMs with a scientific, technological or economic component. It is (for me) unclear if this is mandatory or not.

The *United States, Australia and the OPEC* did not come forward with a proposal so far, but it is clear that they are not in favour of any commitments.

AOSIS. Two lists are tabled.
Parties in Annex X shall adopt and implement PAMs in list A; e.g. efficiency standards and taxation.

Parties shall give high priority to the adaptation and implementation of PAMs in list B; e.g. voluntarily agreements, fuel switching.

The proposals put forward so far vary a lot. Whatever the elaboration of the proposal is in terms of actions to be taken, two main points of discussion are to what extent policies and measures should be binding or voluntary and to what extent they should be common or coordinated. The argument against common and coordinated measures is that the national situations differ, and therefore it is best to choose and implement measures that fit the specific circumstances. However, the EU is of the opinion some measures are only effective if taken in common with other Annex I parties: competition position and synergy effect. The argument in favour of binding measures is to give target credibility and overcome internal resistance. The arguments against are flexibility and sovereignty of Parliaments).

In order to try to get agreement, the chairman of the AGBM group - Estrada - tabled a proposal himself:

The *Estrada proposal* contains of three lists:

A - Policy objectives
B - Possible mechanisms for implementation
C - Menu of Policies and Measures, Parties can choose according to national situation

At this moment, the EU is making an inventory of the reduction and economic potential of PAMs, and where there is room for agreement. If possible, then these issues should become common or co-ordinated.

4. Flexibility

Flexibility in place (JI) and time (banking and borrowing) to fulfil commitments.

4.1. JOINT IMPLEMENTATION

Principle of JI: fulfil part of own commitments in other country - credits can be used to reduce less in own country.

Status: pilot phase - AIJ - no credits - only between Annex I countries.

Opinions:

- Business, electricity and energy sector very much in favour
- US, Australia in favour to start JI world-wide, including credits
- Developing countries: reservation ('carbon colonies'), wait for results pilot phase
- CEEC's: projects including credits welcome, need for capital for investments
- The EU is of the view that Joint Implementation among industrialised countries can be useful even though its share should at this stage be limited to a minor part of the emission reduction, the major part still to be achieved by Policies and Measures implemented "at home". Regarding Joint Implementation with developing countries, we need to wait for the results of the pilot phase in which several 'Activities Implemented Jointly' are implemented and evaluated. During this experimental period, that probably will end in 1999, Parties must draw up fair criteria under which Joint Implementation could take place, and establish a solid institutional framework before a decision about possible crediting of GHG reductions at a global level could start.

4.2. BANKING AND BORROWING

Principle: Cost-effectiveness. Assumption that there is a budget for each Party to the Conference. Banking: save GHG units for future use, or sell them to other participants for future use. Borrowing: allows future allocations of GHG units to be used to offset GHG emissions in excess of the emission limit in the current period. It is said that it could allow participants to mitigate GHG emissions at lower costs if mitigation costs would be lower in the future.

Opinions:

US: Very much in favour of this system.
EU: Borrowing is out of the question, because it is contradictory to the principle of intergenerational equity, which requires a development without harming the quality of life for future generation. Banking is not a preferred option as well. Illogical to put more risk on the environment than necessary, just because the budget is apparently too high.

4.4. EMISSION TRADING

Principle: Cost-effectiveness. In a tradeable 'permits' system, an environmental goal is set and polluters are required to hold a certificate or permit for each unit of pollution they emit (in some systems, polluters must hold a certificate only for the units of pollution they wish to sell). If a participant has permits it does not use, it can sell them to another participant who need extra permits to cover excess pollution (above its limit). Through such trading, a market price for the permits emerges which reflects the cost of pollution reduction. Each participant can then decide whether it is cheaper to reduce pollution or to purchase permits.

Status: Until now not possible. Perhaps after JI pilot phase has positive results, emission trading becomes more acceptable.

Opinions:

US: Very much in favour of this instrument.
EU: Reservation. We need to agree targets first, then see how we can set up some practical mechanisms for emission trading.

5. Conclusions

The AGBM process is difficult considering the many Parties involved. And the enormous and divergent interests that are at stake. However, it is very important to reach agreement in Kyoto on reduction targets and policies and measures. Kyoto is the place to take concrete and measurable steps for combatting climate change.

Before Kyoto, there will be two more sessions of AGBM, in which Parties can try to reach agreement. The EU has so far played a leading role in the negotiations and, considering the fact that most Member States are strongly committed to obtaining credible results, the Union will also play an important role in the outcome of the discussion in Japan.

An important next step for the EU is to find common ground with others that have proposed policies and measures regarding possible mandatory measures and appropriate coordinating mechanisms. We remain convinced that we need to agree on approached in certain areas to exploit the potential in a cost effective manner and avoid distortions of competition.

SECTION III

POLITICAL AND ECONOMICAL INSTRUMENTS - AN OVERVIEW

INTRODUCTION

THE RANGE OF POSSIBLE POLITICAL AND ECONOMICAL INSTRUMENTS FOR THE MITIGATION OF GLOBAL CLIMATE CHANGE

JÜRGEN HACKER
UMB Environmental Management Consultancy Hacker GmbH
Selchowstrasse 1, D-14199 Berlin, Germany

1. Rational Ecological Policy

As in other political fields, a rational ecological policy can be described by a simplified three-step approach:

1st step:	Analysis of the status-quo
2nd step:	Definition of the desired status
3rd step:	Selection of target-oriented reaching strategies

This sequence is important. If you miss one step, or driven, for instance, by the political need for short-term action, do not work it out properly, it is very doubtful whether the target will be reached by the selected measures at all or at least in the fastest or most cost-efficient way.

But even when you have worked out the first two steps properly, as we have done in this book in the first two sections, before you discuss specific actions it is advisable to have a rough overview of the different instruments at ones disposal for tackling environmental problems in general.

2. Policy Instruments for Environmental Protection

The aim of this introductory article is to discuss the different instruments for environmental protection, not in an ambitious scientific way, but as a comprehensible lead-in for non-experts.

There are different ways to structure the wide range of possible instruments depending of the point of view and the interests of the classifier.

J. Hacker and A. Pelchen (eds.), Goals and Economic Instruments for the Achievement of Global Warming Mitigation in Europe, 139–144.

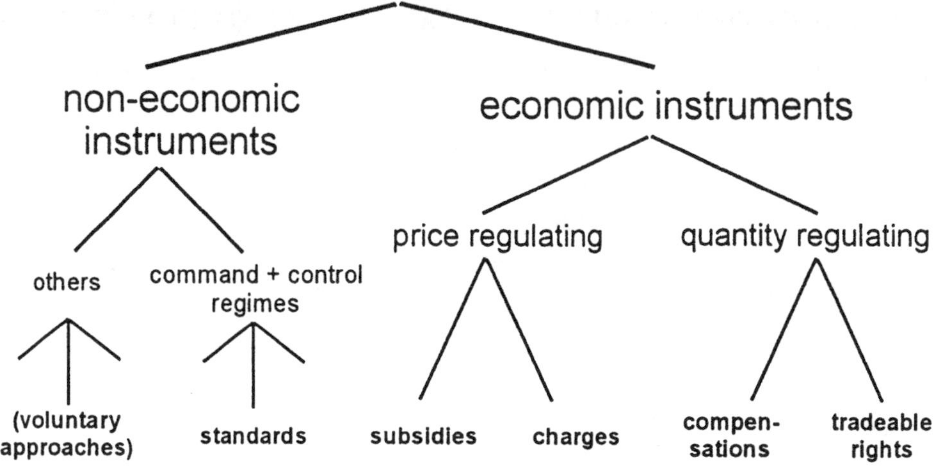

FIGURE 1. Political Instruments for Environmental Protection

Figure 1 shows a structure which is very suitable in the context of this book. As this will concentrate on economic instruments later on, it is useful to differentiate first between economic and non-economic ones.

Economic instruments try to reach the defined environmental targets by setting economic incentives for the users of the natural resources to behave in the political desired direction. The decision whether to pay and use the natural resources or not remains with the individual potential user. Economic instruments can be subdivided according to the type of intervention into price and quantity regulating instruments.

The first type directly influences the market price of goods or services by charges or subsidies. These price variations cause indirect changes in the quantities of natural resources used. However, it is impossible to predict the exact amounts used, because these depend on the individual reactions of the market participants.

The second type of economic instruments intervenes by regulating the quantities of natural resources available on the markets, for instance by limiting the rights to use the atmosphere for the disposal of gaseous wastes. This class of economic instruments can be subdivided into compensation solutions and tradeable rights or permits. In contrast to the first type, these instruments influence market prices indirectly. The price fixing is left to the market search process according to the optimum allocations of the given amount of scarce resources. However, the amount of natural resources used is definitely within the imposed limitation.

The non-economic instruments are subdivided into command and control regimes and in a category of a broad variety of other instruments - "strong" ones like environmental liability regulations and environmental criminal law and "soft" ones like environmental education, consumer information or the EU eco-audit system. These instruments cannot be assigned to any of the other classes and do not justify a class of their own in the context of this book.

Command and control regimes are the most common environmental instruments at present. Most of them are based on standards which the users of natural resources have to comply with. Minor command and control regimes operate with individual injunctions or with usage commitments.

Voluntary agreements are grouped to the other "soft" instruments due to their voluntary character and because they can "contain all kinds of instruments. Goals or technical instructions for a specific industry, but also scales of charges and information system are conceivable. That is why the agreements do not have any working mechanism in their own right, rather the latter depends on the instruments provided in each specific case" (Rennings et al., see page 185). It is characteristic that they are achieved by political negotiations, mainly between governments and business sectors, partly also between different public bodies. In most cases voluntary agreements are hardly voluntary, but agreements to avoid more stringent or costly instruments like taxes or standards. As part of these agreements is similar in their function to economic instruments they are discussed separately in detail in this section (see pp. 159-204).

All of these different classes of instruments can be designed as

- product based
- technology based, or
- emission (reduction) based.

More details of the implications of these bases are described by Pelchen in the next chapter (see pp. 152-155).

3. Types of Price-Regulating Instruments

The figures 2 and 3 structure the different types of price-regulating instruments in more detail. They are broadly self-explanatory.

Worth mentioning are the different functions of environmental charges: Besides the function as ecological incentive already mentioned, they produce financial income at the same time. It depends on the design of a charge which function dominates. In general, the financial function dominates with taxes and the ecological incentive dominates with fees. Though both functions are effective in most environmental charges, it is possible to design charges with only one function. For instance, a surcharge on a trade tax to enable a government to finance certain public environmental damage can give no ecological incentives to the tax payer.

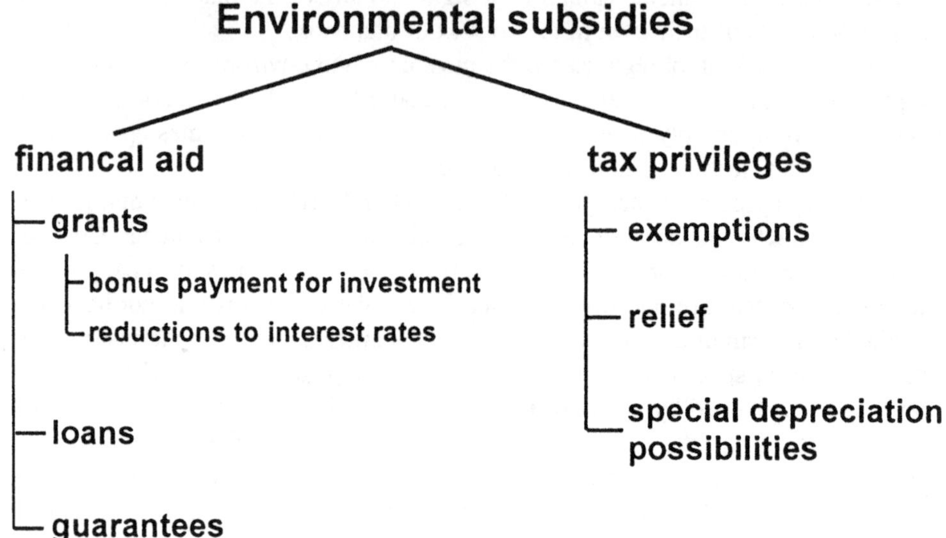

FIGURE 2. Environmental subsidies

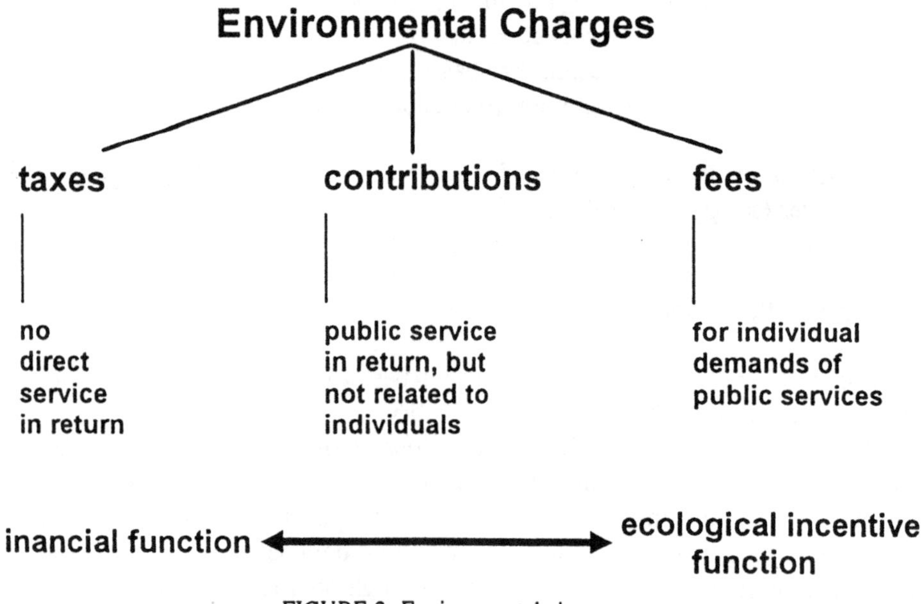

FIGURE 3. Environmental charges

On the other hand if the total income of a fee is given back to consumers, for instance by reduced general income tax, this fee has no financial function any more, only a financial redistribution remains.

4. Types of Quantity Regulating Instruments

Figures 4 and 5 give an overview of the two groups of quantity regulating instruments.

Compensation Systems/Solutions

- **basis : commitments by other environmental instrument**
  (mostly command & control regimes)
- **compensation defined for individual source/project**
  (lives & dies with sources/projects)

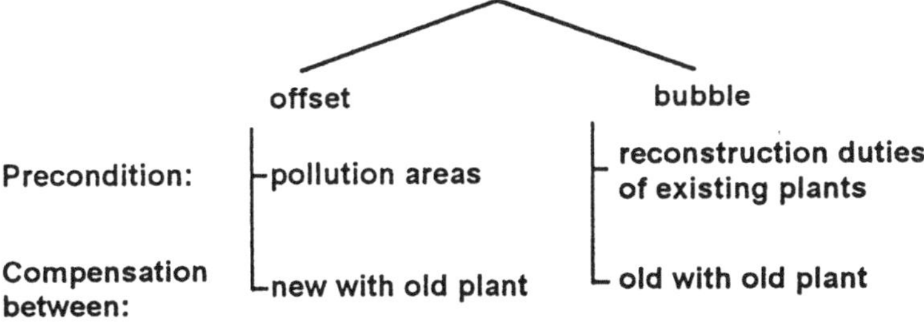

FIGURE 4. Compensation Systems/Solutions

Compensation systems or solutions allow compensation for a given environmental commitment by a more cost-efficient alternative. The economic principle behind this is explained in detail by Jepma (see pp. 260-261).

To prove that the compensation is at least environmentally neutral the impacts of both alternatives on the environment have tc be evaluated on a quantitative basis, which justifies this classification.

Such solutions can be implemented within one type or across different types of environmental instruments.

For instance in Germany, there is a compensation solution in the command and control clean air regime which make it possible under certain criteria to offset the failure to fulfill a certain standard in one plant by a higher standard in a neighbouring plant. But it is also possible to design a compensation regime which allows the exemption of a tax by the fulfilment of a more ambitious standard. Most experience with compensation solutions has been gathered in the Clean Air Act of the USA with its options for offsets and bubbles.

Tradeable Rights/Permits

- **basis : total desired use of nature / emissions budget**
- **individual, autonomous rights (independent of sources/plants)**

different design points of rights/permits

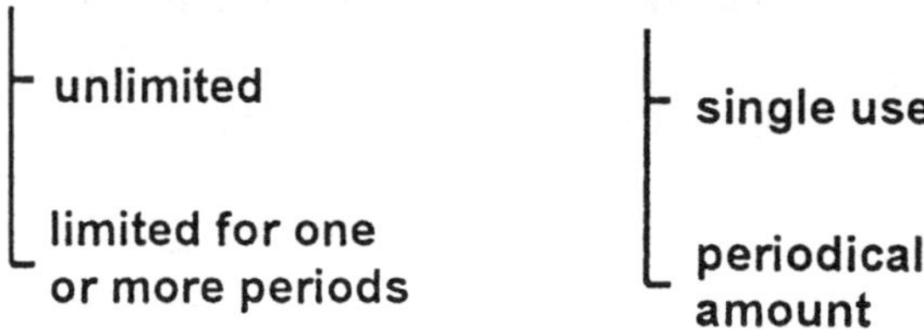

FIGURE 5. Compensation Systems/Solutions

In contrast to the compensation instrument tradeable rights or permits need their own definition of the desired maximum use of nature. The total use of nature is separated in autonomous individual rights, which are independent of sources or plants. Therefore they can be traded without restrictions within the boundaries of the system.

The tradeable rights system can be designed in many different ways which will be the main focus of this book and be discussed in detail in Section IV (see pp. 293-406).

A FRAMEWORK FOR THE EVALUATION OF POLITICAL AND ECONOMICAL INSTRUMENTS FOR GLOBAL WARMING MITIGATION

Dr. ARTHUR PELCHEN
UMB UmweltManagementBeratung Hacker GmbH
Selchowstr. 1, D-14199 Berlin, Germany

Abstract. This paper introduces a possible general framework for the evaluation of different political and economic instruments in a transparent and comprehensible way. For illustration exemplary results of an evaluation are presented as well. Therefore possible criteria for the evaluation are explained and analysed. The most important criteria are the accuracy and the speed of achieving the environmental target, the economic efficiency, the compatibility with the market economy and the compatibility with the polluter-pays-principle. Additional criteria are the political acceptability, the distributional effects and the incentives for additional environmental improvements. In the first step of the exemplary evaluation it is shown, that independent of the chosen type of instrument, i.e. standards, charges or subsidies, the best are emission-based instruments followed by technology-based and product-based instruments. The second step compares the different emission based standards, charges and subsidies with tradeable permits. The result is a ranking with tradeable permits coming off best, followed by emission charges (taxes), emission standards and emission reduction subsidies.

1. Introduction

The aim of this article is to present a general, comprehensible and transparent framework for the evaluation of political and economic instruments in environmental protection before starting the exposition and discussion of the individual instruments in the subsequent papers. This qualifies the reader to use this framework for their own conclusions concerning the suitability of the different instruments for the mitigation of global warming. It is demonstrated in this paper that it is possible to assess instruments according to a range of criteria in order to yield less subjective and more comprehensible results. The presented framework is based on a dissertation by Knüppel (1988).

In view of this aim, the possible criteria for evaluation are explained and analysed, followed by a brief description of the mechanism to weigh the marks for the individual criteria and to combine them to an overall assessment. Additionally the results of an evaluation using this framework are given for illustration, based on the examples by Knüppel (1988).

J. Hacker and A. Pelchen (eds.), Goals and Economic Instruments for the Achievement of Global Warming Mitigation in Europe, 145–157.

2. Possible criteria for the evaluation

The following paragraphs briefly outline the criteria, that are commonly used in mainstream economic assessments of environmental policy instruments. In order to be operational in the framework for the evaluation and to be able to calculate a single value for the overall comparison of the instruments, the criteria should be quantifiable and independent of each other.

2.1. ACCURACY OF ACHIEVING THE ENVIRONMENTAL QUALITY TARGET

Political and economical instruments are usually designed to achieve a certain goal. These goals should be specified in the political decision making process for example in laws or international agreements. It is obvious that the ability of instruments to achieve these given goals accurately, is one of the most important factors in assessing the quality of political and economical instruments. To be able to assess a certain set of instruments according to the accuracy, with that they achieve a given environmental goal, three conditions have to be fulfilled:

1. the environmental goal must be defined in a quantifiable way,
2. there must be a known cause-effect-relationship between the instruments and the environmental goal and
3. the degree of achieving the goal must be measurable.

In the literature this criterion is sometimes referred to as the ecological efficiency as opposed to the economic efficiency, which is treated below (Rat der Sachverständigen für Umweltfragen, 1978; Wicke, 1982).

2.2. SPEED OF ACHIEVING THE ENVIRONMENTAL QUALITY TARGET

Similar to the first criterion it is desirable that a political or economical instrument achieves the defined goal quickly and without a lengthy trial and error process. The speed of achieving this goal depends mainly on two factors:

- On the one hand on the time that is needed to implement the instrument: this in turn is mainly influenced by the fact how easily the chosen instrument can be included in the existing legal framework: in general an instrument, which is similar to existing legal instruments, is easier to codify in legal language than an instrument, that has never been used before and necessitates extensive negotiations and alterations in the legal system. In those cases, where the emission of an environmental pollutant is not yet legally regulated at all, this general rule might not be valid.

- on the other hand on the time that is needed in the application phase of the instrument to actually achieve the given environmental goal: this in turn is influenced by the speed that the economic subjects are able to adapt to the new situation. An instrument that allows for an increased flexibility in the decisions of the economic subjects in the adaptation process will achieve a given goal quicker than an instrument giving rigid instructions for the adaptation.

This instrument requires the same preconditions as the first instrument.

2.3. ECONOMIC EFFICIENCY

Besides the ecological efficiency dealt with above, the assessment of the economic efficiency of an instrument is of great importance. The evaluation is based on the economic rationality principle, which can be defined in two different ways for the specific conditions of the environment:

- Minimum principle: Achievement of a given environmental effect with lowest possible macro-economic cost
- Maximum principle: Achievement of the greatest possible environmental effect with a given macro-economic cost

Since the behaviour of the economic subjects is in general based on these economic principles and since the cost of a certain measure is decisive for its acceptance, they should also be applied to political and economic instruments for environmental protection as well. Hence a political or economic instrument for the achievement of global warming mitigation is the better the lower its related macro-economical costs for the achievement of the environmental goals. In the context of political instruments two groups of costs have to be considered:

- The first group comprises the costs related to the implementation and day-to-day operation of the chosen instrument like for example costs for monitoring. Hence the more bureaucratic instruments are more costly in respect to the operational costs, whereas those featuring new regulatory designs, that necessitate more alterations to the existing legal framework, are more costly in respect to their first time implementation. Again this might not be valid for pollutants that are legally unregulated so fare.
- The other group of costs is related directly to the abatement, reduction or sequestration of the emissions. Here again the instruments allowing some extra choice and flexibility to the economic subject are less costly.

Because of the fundamental importance of the criterion for the acceptance of an instrument it is included in nearly all assessments of political and economical instruments for environmental protection found in the literature.

2.4. COMPATIBILITY WITH THE MARKET ECONOMY

The criterion of compatibility with the market economy is obviously only of importance, if the market system is regarded as a desirable feature for the future. Given this assumption all political and economical instruments for the mitigation of global warming should be compatible with the market economy and should fit in the economical framework without any problems. For the evaluation of market compatibility Knüppel (1988) uses six subcriteria which are visualised in Figure 1 and explained in the following paragraphs. From these subcriteria the compatibility with the market economy is calculated and expressed on a scale between market based and centrally planned.

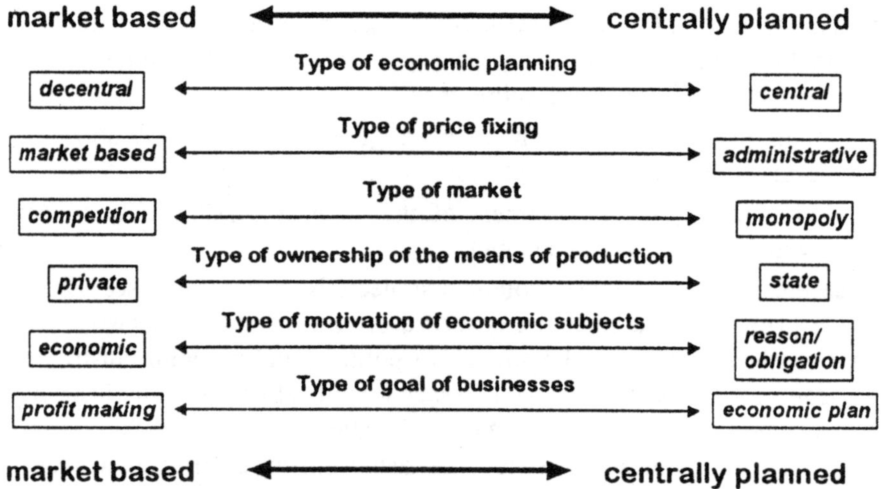

Figure 1. Criteria for the compatibility with the market economy

2.4.1. *Type of economic planning*

Every economic system necessitates decisions on the type, amount and quality of products and services to be produced. These allocation and distribution problems are either solved centrally or decentrally. While in the centrally planned economies like the former Soviet Union or East Germany central governmental institutions decided on economic 5-year-plans, it is typical for the market-based systems, that every consumer and producer decides decentrally on the production and distribution of goods and services. Supply and demand are co-ordinated by the equilibrium price in the relevant markets. Hence a political or economical instrument is the more market-based and therefore better the less it interferes with the market mechanism and the more it leaves the decisions on allocation and distribution to the individual producer and consumer.

2.4.2. *Type of price fixing*

Similar to the first subcriterion the prices for goods and services can either be centrally determined by government institutions or they can be regulated on markets by supply and demand. Related to environmental resources an instrument is considered more market based and therefore better, if it does encourage market-based price fixing to express the true scarcity of the resource. In this context an emission charge does not lead to a market based and scarcity indicating price for the environmental resource.

2.4.3. *Type of market*

In economic theory market types are distinguished by the number of participants both for supply and demand. Eucken (1965) mentions five market types between full competition with an unlimited number of participants on both sides of the market and the monopoly, where either supply or demand is dominated by a single participant. Since competition in a market-based system is generally improving the overall performance of the system in terms of efficiency and quality of supply, it is generally regarded as superior. Applied to economic and political instruments for the mitigation of global warming this means that those instruments are better that allow for as much competition among the different economic subjects as possible. Another aspect is whether or not the instruments influences the access of new producers on the market.

2.4.4. *Type of ownership of the means of production*

The main distinction is between private and state property. Private property is the dominating type of ownership in market-based systems and has the advantage, that decisions on the use of the property are taken decentrally and flexibly by the owner. In opposition state property is commonly related to state bureaucracies centrally taking the decisions on the use of the means of production. Since the atmosphere is a part of the means of production, economic and political instruments can be evaluated on the basis of how they treat the property on the right to emit a certain amount of GHG.

2.4.5. *Type of motivation of the economic subjects*

According to the theories of Marx the economic subjects in a market-based system are motivated by the maximisation of their private economic benefit, whereas in the centrally planned economies, since there is no divergence between the private and public benefit, they are motivated by reason and obligation to fulfil the social goals. For the evaluation of economic instruments in the framework of a market-based system this means that they are rated the better the less they obstruct the motivation of the economic subjects to maximise their private benefits. Hence an instrument that rewards the overfulfilment of the environmental obligations by increasing the benefits for the economic subjects is also rated favourably.

2.4.6. *Type of goal of businesses*
Similar to the motivations of the individual economic subjects the goal of enterprises in a market-based economy is to maximise their profits. In opposition the enterprises in centrally planned economies mainly strive to fulfil the 5-year-plan decided by the government. Again this means in the context of a market-based system that those instruments for global warming mitigation are rated better that do not hinder the goal of profit maximisation.

2.5. COMPATIBILITY WITH THE POLLUTER-PAYS-PRINCIPLE

It is a common principle in our legal system (and our moral values) that the party that is responsible for a damage is also liable for the compensation. This principle applied to the environment is the polluter-pays-principle (PPP). Hence all instruments for the mitigation of global warming can be evaluated against this principle by checking in how fare they make sure, that the parties responsible actually pay for the damages. One problem with the PPP is to decide, whether the producer or the consumer of an environmentally damaging product is responsible for the pollution.

2.6. POLITICAL ACCEPTABILITY

This criterion is certainly an important one for the practical implementation of a new instrument. On the other hand political acceptability is a phenomenon that is changing over time and that depends very much on particular interests of certain pressure groups. Many new developments in politics were not acceptable when brought to the political agenda for the first time and are now common practice in many parts of the world. This criterion is therefore rejected as being prejudiced by Knüppel (1988).

2.7. DISTRIBUTIONAL EFFECTS

Another criterion often found in the literature is that of distributional effects. Commonly five groups of distributional effects are distinguished related to the environment (Jarre, 1975; Osterkamp, 1975; Zimmermann, 1982):

- Distribution of environmental damage,
- Distribution of environmental assets,
- Distribution of environmental improvements due to policy,
- Distribution of costs for environmental improvement and
- Distribution of losses in other social areas due to the achievement of the environmental improvements.

Because it is methodologically very difficult to attribute these effects to the use of different instruments, this criterion is seldom used in assessments.

2.8. INCENTIVES FOR FURTHER EMISSION REDUCTION

Political and economical instruments can also be evaluated whether or not they set incentives for further emission reductions beyond their immediate goals. On the one hand this aspect is already included in the subcriteria 'type of motivation of economic subjects' and 'type of goal of businesses' (see 2.4). On the other hand there is no need for further emission reductions, if the environmental goal is defined correctly and achieved precisely with the chosen instrument. In the case of carbon dioxide zero emissions are environmentally not required. Therefore it is not necessary to include this aspect as an additional criterion.

2.9. COST OF MONITORING

Some authors also mention the cost of monitoring as an criterion for the assessment of instruments. This is certainly an important aspect, but according to the definition of economic efficiency used in this paper (see 2.3.) these costs are already included there. Hence an additional consideration is not necessary.

3. Weighting and combining the criteria

One of the goals of this evaluation is to calculate a single value to rate the overall quality of the different political and economic instruments in a transparent and comprehensive way. Since it was one of the preconditions for the choice of the criteria, that they were quantifiable, it is possible to determine a mark for each chosen criterion and each instrument. These values are then combined to represent the overall result of the evaluation for the given instrument. Since not all of the above mentioned criteria are of equal importance, it should be possible to assign different weights to the individual criteria. Mathematically the calculation of the overall evaluation with different weights is equivalent to the following equation:

$$E_Q = \frac{1}{\sum_{i=1}^{n} w_i} \cdot [w_1 \ w_2 \ w_3 \ \dots \ w_n] \cdot \begin{bmatrix} m_1 \\ m_2 \\ m_3 \\ \vdots \\ m_n \end{bmatrix}$$

subject to: E_Q = Overall Quality Evaluation n = Number of chosen instruments
w_i = weight for criterion i m_i = mark for criterion i

For the special case, that all criteria are weighted equally the overall evaluation is equal to the arithmetic mean of the marks for the individual criteria. This method is also used to aggregate the subcriteria for the criterion 'compatibility with the market economy', the aggregate results being rounded for the inclusion in the overall evaluation.

4. Example for evaluation

4.1. INTRODUCTION

Despite the subjectivity of the marks assigned to the individual instruments for the different criteria, the results of the evaluation by Knüppel (1988) are given here for illustration. In this evaluation Knüppel (1988) only uses the first five criteria discussed above and weighs all criteria (and subcriteria) equally. Knüppel (1988) has carried out the evaluation for the following instruments:

- Product subsidies
- Technology subsidies
- Emission subsidies
- Emission tradeable permits
- Product standards
- Technology standards
- Emission standards
- Product charges
- Technology charges
- Emission charges

It has to be added that tradeable permits can also be designed as an technology- or product-based instrument. In the light of the advantages of the emission-based instruments (see below) the other two types of tradeable permits are neglected in the assessment by Knüppel (1988).

4.2. RESULTS

In a first step the results of the comparison of the product-, technology- or emission-based types of subsidies, standards and charges are presented in Figure 2, 3 and 4. According to these figures calculated by Knüppel (1988), there is the same hierarchy in the total assessments for subsidies, standards and charges with the emission-based instruments rated best followed by the technology- and product-based instruments. This is mainly explained by differences in their accuracy and speed of achieving a given environmental target. For illustration, let us assume that the environmental target was to reduce the emissions of a certain harmful gas that is emitted using a certain technology to produce a certain product. Only emission-based instruments, like for example an emission tax, directly influence the rate of emission by attaching right at the cause of the environmental problem and increasing the production cost exactly proportional to the emissions of the harmful gas.

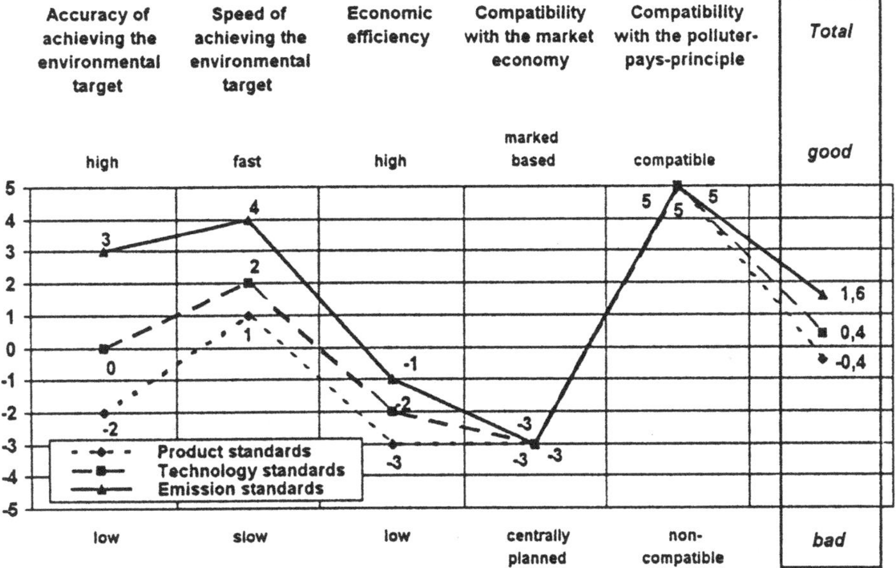

Figure 2. Evaluation of the different types of standards

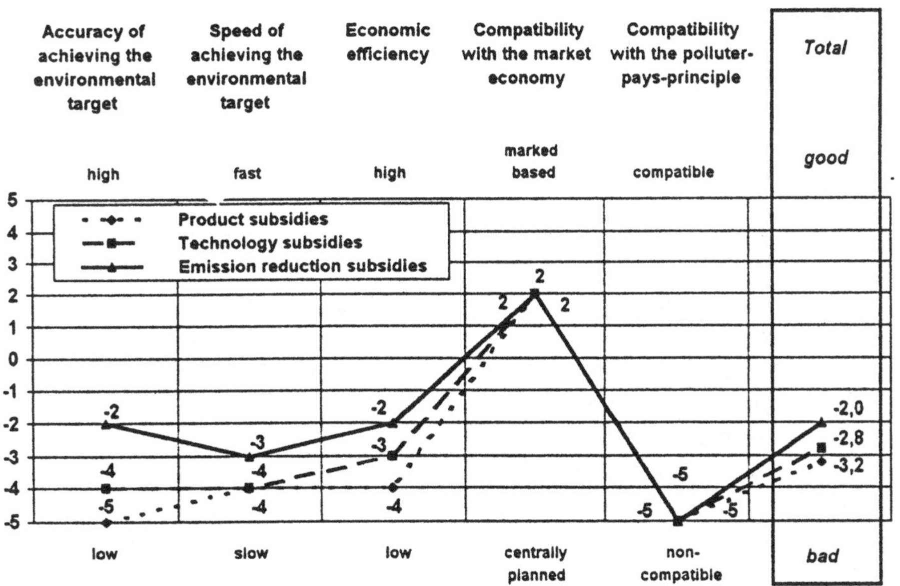

Figure 3. Evaluation of the different types of subsidies

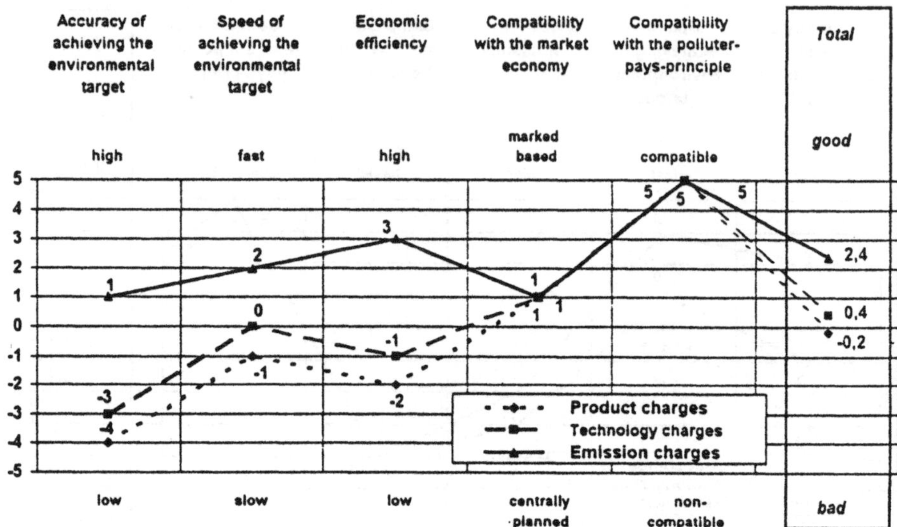

Figure 4. Evaluation of the different types of charges

On the other hand technology-based instruments might cause a shift towards technologies, that reduce the emissions of the harmful gas; nevertheless the actual emissions do not only depend on the technology used, but also on the scale of production. Hence it is less certain that a given emission reduction goal is achieved by a technology-based instruments than by an emission-based one. In the case of product-based instruments, these regulate the use or production of a certain good, if this is associated with the emission of the harmful gas. Since this neither takes the technology used nor the production (or usage) volume into account, there are again determinants for the emissions, that are not covered by the use of product based instrument.

Only in the case of a one-dimensional and linear relationship between a product and the related emissions and vice versa between the emissions and a single product, i. e. if a constant emission factor can be calculated for the single product, a product-based instrument will yield the same results as an emission-based instrument. In the case of petrol and the related carbon dioxide emissions it is possible to calculate an emission factor, but there are other products like coal or natural gas that are also responsible for carbon emissions and that reveal different emission factors. Hence it is again necessary to use an emission based instrument.

In summary it can be stated, that the closer an instrument is attached to the relevant cause of an environmental problem the more accurate and the quicker the given environmental goal is achieved. Consequently emission-based instruments are preferable compared to technology- and product-based instruments independent of whether they come along as emission-subsidies, -standards or -charges.

Figure 5 summarises the hierarchy for the base of the instruments.

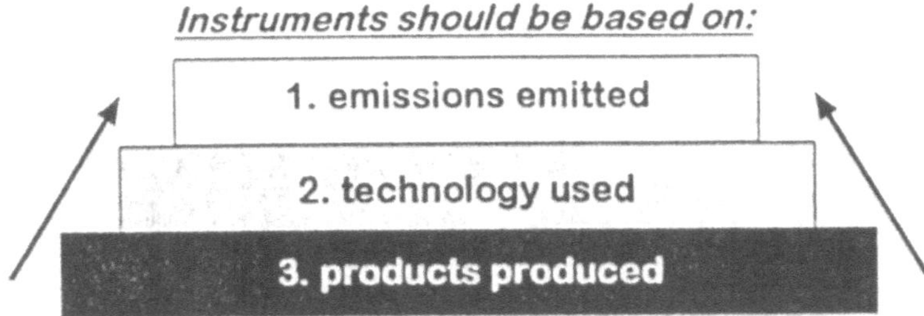

Figure 5. Preliminary hierarchy for the base of environmental instruments

On the basis of the favourable outcome of the first step of the assessment for the emission-based instruments, i. e. the emission-based subsidies, standards and charges, are compared to tradeable permits in a second step. In addition to the evaluation according to the main criteria (Figure 7), the results of the subcriteria for the compatibility with the market economy are presented as well (Figure 6).

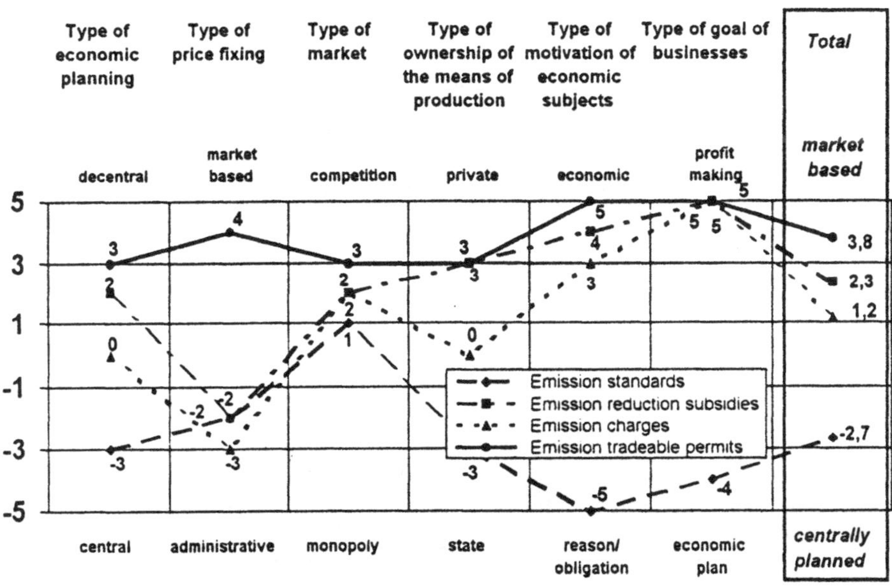

Figure 6. Market compatibility of the different emission based instruments

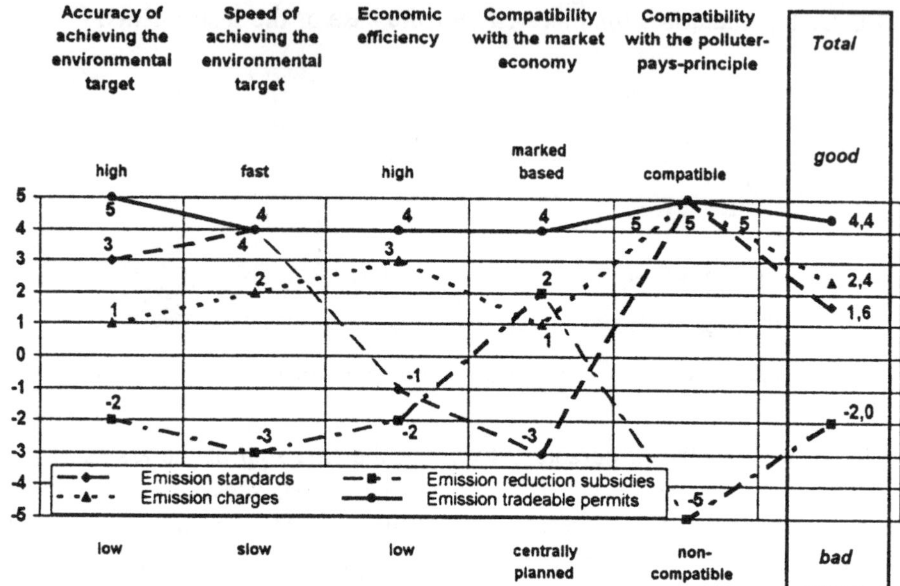

Figure 7. Evaluation of the different emission based instruments

Although the marks given by Knüppel (1988) are somewhat subjective and therefore debatable, the ranking in the overall results can be found in many different studies. Hence the evaluation yields the following overall hierarchy of the emission-based instruments: the best instrument are the tradable emission permits, followed by the emission charges, the emission standards and the emission reduction subsidies (Figure 8). This result is in line with the results presented by Endres in this volume (see pp. 409-425).

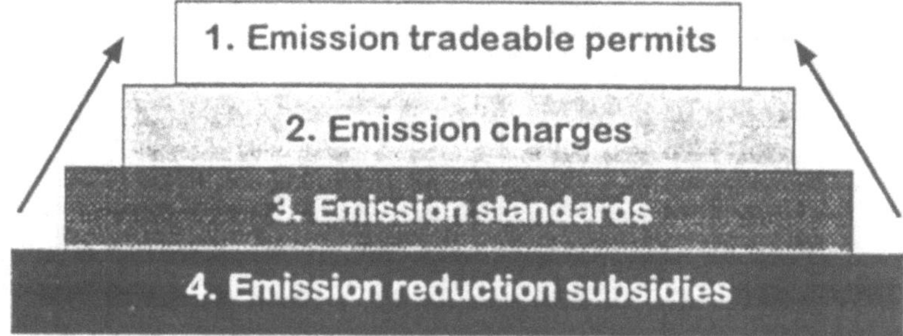

Figure 8. Hierarchy of instruments

5. Conclusion

Although not all readers will agree with the specific outcome of the evaluation by Knüppel (1988), maybe not even with the ranking of the instruments found in his evaluation, the general framework for the evaluation, possibly extended by additional criteria, that are not used by Knüppel (1988) or not even mentioned in this paper, can be used by everyone to come to other conclusions. The results of such an evaluation performed by the participants of the Advanced Study Course are presented in this volume (pp. 427-435).

The specific results of Knüppel (1988) are mainly presented as an example to illustrate the methodology and to present a possible ranking of the instruments, that can be, at least qualitatively, explained and that is in line with the opinion of the author and other evaluations both in the literature and in this volume (see article by Endres, pp. 409-425).

Another reservation has to be made with regard to Knüppel's results: He evaluates the instruments on a very general, theoretical and abstract level without taking specific designs of the various instruments into account. It is therefore possible that the results of this evaluation will differ for instruments used for the mitigation of global warming as exposed and discussed in the following papers in general and even for different designs of the same type of instrument.

References

Eucken, W. (1965) *Die Grundlagen der Nationalökonomie [The basics of the national economy]*, Berlin.

Jarre, J. (1975) *Umweltbelastung und ihre Verteilung auf soziale Schichten [Environmental effects and its distribution on social strata]*, Göttingen.

Knüppel, H. (1988) *Umweltpolitische Instrumente - Analyse der Bewertungskriterien und Aspekte einer Bewertung [Environmental policy instruments - Analysis of the criteria for evaluation and aspects of the evaluation]*, Dissertation, Universität Fredericiana, Karlsruhe.

Osterkamp, R. (1975) *Standards und Steuern als alternative Instrumente der Umweltpolitik - Ein theoretischer Vergleich [Standards and taxes as alternative instruments in environmental policy - a theoretical comparison]*, Dissertation, München.

Rat der Sachverständigen für Umweltfragen (1978) *Umweltgutachten 1978 [Environmental Assessment Study 1978]*, Bundestagsdrucksache 8/1938, Bonn.

Wicke, L. (1982) *Umweltökonomie - Eine praxisorientierte Einführung [Environmental economics - a practical introduction]*, Munich.

Zimmermann, K. (1982) Umweltpolitik und Verteilung - Sozioökonomische Hintergründe einer modernen Verteilungsfrage [Environmental politics and distribution - Socioeconomic background of a modern distributional issue], in H. Möller et al. (Eds.), *Umweltökonomie [Environmental economics]*, Königstein.

SECTION III

POLITICAL AND ECONOMICAL INSTRUMENTS - AN OVERVIEW

VOLUNTARY APPROACHES

GLOBAL AND EUROPEAN VOLUNTARY APPROACHES IN MITIGATION OF GREENHOUSE GAS EMISSIONS ON MUNICIPAL LEVEL AND THEIR RESULTS

Dr. KLAUS MÜSCHEN
Energy Planning Department
Ministry of Urban Development,
Environmental Protection and Technology
Am Köllnischen Park 3, D-10173 Berlin, Germany

1. Introduction

In the beginning of the 90ths many cities and local governments all over the world start to force actions in mitigation of greenhouse gas emissions. There are two networks which concentrate the cooperation of these local actors: The Climate Alliance and the Cities for Climate Protection Campaign.

The Climate Alliance was founded in 1990 by European local authorities together with delegates of the indigenous peoples of Latin America. The objectives of the Climate Alliance are the reduction of greenhouse gas emissions and the preservation of the world's rainforest. In 1997 already 650 European cities are members of the Alliance.

The Cities for Climate Protection Campaign was conceived at the first Municipal Leaders' Summit on Climate Change held at the UN in New York in January 1993. The more than 60 cities in Europe which has joined the campaign have pledged to develop and implement a local action plan to reduce greenhouse gas emissions on the local level.

Two years ago the Heidelberg Conference on „How to Combat Global Warming at the Local Level" took place. After discussing creative approaches to deal with global and local environmental problems on a local level the participants, Mayors and authorized representatives of local authorities, signed the Heidelberg Declaration. The signatories plegde to take actions necesary to reduce CO_2 emissions by 20% from 1987 levels by the year 2005.

In March 1995 the Second Municipal Leaders'Summit on Climate Change was

J. Hacker and A. Pelchen (eds.), Goals and Economic Instruments for the Achievement of Global Warming Mitigation in Europe, 161–167.

held in Berlin parallel to the Conference of the Parties (COP) of the UN Framework Convention on Climate Change, and in October 1995 the Third Local Government Leaders 'Summit on Climate Change in Saitama/Japan. The next summit will be held in Nagoya/Japan one week before the COP 3 starts in Kyoto. The summits gave presentations of the climate-protection initiatives and discussions on local strategies for combatting the problem of global warming. In Berlin the Summit also presented an Action Plan to the UN Conference. Key elements included:

- strong support for the Alliance of Small Island States (AOSIS) protocol calling on developed nations to commit themselves to a 20% reduction in CO_2-emissions;
- formal COP recognition of the municipal sector for the purpose of future consultation through the Subsidiary Bodies;
- a call to the COP to establish jointly with minicipalities worldwide the Local Authority Climate Assembly, which would seek to coordinate and facilitate municipal action on climate change in concert with national governments' national action plans.

In autumn 1996 the Heidelberg Declaration Follow-Up Workshop was held in Heidelberg again. The leading questions were as follows:

- What progress have the cities made to meet the commitments they made by signing the Heidelberg Declaration and by joining the CCP?
- Which programmes have been successfully implemented to reduce greenhouse-gas emissions?
- Which approaches were less successful and why?

In Heidelberg cities of Japan, USA, Peru, Uganda, Canada and Europe discussed their experiences. The following European cities gave reports on their activities at the Heidelberg Workshop:

- Barcelona, Spain
- Berlin, Germany
- Freiburg, Germany
- Gdansk, Poland
- Graz, Austria
- Hanover, Germany
- Helsinki, Finland
- Heidelberg, Germany
- Karlsruhe, Germany
- Katowice, Poland
- Leipzig, Germany
- Lucerne, Switzerland
- Lviv, Ucraine
- Montpellier, France
- Newcastle upon Tyne, UK
- Palermo, Italy
- Rome, Italy
- Saarbrücken, Germany
- Viernheim, Germany
- Warsaw-Mokotóv, Poland
- Zurich, Switzerland

2. Local Action Plan

All cities already established local action plans to increase greenhouse-gas emissions. The main problem still is the evaluation of the enforced measures. The databases for evaluation must be improved to ensure a detailed monitoring of the action plan. The available instruments of the action plan can be classified in regulation and standards, funding, self-obligation, education and information.

2.1. BUILDING STOCK

The projects *Thermo Profit* (Graz) and Energy Saving Partnership (Berlin) started to finance retrofitting and energy efficency measures for estate administration in a contracting model. A number of houses were offert to private energy service companies for a 10 (ore more) years contract. The investment for energy efficiency will be financed by the energy cost savings.

Many cities started to increase the share of *district heating* and natural gas for the housing stock, so replacing lignite (Heidelberg, Katowice, Berlin, Saarbrücken, Leipzig).

At the local level there is an increase in the use of *small combined heat and power plants* (Berlin, Leipzig, Saarbrücken). Typical applications are hospitals, school or sporting centers, industrial use of heat and power, new housing areas.

2.2. MUNICIPAL BUILDINGS

The key to reduce energy consumption in the public sector is to develop and implement a centralised *energy management*. Many cities have already established a Municipal Energy Management with special responsibilities in controlling the energy consumption and in coordinating the measures (i.e. Graz, Heidelberg, Saarbrücken, Berlin).

Many cities set *low energy standards* for new public buildings because national standards often miss the state of the art of energy conservation. (Heidelberg, Berlin).

The City of Berlin operates approximately 6000 buildings of various sizes and conditions. The energy-costs of of the real estate amount to about DM 500 million annually. According to a number of studies, the cost-cutting and energy-saving potential of these buildings is on average 30%. So we initiated a new model: *Energy Savings Partnership Berlin* (ESP). The objective is to transfer the energy management of a pool of buildings to an energy saving partner. This private energy service company is responsible for the planning, constuction, financing, operation as well as the servicing of all energy-relevant installation components of the buildings. The pool of buildings presents a various cost cutting potential. This mix minimizes the risk for the energy saving partner in predicting the energy conservation. The energy saving partners will be required to reach a minimal energy cost savings of 6% annually. Any surplus coming from additional saving measures is to be shared with the City of Berlin.

Last summer we start the bidding for the first pools, 50 buildings. The procedure comprehends the following steps:

- Public announcement for bidding firms and consortia
- Initial selection of 10 bidders
- Negotiation, comparison and optimization of bids
- award of contract

The results of the bidding are a great challenge for both the city of Berlin and the energy saving partners. We stipulated two 12 years contracts - 50 buildings each - with guaranteed 9% and 11.25% reduction of energy costs for the Berlin budget. Because the real cuts in energy costs are approximately 25 - 30%, the difference flows in the budget of the energy saving partners to refinance their investment of the energy saving measures. This underlines the cooperative idea of the model: „The best way to save is to share“. Now we are preparing the next pools for bidding.

2.3. LOW ENERGY HOUSES

Pilotprojects in many cities demonstrate the use of solar energy and the state of the art of low (or Zero) energy houses (Graz, Heidelberg, Berlin). The biggest problem is still to reduce the additional cost of the advanced energy standards of these buildings. However there are examples now to build low energy houses with a requirement of less than 40 kWh/m^2*a at the same costs as normal standard houses with more than the double energy consumption.

2.4. SOLAR ENERGY AND RENEWABLES

Most of the CCP cities started *promotion campaigns* on Solar Energy and Renewables. This includes funding and financial support for these energy systems. A very interesting action is shown by the city of Graz, where prominent people build their own solar energy system.

In Berlin we initiated a *Solar Forum* together with the Technical University Berlin and the Berlin Technology Foundation. The idea is to promote solar energy systems among different actors in this fields like companies, architecs, planners, associations, administrations. The issues of the last meetings were the Solar Ordinance, Cost Covering Compensation for solar power and the use of daylightning.

The Berlin Parliament (Berliner Abgeordnetenhaus) approved a law in 1995 allowing the governing Senat of Berlin to enforce a *Solar Ordinance*, which foresees a 60% use of solar warm water für new houses. The discussion on the Solar Ordinance is still going on because the resistence of the association of trade and industry, which prefers a self-obligation instead the Solar Ordinance to promote the use of renewables.

More than 10 cities in Germany deal with different models of *cost covering compensation* for feeding solar power to the grid. Because the cost of solar power is still more than 1$/kWh, a cost covering compensation can cause a growth in the market without burdening the public budget. The result is a very low increase of the price for electricity for all consumergroups in the city.

2.5. INDUSTRIAL AND COMMERCIAL SECTOR

The City of Graz starts a consulting initiative for energy efficiency in the industrial and commercial sector named *EcoProfit*. This includes consulting about funding and financing of energy concepts especially for small and medium enterprises (SME).

In many regions, *energy agencies* were started to support this consulting process. A priority is given to measures which do not need extra financial support like contracting and third-party-financing for the diffusion af small cogeneration systems and the use of waste heat.

2.6. TRANSPORT SECTOR

The transport sector presents the biggest problems in reaching the goals on reducing greenhouse-gas emissions. The experiences of many cities are hopeful:

The City of Graz (Austria) established a plan called „*EcoDrive*" promoting Ultra-Low Emission Vehicles (e.g. electric and solar vehicles) through pilot projects and incentive systems.

In Katowice (Poland) the local government *rebuild the public transport sector*, to cut the increase in the use of private cars especially in the last years.

In the City of Heidelberg the new traffic plan was coordinated by the *Traffic Forum* Heidelberg with the participation of the citizens. Representatives from all groups, associations and institutions interested in the traffic situation of the city discussed the objectives of an ecological and economic mobility. The priority list of the plan includes the construction of new tram routes, increasing the frequency for public transport, introduction of local trains, expansion of traffic networks and a parking management.

Different public transport companies offered *job tickets* and student tickets for public transport.

A special new measure is the Karlsruhe (Germany) *Track-Sharing Model*. After investigation the model consists of three components:

- a vehicle able to use regional German Railway (DB) tracks and light rail tracks in the city center;
- connecting DB tracks to the already existing tramway system;
- building new stops on existing heavy-rail lines which can be served without lengthening travel times because the improved acceleration of light rail vehicles.

Since 1992 the first line is in operation. The number of users have more than doubled compared with the former heavy-rail trains. Corresponding, the use of cars has decreased. The model is particularly suited for medium-sized cities and regions with 200.000 - 500.000 inhabitants. Other cities - Geneva, Saarbrücken, Kassel, Ulm - have or will adopt the track-sharing model too.

2.7. UTILITIES

Many cities are promoting *energy service* companies or local energy agencies. Especially the public utilities owned by the city could be forced to offer new energy services such as demand-side-management, contracting models, premium programmes, cogeneration and district heating.

2.8. OTHER ACTIVITIES

- Energy Efficient Student Hotels (Graz)
- Bonus-Systems for schools participating in energy and cost savings (Hannover, Hamburg, Berlin)
- Local Agenda 21 (Heidelberg, Saarbrücken, Helsinki, Berlin, etc.)
- Public-Awareness / Climate Campaigns and Energy Dialogue (Heidelberg, Berlin)
- Contracting, Third Party Financing

3. Conclusion

Although I was able to describe only a small part of the activities of cities in the Climate Protection campaign in Europe, there are some conclusions at the Heidelberg Workshop. Two other networks of European cities - the Climate Alliance and the Energie-Cités attending the Heidelberg Workshop aggreed with these conclusions:

- In evaluation of local action plans is a gap to the evaluation of different measures. On particular measures we have well known results like the energy savings or reduction in greenhouse gas emissions. But we need calculated data bases and balances for the hole action plans of cities.

- We have to work harder on the transport sector. The reduction in greenhouse-gas emissions in other sectors will be (over-)compensated by the increase in transport and car use. In Germany the free time activities are responsible for more than 50% of the car use. Normal cars become more efficient and we are talking about the 3-liter-car. In USA the most popular off-road cars still need 20 l per 100 km.

- We also have to take more effort at the national and international level. The Berlin Summit had presented an Action Plan to the UN Conference demanding a national and international support for the municipal climate activities. We need higher and effective energy taxes to internalize the social and ecological costs of the use of energy. Higher energy prices affect a higher public awareness on energy conservation and lead the actors towards economic efficiency and energy saving measures.

- We need more cooperation and coordination among the city networks in Europe. The Cities Climate Protection campaign, the Climate Alliance of European Cities and the Energie-Cités are acting on the same issues.

- Last but not least we have to awake the interest of other cities and convince them to join the Cities for Climate Protection campaign. Together we can compare different experiences and learn from each other about new solutions and models of saving energy and reducing greenhouse gas emissions.

VOLUNTARY COMMITMENTS TO MITIGATE GREENHOUSE GAS EMISSIONS - THE EXAMPLE OF GERMAN INDUSTRY AND TRADE*

FRANZJOSEF SCHAFHAUSEN
Federal Ministry for the Environment, Nature Conservation and Nuclear Safety
Postfach 12 06 29, D-53048 Bonn, Germany

* This contribution is an updated version of the presentation to the Advanced Study Course, including especially the results of the First Monitoring Report which was not yet available at the time of the course. This contribution is dated 29th December 1997.

1. The role of voluntary commitments in the Environmental Policy in Germany

Voluntary agreements, voluntary declarations and voluntary commitments are one category of numerous categories of instruments used by the German environmental policy since more than two decades. You can find first discussions and analyses at the beginning of the seventies.

Looking at the different categories of tools

- regulations/command and control approaches,
- economic incentives/instruments,
- complementary mechanisms like education, advice and information

voluntary agreements are without doubt part of the second category.

In the German Environmental Policy voluntary commitments are relatively often used and therefore very well known and approved approaches. Between the early seventies and today you can find approximately 80 - 100 different commitments covering various issues with different outcome.

The main fields of such commitments were

- waste reduction,
- reduction of product related emissions,
- substitution of products (for instance the CFC-reduction in the early nineties),

J. Hacker and A. Pelchen (eds.), Goals and Economic Instruments for the Achievement of Global Warming Mitigation in Europe, 169–182.

But never before German Environmental Policy had used such far-reaching, all branches embracing voluntary commitments like the Declaration by German Industry and Trade on Global Climate Prevention.

Obviously, there has been a similar recognition in other OECD countries and a large number of voluntary agreements have been made. For example, a workshop on voluntary agreements on climate protection prepared by the Federal Ministry for Environment, Nature Conservation and Nuclear Safety (BMU) and the International Energy Agency (IEA) held in October 1995 in Bonn showed that in the last two years voluntary agreements have been reached in nearly all OECD countries (IEA, 1997). Naturally, these voluntary agreements have all quite different specific features, e.g. with regard to their binding nature, the targets set, the terms of the agreements, reciprocity etc. However, there is no doubt that voluntary agreements are regarded and used as a serious and real alternative to other instruments and strategies in climate protection.

The philosophy behind voluntary agreements/commitments has not changed since the early seventies:

- building of consensus by participatory processes,
- complementation and substitution of traditional command and control approaches and/or economic mechanisms,
- decentralisation of decisions,
- creation of flexibility to support an cost efficient and sustainable environmental policy.

During the seventies, eighties and nineties German Environmental Policy creates a very close network of regulations, economic incentives and complementary mechanisms to prevent environmental degradation. Water and waste management, Clean Air Act with it's ordinances, Soil Conservation, Chemical Safety and Nature Conservation are some of the fields of action. During the eighties the issue of prevention and co-operation gains more and more importance. This offers the chance to use more flexible approaches and to decentralise decisions on environmental protection. Use of standardisation, voluntary actions, emissions trading and joint implementation were some main fields of discussion and implementation.

The question was not: Implementation of traditional command and control measures in every case. The question was: How to meet a clearly defined target in a cost-effective manner.

2. The history of voluntary commitments on Climate Change by the German Industry and Trade

The presentation and implementation of the Declaration by German Industry and Trade on Global Warming Prevention constitutes a new element in the climate Change protection policy of the Federal Government. In its climate protection policy the Federal Government is applying not only the polluter-pays principle and the precautionary principle but also the principle of co-operation between all important actors. This development in environmental policy is indicative of a high level of environmental awareness in our society, on the basis of which calls can be made for individual responsibility and initiative.

Only a few weeks before the Conference of the Parties to the United Nations Framework Convention on Climate Change (UN-FCCC) in spring 1995 in Berlin, German Industry and Trade entered into a commitment, „to make special efforts on a voluntary basis to reduce its specific CO_2 emissions and its specific energy consumption by up to 20% by the year 2005 (base year 1987)" (BMU et al., 1995). The general commitment to reduce greenhouse gas emissions was based on declarations by 15 industrial associations, which had formulated different targets in specific individual commitments. This voluntary obligation was subject to criticism from various sides. Not only the declared aim, but also the monitoring of the obligations were considered in need of improvement.

On the anniversary of the First Conference of the Parties to the United Nations Framework Convention on Climate Change on 27 March, 1996, German Industry and Trade presented an updated and extended declaration which was intended to take account of that criticism (BMU et al., 1996). The objective was modified in such a way that the obligations to reduce CO_2-emissions were now set in relation to the international customary year of 1990 and the wording „up to" was dropped. The decision for the year 1990 took account of the political requirements and declarations: It is associated with the considerable empirical, statistical and conceptual problems. Over and above this, the individual declarations were firmed up and in some cases redrafted absolute obligations to reduce CO_2 emissions. As a result, the intended reduction target in some sectors were nor only significantly more demanding than the so-called „business as usual" scenario, but in some cases also exceeded the results of a potential waste heat ordinance. In order to counter the critical objections with regard to the verification of the obligation taken on, it was agreed that there should be accompanying CO_2 monitoring for which the reduction results would be presented at yearly intervals and verified for conformity with the targets.

DIAGRAM 1. List of participating Associations of German Industry and Trade

Federation of German Industries

- Bundesverband Steine und Erden - Cement Industry
- Bundesverband Steine und Erden - Brickworks Industry
- Bundesverband Steine und Erden - Limestone Industry
- Bundesverband Steine und Erden - Fire-Proofing Industry
- Bundesverband Steine und Erden - Ceramic Tiles and Panels Industry
- Bundesverband Glas- und Mineralfaserindustrie - Glass and Mineral Fibers Industry
- Kaliverein - Potassium Association
- Verband Deutscher Papierfabriken - Paper and Pulp Industry
- Verband der Chemischen Industrie - Chemicals Industry
- Wirtschaftsverband Metalle - Non-Ferrous Metal Industry
- Wirtschaftsverband Stahl - Steel Industry
- Wirtschaftsverband Zucker/Verein der Zuckerindustrie - Sugar Industry
- Gesamtverband der Textilindustrie - Textile Industry
- Mineralölwirtschaftsverband - Oil Refining Industry

Bundesverband der deutschen Gas- und Wasserwirtschaft (BGW) - Federal Association of Gas and Water utilities

Vereinigung deutscher Elektrizitätswerke (VDEW) - Association of German Electricity Suppliers

Verband der Industriellen Energie- und Kraftwirtschaft (VIK) - Industrial Energy Consumers and Self-Producers

Verband kommunaler Unternehmen (VkU) - Association of municipal enterprises

The fundamentally new element in the Declaration of the German business community on Global Warming Prevention is the extraordinarily broad scope of application. Altogether, the declaration covers an approximate 71% of the total energy consumption of German Industry and 99% of public electricity production. The commitments cover entire production processes and - in some cases - also individual products.

DIAGRAM 2. The key element of the voluntary commitment by German Industry and Trade

„German Industry and Trade are prepared to make a special effort on a voluntary basis to reduce their specific CO_2 emissions or their specific energy consumption by 20% in the period up to the year 2005 (base year 1990)."

3. The main elements of voluntary commitments on Climate Change in Germany - a step by step approach -

On 27 March 1996 roughly 2/3 of the German industry and trade gave an undertaking to the Federal Government to make every effort on a voluntary basis to reduce specific CO_2 emissions and specific energy consumption by 20% by the year 2005, compared to 1990th emission level.

This pledge made by German industry is based on voluntary commitments entered into by the various sectors of industry and trade, including the energy sector. Each of these voluntary commitments has its own special features which take into account the current conditions in the sector and the different potential to contribute to climate change prevention.

In contrast to the situation in the Netherlands or in the United States of America for example, in Germany whole sectors of trade and industry - represented by their associations - have entered into such voluntary commitments with the government and not individual companies.

These commitments are *politically binding* undertakings (in the sense of unilateral, *not legally binding* declarations) but not formal agreements or contracts for instance between individual companies and the government (as usual in the Netherlands).

With voluntary commitments the emphasis is on the word „voluntary" and it was indeed a voluntary pledge made by German Industry and Trade to the government. The Federal Government has responded positively to these commitments and in turn made a political declaration that it will - for the time being - dispense with further command and control measures like the so-called „waste heat ordinance" and will take into account the CO_2 reduction reached under the regime of voluntary commitments in the case of the introduction of an EU-wide CO_2-/energy tax.[1]

In contrast to the Dutch case, the German Government has

- provided no financial support for the implementation of the voluntary commitments on Climate Change Prevention,
- offered no easing of approval procedures required by environmental law,
- pledged no consistent interlinkage of the content of environmental policy with other policies, e.g. energy policy or employment policy.

DIAGRAM 3. Size of the voluntary commitment on Global Warming Prevention by German Industry and Trade

71%	Total energy end use of German Industry
99%	public electricity production

These differences between the German and - for example - the Dutch voluntary agreements is also a consequence of the fact that in Germany there has certainly never before been such a far-reaching, all-embracing voluntary commitment covering nearly the entire German industry. The high political importance which the German population attach to climate protection also plays an important role. It does not make it easy for politicians to entrust this issue to a new instrument which they only partly have the ability to shape.

The monitoring process must - on the one hand - take into consideration the branch-specific differences and - on the other hand - guarantee clear, plausible and comparable results as well as for the politicians and the German Industry.

If you take an objective look at the climate protection „pact" between German Industry and the Federal Government, you can see that considerable progress on climate protection can be made with this new political approach which can lend a new dimension to environmental and climate protection. However, anyone who believes that this approach had met with undivided approval is quite wrong. Science, the Greens and the Ecology Movement have severely criticised both the government and industry (Rennings et al., 1997; Fischedick et al., 1995; Deutsches Institut für Wirtschaftsforschung, 1995; Wuppertal-Institut, Erster Monitoring Bericht, 1997).

[1] E.g. based on the Proposal of the EU-Commission „ Proposal for Council Directive „Restructuring the Community Framework for the taxation of energy products", (COM(97)30 final).

4. Conclusions

As mentioned before the Federal Government and the German Industry and Trade agreed in February 1996 on a comprehensive and transparent monitoring system (BDI, 1996). The first monitoring report was presented by the Rheinisch-Westfälisches Institut für Wirtschaftsforschung, Essen as the external and neutral monitor on 18 November 1997 (Hillebrand et al., 1997). In the Conclusions of the First Monitoring Report is stated on pages 48-51:

„Even if a period of not quite two years experience with the Voluntary commitment Declaration from German Industry and the monitoring report is relatively short, the report nevertheless permits a number of conclusions which may be of significance as to the final success of this instrument.

It is first to be noted that the report from the associations contains a large number of actions which document the special effort to reduce CO_2-emissions. Even if the actions cannot be attributed to the voluntary obligation alone, they still document the intensive efforts to achieve a more rational use of energy and a reduction in CO_2-emissions. The efforts are directed both at the optimisation of individual production processes, and at energy supply in general. It becomes clear that isolated improvements of individual production processes are increasingly running up against technical and physical limits, and a further increase in energy efficiency therefore hardly appears possible or is associated with unreasonably high costs. It can be deduced from this finding that integrated supply strategies are an attractive opportunity to use energy as rationally as possible and jointly. Examples of this are provided not only by the steel industry, the cement industry or the chemicals industry, but above all by the consultancy initiative of the VIK (Verband der Industriellen Energie- und Kraftwirtschaft - Association of Industrial Energy Consumers and Self-Producers). Even if this initiative was only called into being last year, the results already achieved indicate that co-operation between various associations and companies will be able to bring about a tangible increase in the use of residual heat from industrial plant and significantly improve the CO_2 balance, without the involvement of legislation or additional taxes.

A large number of examples also demonstrate that increased energy efficiency can also reduce the input of other raw materials. To this extent, the actions not only induce a reduction in CO_2-emissions, but contribute in general to a more sustainable use of resources. Examples of this can be found, for instance, in the steel, cement and paper industries.

TABLE 1. Declaration by German Industry and Trade on Global Warming Prevention - Synopsis of the Declarations of 10 March 1995 and of 27 March 1996

10 March 1995	27 March 1996
15 Associations of German Industry and Trade	19 Associations of German Industry and Trade
Base Year: 1987	Base Year: 1990
Target Year: 2005	Target Year: 2005
„...reduction of specific CO_2-emissions or specific energy consumption up to 20% ..."	„...reduction of specific CO_2-emissions or specific energy consumption by 20%..."
Monitoring-concept announced	annual Monitoring-System by an external Monitor implemented
------------	absolute reduction of CO_2-emissions by a total of 170 Mio t (base year 1990/target year 2005)

The voluntary commitment and CO_2 monitoring are still in the experimental phase. It can not therefore be regarded as surprising that the monitoring process itself and its results are still in need of improvement. This concerns, among other factors, the reports from the associations involved in monitoring. In comparison, for example, with taxes or levies, this instrument demands a higher degree of preparedness to disclose information on actions intended to increase energy efficiency. This requires not only staff capacities to perform this function, but also as a rule demands access to internal data and information from within companies. This could explain, at least in part, the heterogeneous nature of the descriptions of actions on which this monitoring report is based. In spite of this information problems, it is surely indisputable that the presentation of the actions is an essential part of the monitoring process, and the voluntary obligation. Only with transparent presentation of the actions taken, the objective of monitoring to verify the reduction obligations received and also constitute a firm basis for the removal of information deficits in the field of use and conversion of energy, can be fulfilled.

....

DIAGRAM 4. Characteristics of volunatry commitments in Germany

- Clear Commitments
- No contracts between companies/associations and the government (unilateral, legally not binding declarations)
- Transparent Monitoring-System (periodical verification by a third party - RWI, Essen)
- Publication
- No Sanctions - but implementation of regulatory or economic mechanisms/instruments if the commitments will not be fulfilled

The comparison of the declared objectives with the efficiency improvements or CO_2 reductions already achieved shows that in some industries the target was almost achieved as early as 1995. This should be an incentive to reconsider the targets. A linear extrapolation of the CO_2 reductions achieved between 1990 and 1995 would not be appropriate, as the savings achieved in East Germany certainly cannot be repeated to the same extent. It is also to be taken into account that, when one remembers that energy consumption and capital input are complementary, the increases in efficiency are tied to investment cycles and can therefore take place intermittently. Nevertheless, degrees of target achievement of 85 or 90% point to additional savings potentials, which should lead to a corresponding redefinition in order to document the effectiveness of this instrument."

It should be recognized that the Declaration by German Industry and Trade on Global Warming Prevention has to be implemented within a dynamic process. Review phases and dialogues between the Federal Government and the various associations on the basis of the yearly CO_2-Monitoring reports are essential elements of this process.

TABLE 2. Absolute CO_2 reduction under the Declaration by German Industry and Trade on Global Warming Prevention between 1990 and 1996

Association	Reduction between 1990 and 1996 in %	
Potash Industry	77,0	
Cement Industry *	22,5	
Lime Industry * **	11,1	
Ceramic Tiles and Slabs	32,5	
Brick Industry	5,6	
Refractory Industry * **	12,8	
Iron and Steel Industry	17,1	
Non-ferrous Metals Industry	13,2	
Chemicals Industry	24,2	
Paper Industry	8,2	
Glass Industry *	5,2	
Textile Industry	30,0	
Sugar Industry	42,8	
Total Industrial Associations	20,6	
Public Electricity Supply		9,7
Gas Industry		21,4
Petroleum Industry		20,3
Municipal Energy Supply		-----

* Base year 1987 ** West Germany

To give only two examples for the different reasons behind the development presented in TABLE 2:

- It can be seen that the clear reduction of CO_2-emissions in the textiles industry was brought about by relocating companies outside Germany.
- In the steel industry the monitoring report reveals the effects of the changeover from top-blown steel making to electric steel making, which allows a greater use of scrap.

5. A first outlook - lessons learned from the First Monitoring Report

From the politicians point of view, a major prerequisite for the success and therefore for the acceptance and sustainibility of voluntary commitments is that - in principle - no one is allowed a free ride.

The occurrence of this problem cannot be entirely prevented but with the German solution it can be minimised on the side of industry for the following reasons. Firstly, the branch associations are characterised by an more or less streamlined and transparent organisation which can be expected to take successful and disciplined action in such matters. Secondly, the fact that two thirds of German industry has signed these voluntary commitments clearly restricts the potential for free riders. Thirdly the Federal Government has only announced that in the event of an EU-wide tax on CO_2/energy it „will make every effort to ensure that those sectors of industry involved in the voluntary commitment campaign are exempt from such a tax or that full credit is granted for the CO_2 reductions they have achieved.“ (Press and Information Office of the Federal Government, 1996)

Another major and indispensable criterion is the comprehensive, ambitious and transparent monitoring once a year. Monitoring ensures, on the one hand, that free riders are recognised relatively quickly and that their potential advantage is kept within narrow defined limits.

But monitoring also serves other functions.

It ensures firstly that government, science and public can - at reasonable intervals - check whether industry is keeping the pledges it makes.

It ensures secondly that the process of implementing and developing the voluntary commitments step by step is based on an ambitious review concept using the steps:

- development and implementation of first commitments,
- review of the progress and analyse of the options,
- start of a dialogue between government and industry and trade associations,
- adjustment of the origin commitments.

In this process the quality of the pledges will be repeatedly reviewed and the pledges of 1995 and 1996 will be further developed in a dialogue between the government an the business community.

The momentum of this process has already been seen in the further development of the „Declaration by German Industry and Trade on Global Warming Prevention" between 1995 and 1996.

For the present situation this means:

1. New negotiations with industry and trade associations recognising the fact that the First Monitoring Report has shown that there are in different branches additional CO_2 reduction potentials because the degree of the target achievements of these branches between 1990 and 1995 is between 85 and 90%. On the basis of this results of the First Monitoring Report the Federal Cabinet decision on Climate Change dated on the 6 November 1997 argue that under the voluntary commitment by German Industry and Trade on Global Climate Prevention is an additional CO_2 reduction potential of 10 - 20 million tons (target year 2005) (Federal Ministry for the Environment, 1997).
2. Starting negotiations with additional associations which have not entered the Declaration of German Industry and Trade on Global Climate Prevention.
3. Covering not only CO_2 but also other greenhouse gases like N_2O (Chemicals Industry), CH_4 (Gas-, Mineral Oil- and Coal -Industry), CF_4 and C_2F_6 (Aluminium Industry), PFC's, HFC's and SF_6 (Chemicals Industry) under the Declaration by German Industry and Trade.
4. Extending the present voluntary commitments from production processes to products as actually done by Glass Industry (Isolation windows) and Chemicals Industry (Isolation material). There is additional potential for instance in the fields of household appliances and cars.
5. Combining voluntary commitments with mechanisms like „ joint implementation" or „emissions trading". Some associations like the Association of Chemicals Industry, the Association of Public Electricity Suppliers, The Association of Gas Companies are interested in such a innovative combination of different mechanisms.
6. Combining the voluntary commitments with the Proposal of the EU-Commission on Restructuring the Excise Duties on Energy in the European Union.

6. Final Remarks and Conclusions

Voluntary commitments have a long tradition in environmental policy in Germany. Everyone involved is aware that a failure will lead to government regulation. To this extend a great deal of political and moral pressure is exerted by such approaches.

Voluntary approaches can achieve policy objectives and help integrate economic and environmental goals, but their effectiveness may vary according to how structured or how closely monitored they are. They can provide a useful long-term signal to participants, assuring greater continuity, specificity, flexibility and capturing a shared perception of solutions. Their success will be influenced by various factors including socio-economic factors, technological progress and market developments (business and investment cycles). In every case these approaches can yield some benefits in terms of greater understanding and co-operation between industry and government.

Having a look back over the past three decades the Federal Environment Agency (Umweltbundesamt, Berlin) in a detailed analysis has established that

DIAGRAM 5. Benefits of Voluntary Commitments

- The polluter (company) can choose the most effective means to met the target(s)
- Increased Flexibility
- Fast results
- Lower costs for the government for administration and implementation
- Objectives can go beyond the targets for regulations
- Business community can use the „Green-Image"
- Increased trust among government and business community
- Regulation or other economic incentives are not appropriate in some cases

the voluntary commitments can be regarded as a convincing success. This conclusion is more or less in line with the results of the First Monitoring Report presented by the Rheinisch-Westfälisches Institut für Wirtschaftsforschung, Essen in November 1997. I believe that voluntary commitments are a new path in environmental policy. This path is based on the principle of co-operation. It needs mutual trust, as well as controls.

References

Bundesverband der Deutschen Industrie e.V. - BDI - (1996) CO_2-Monitoring, Konzept für die Erstellung von regelmäßigen Fortschrittsberichten zur transparenten und nachvollziehbaren Verifikation der „Erklärung der deutschen Wirtschaft zur Klimavorsorge", Cologne.

Deutsches Institut für Wirtschaftsforschung (1995)„Selbstverpflichtung" der Wirtschaft zu CO_2-Reduktion: Kein Ersatz für aktive Klimapolitik, DIW-Wochenbericht **14**, Berlin.

Federal Ministry for the Environment, Nature Conservation and Nuclear Safety - BMU - (Ed.) (1994) The Federal Government's Decision of 29 September 1994 on Reduction Emissions of CO_2 and Emissions of other Greenhouse Gases in the Federal Republic of Germany , Bonn.

Federal Ministry for the Environment, Nature Conservation and Nuclear Safety - BMU -, Federal Ministry for Economics, Federation of German Industries e.V. (Ed.) (1995) Declaration by German Industry and Trade on Global Warming Prevention, Cologne.

Federal Ministry for the Environment, Nature Conservation and Nuclear Safety - BMU -, Federal Ministry for Economics, Federation of German Industries e.V. (Ed.) (1996) Updated and extended Declaration by German Industry and Trade on Global Warming Prevention, Cologne.

Federal Ministry for the Environment, Nature Conservation and Nuclear Safety - BMU - (Ed.) (1997) Decision of the Federal Government on the Climate Protection Programme of the Federal Republic of Germany on the Basis of the Fourth Report of the CO_2 Reduction Interministerial Working Group (CO_2 Reduction IWG), Bonn.

Fischedick, M.; Kristof, K.; Ramesohl, S. and Thomas, S. (1995) „Erklärung der Deutschen Wirtschaft zur Klimavorsorge": Königsweg oder Mogelpackung ?, Wuppertal Papers No. 39, Wuppertal.

Hillebrand, B.; Buttermann, H.-G. and Oberheitmann, A. (1997) First Monitoring Report: CO_2-Emissions in German Industry 1995 - 1996, RWI-Papiere, No. 50, Rheinisch-Westfälisches Institut für Wirtschaftsforschung, Essen.

International Energy Agency - IEA - (Ed.) (1997) Voluntary Approaches for Mitigating Greenhouse Gas Emissions, Conference Proceedings, Bonn.

Press and Information Office of the Federal Government (Presse- und Informationsamt der Bundesregierung), Press Bulletin No. 118/96 , 27 March 1996, „Declaration by German Industry on Global Warming Prevention, Encouraging annual assessment of the second round of CO_2 voluntary commitment, German Industry specifies and extends its commitment of 10 March 1995".

Rennings, K.; Brockmann, K.-L. and Bergmann, H.(1997) Voluntary Agreements in Environmental Protection - Experiences in Germany and Future Perspectives, Discussion Paper No. 97-04 E Zentrum für Europäische Wirtschaftsforschung (ZEW), Mannheim.

Wuppertal-Institut (1997) Erster Monitoring-Bericht zur Erklärung der deutschen Wirtschaft zur Klimavorsorge, Nachbesserung der Klimaschutzerklärung und des Monitoringverfahrens dringend notwendig, Wuppertal/Kyoto, 9.12.1997.

VOLUNTARY AGREEMENTS IN CLIMATE PROTECTION - EXPERIENCES IN GERMANY AND FUTURE PERSPECTIVES

Dr. K. RENNINGS, K. L. BROCKMANN, H. BERGMANN
Zentrum für Europäische Wirtschaftsforschung (ZEW)
Postfach 10 34 43, D-68034 Mannheim, Germany

Abstract

A trend towards „softer" regulation, especially in the form of negotiated environmental agreements, is observable in national and international environmental policies. Such agreements are controversial, because there are fears that government will relinquish its responsibility for environmental protection. This paper analyses experiences with voluntary agreements in German climate policy. Especially the voluntary agreement made by a number of industries on a CO_2 reduction by the year 2005. The paper is based on a study commissioned by the German Federal Ministry of Economics (Rennings et al. 1996).

Proponents of voluntary agreements argue that this instrument provides incentives to the business sector for the development of efficient, innovative and environmentally-friendly solutions. Analysing voluntary agreements in German climate policy, we conclude that it is hard to detect solutions derserving such attributes. These agreements are unlikely to produce results that go beyond what industry would have done in any case and they avoid using economic incentives. The agreements are non-binding and unenforceable, with the negotiating process leading to a watering down of the environmental goals government had originally aimed at. A preference for negotiated solutions on principle, as espoused by the Federal Government in Germany, seems to be „counterproductive". If the government clearly signals its willingness to refrain from using regulatory or economic instruments in favour of industry agreements, it weakens its negotiating position. The government also limits its options should the implementation of the agreement prove unsatisfactory. Government needs to be „in control" in order to leave its choice of policy instruments open and to be flexible. In a last step, we derive some general conlusions concerning reasonable strategies and applications of voluntary agreements within the European Union.

J. Hacker and A. Pelchen (eds.), Goals and Economic Instruments for the Achievement of Global Warming Mitigation in Europe, 183–204.

1. Introduction

By reverting to common command and control measures preventive environmental protection can entail a considerable extension of government intervention. The German Federal Government believes that this should be increasingly counteracted through cooperation between government and the business community, amongst other things through voluntary environmental protection measures. However, these „soft" instruments involve the risk of government relinquishing its responsibility for the environment. Negotiated agreements made by the business community, like the pledges to reduce carbon dioxide emissions, are therefore controversial from an ecological and economic point of view.

The economic perspective is specified here from a neo-liberal perspective. In a liberal approach, markets should coordinate the allocation of goods as far as no market failure can be observed. But when external effects, information deficiencies or inflexibilities exist, the state is responsible for correcting these kinds of market failure. However, the state has to choose the option which minimises market distortions. In other words: Market-based instruments like taxes or tradeable emission permits are generally preferable. Nevertheless, other types of instruments like voluntary agreements can be used alternatively or additionally when they have advantages with regard to certain criteria (e.g. efficiency, institutional controllability, minimisation of unwanted side-effects).

Especially the German Ordo-Liberal-School in the tradition of Eucken argues in favour of a government correcting market failure, and against an interventionist state. Principles and criteria of the ordo-liberal school will be specified in chapter 3 where we in general develop an analytical framework for the valuation of environmental policy measures with special interest in the application to voluntary agreements in climate policy. Since ordoliberalism is mainly a German economic school, it should be mentioned that similar ideas have been established in other countries. For example, the so-called Oxford-Liberals have created the guideline „as much competition as possible, as much planning as necessary" (GROSSEKETTLER 1991:106).[1]

---

[1] However, the ordoliberal school focuses more on the responsibility of the state to establish basic rules for a regulatory framework within long-term oriented economic policy, while the Oxford-Liberals give higher weight to policy measures in the short run.

2. Attributes of voluntary agreements in Germany

2.1. NEGOTIATED AGREEMENTS: SOFT, NON-VOLUNTARY AND OUT OF KEEPING WITH A MARKET ECONOMY

Negotiated agreements made by the business community are generally called voluntary. However, what government does is more like showing the instruments of torture to the victim as a first step of torture. Comparable to what, in former times, torturers would do with their instruments, nowadays the Minister for the Environment presents a draft for an ordinance so as to achieve „voluntary" concessions (MURSWIEK 1988:985). Hence, in principle such cooperative solutions can be interpreted as barter transactions in which the business community imposes an obligation on itself to act in a certain manner, and government in return refrains from enforcing the desired conduct.

In the debate on environmental policy, voluntary agreements are sometimes pictured as „a free-market instrument". This description is only justified in the case of instruments that use the price mechanism. There are, however, only a few exceptional voluntary agreements which fall back on this fundamental free-market principle. Yet, as a rule negotiated agreements shy away from such „tough" economic instruments and do not touch the structure of relative prices; after all, negotiated agreements do not spring from the market system, but rather from political negotiations between government and trade associations. However, an approach that is essentially based on negotiated solutions should not be characterized as market-based, but as a corporatist approach (HOLZHEY/TEGNER 1996:426-427). The main difference is that consumers have commonly no possibility to participate in the negotiation process, although the consumer surplus is highly influenced by the result of the negotiations (improvement of environmental quality).

2.2. SOVEREIGN VERSUS CORPORATIST APPROACH IN ENVIRONMENTAL POLICY

Voluntary agreements can contain all kinds of instruments. Goals or technical instructions for a specific industry, but also scales of charges and information systems are conceivable. That is why the agreements do not have any working mechanism in their own right, rather the latter depends on the instruments provided in each specific case. Having said this, there are typical characteristics of negotiated agreements concerning political decision-making and enforcement of agreements.

Voluntary environmental protection measures differ markedly from the classic model of a rule-issuing government trying to achieve its environmental policy goals through commands and prohibitions. Government refrains from using any instruments of formal power, i.e. from enacting legal norms, and instead enters into negotiations on the realization of the environmental goals with the groups affected. In doing so, government uses the „threat" of enacting restrictive legal norms as a starting point for negotiating a voluntary agreement and thus exerts a guiding influence on the business community's conduct (HOFFMANN-RIEM 1990:400, 426; BROHM 1992:1025, 1027). In contrast to traditional sovereign actions, government seeks a solution via consensus-building. However, this bargaining process holds dangers. Unlike any cooperation in the form of the right to be heard, the right of participation or of involvement, agreements are based on a system of service-and-service-in-return.

Concerning the enforcement of an agreement, what has to be stressed is that a mere promise still fails to guarantee its actual implementation. Companies not complying with the pledges (given by their association) do not risk any fines, penalties or other coercive measures (cf. HOFFMANN-RIEM 1990:400, 438; MURSWIEK 1988:985, 988). There is no claim to performance in the case of a voluntary agreement; indeed, what characterizes agreements is the lack of any binding force and of enforceability. In contrast to a contractual relationship, the party pledging voluntary measures never enters into any legal obligations. Contracts and agreements are similar in that both have an exchange of services in common. The relationship „do ut des", in other words a mutual give-and-take, is typical of contractual relations and agreements alike. However, the binding force of a contract - „pacta sunt servanda" - is completely foreign to negotiated agreements. Agreements are more likely to be categorized as a gentlemen's agreement, which does not entail any legal consequences.

3. Analytical roster for assessing voluntary agreements

The Freiburg ordoliberal school holds that the absence of markets, and any functional defects of existing markets, will result in corrective requirements which cannot be covered solely by an evolutionary competition of institutions - a view which will be pursued here below. This school regards as inadequate the process of natural selection emerging from international competition between institutions evolving as a quasi-random process, but not specified by the state, and the resultant elimination of inefficient institutional bodies. On the contrary, the Freiburg neo-liberals' concept of a free-enterprise economy presupposes that the state must selectively create institutions for countering existing defects and challenges (GROSSEKETTLER 1991:104-106). The efficiency of such institutions must be measured in terms of how far the principles of free enterprise are respected.

Seen from the perspective of ordoliberalism voluntary agreements are regarded with scepticism. Government no longer - as postulated in the ordoliberal model - stakes out the regulatory framework within which entrepreneurs can dedicate themselves to profit-making. Instead government delegates this responsibility to the businesses themselves. However, in the opinion of ordoliberals a laissez-faire policy with a market economy left to its own devices tends to destroy itself. Failure to protect and promote competition may lead to closed-off markets, cartels, price fixing and ultimately to a distorted price structure. A lack of competition and the wrong price signals in turn may entail a spate of government interventions, e.g. measures to monitor prices and regulatory requirements concerning the application of given environmental technologies (due to a lack of price incentives). That is why those economists who consider government to be a powerful custodian of the system in the tradition of Eucken's ordoliberal school view the trend in environmental policy towards voluntary agreements with great concern (MAIER-RIGAUD 1995).

However, the following assessment will test these hypotheses by applying the ordoliberal principles to concrete case studies of voluntary agreements. The generalized analytical roster in TABLE 1 comprises a catalogue of check criteria applying to the selection of instruments for eliminating deficits in a free-enterprise economy (cf. RENNINGS et al., 1996 for a detailed explanation of the criteria involved).

TABLE 1. Systematized check criteria for evaluating economic policy measures in terms of regulatory efficacy

Step 1: Goal formulation and operationalization:

- Formulating the targeted goal system
- Indicators
- Assignment of goals, means and implementing agencies

Step 2: Legitimization of the action's goal in terms of contract theory:

- Hypothetical justification (Rawls)
- Reference to concludent action

Step 3: Selection of the decision-making level/process:

- Subsidiarity principle
- Congruence principle:
  - Equivalence: the user group must coincide with the payer group for a collective good
  - Democratic monitoring: the group of the decision-impacted must coincide with the group of entitled monitorers

Step 4: Economic legitimization of actions formulated:

- **Choice of instruments**: selection of conceivable instruments for goal implementation
- **Effectiveness (goal-conformity)**:
  - Degree of goal attainment (direction and dosage)
  - Speed of goal attainment
  - Invariance against changes in the macro-economic boundary conditions
- Necessity (system-conformity):
  - Market-conformity:
    - Instrumental subsidiarity: designing measures with minimized impact on individuals' powers of decision-making (centralized/decentralized)
    - creation of fully functional markets (free price formation, fully functional competition)
    - Minimizing intervention into the functioning of existing markets
    - Priority of regulatory before process policy: formulation of a long-term orientation framework, and avoidance of stop-and-go measures
  - Minimization of detectable
  - unwanted side-effects:
    - Stability-policy goals (economic compatibility)
    - Distribution-policy goals (social compatibility)
- **Economic efficiency**:
  - Static economic efficiency (cost-efficiency):
    - Purpose/avoidance costs
    - Transaction costs
  - Dynamic economic efficiency (innovation efficiency)
- Institutional controllability: implementability in the political process and allowance for the possibilities for abuse in the political/administrative apparatus

Source: In broad conformity with GROSSEKETTLER (1991: p. 114).

With regard to voluntary agreements, these criteria can be interpreted and applied by the following questions:

Step 1 and 2: Goal operationalisation and justification

- What environmental goal does the measure examined in a given case study refer to?
- How can the targets be measured?
- Can this goal be justified hypothetically (Rawls theory of justice) or empirically (democratic decisions, concludent action)?
- Was the goal originally pursued watered down as early as during the negotiation process or were there any delays?

Step 3: Selection of the decision-making level/process

- Are decisions taken at an appropriate level with regard to the principles of subsidiarity and congruence?

- Is any environmental responsibility delegated and, if so, how is the delegation to be assessed?

Step 4: Economic legitimization of actions formulated

- Choice of instruments: what are the relevant instruments agreements have to be compared with?
- Goal conformity: Is it possible to achieve the specific underlying environmental policy goal with the help of the given agreement?
- System conformity: Is the measure in keeping with the system of a social market economy? Is the system strengthened, does it remain unchanged or is it weakened? Are detectable unwanted side-effects minimized?
- Economic efficiency: Are the costs higher or lower than when using other instruments? What incentives are provided to achieve technological progress?
- Institutional controllability: How immune are voluntary agreements to political influences that can dilute their desired effect? How do the associations deal with the problem of free riders?

4. Goal operationalisation and justification

The Chlorofluorcarbon (CFC)-phase-out and the reduction in carbon dioxide emissions will serve as case studies from the area of climate protection. The CFC-case study derives its relevance from the fact that the progress made in phasing out CFCs is often cited as a case for the economic and ecological advantageousness of using voluntary agreements. Voluntary agreements, the argument goes, result in the ecological goal being attained more quickly, trigger off an innovation drive in industry and safeguard industry's export capability. The carbon dioxide case study is the obvious choice, not only because of its particularly prominent role in the global warming problem. Accounting for 60% of the damage, anthropogenic carbon dioxide emissions are the largest cause of the greenhouse effect. Moreover, with regard to the carbon dioxide problem the German business community presented an updated voluntary agreement in March 1996. This German negotiated agreement has come to be considered a model for the European Union's climate policy.

To start off with, one has to examine whether in environmental policy areas operational goals have been defined which could serve as a yardstick for the evaluation of voluntary agreements. For climate protection there are specific national and international reduction goals for greenhouse gases and ozone-depleting substances to which one can revert without any reservations. The goals are operational and can be justified by theoretical reasoning and concludent actions.

5. Choosing the decision-making level and process

5.1. INTERNATIONAL FREE RIDERS PUSH NEGOTIATED AGREEMENTS IN CLIMATE POLICY

According to the principle of congruence included in the theory of public goods, the club deciding on the provision and funding of a good should ideally be identical to the club benefiting from the good. A stable global climate benefits the entire world population, consequently, a global institution would have to be founded to decide on the provision and funding of the good „climate protection“ in an economically optimal manner. However, as long as there is neither an „Environment Security Council“ nor a comparable institution, the road via international agreements that has been used so far, has to be followed further. In that connection one has to keep in mind that any solution below the global level entails serious additional problems of free rider-behaviour.[2] Amongst other things, the implications of this free rider-behaviour have so far been seen in the fact that none of the leading industrialized nations is willing to play a pioneering role in introducing climate taxes (MULLER 1995). This national wait-and-see attitude in the follow-up process to Rio has led to an environmental policy standstill in the field of global climate protection. Precisely in view of this standstill in international climate policies, voluntary agreements seem to be an instrument politicians and business representatives are taking up readily, because it allows a certain degree of national activities without businesses having to accept serious cost disadvantages in international competition. Voluntary agreements in the field of climate protection have become common all over the world.

---

2 which, admittedly, a global club does not necessarily abolish either, if, say, decisions relating to the funding of climate protection programmes and to the allocation of these funds to the beneficiaries are taken independently of one another.

5.2 NEGOTIATED AGREEMENTS IN CLIMATE POLICY ARISE FROM A „NO REGRETS" APPROACH

Originally, the policy of voluntary climate protection measures was pushed by the US administration in particular, which had committed itself at a very early stage to a „no regrets" strategy concerning climate protection (KRAUSE/ KOOMEY/OLIVIER 1994; RENNINGS 1994:83-86). Unsure about possible climatic damage, the administration concluded that to be on the safe side, it would only order measures to reduce greenhouse gases that even from a managerial point of view were at least cost covering. To date the „no regrets" strategy has contrasted with the „insurance buying" policy of some Western European states like Germany. According to the latter policy, the climate protection standards that are fixed are arrived at on the basis of a global warming deemed just about acceptable and on the basis of reduction goals derived from the latter. In order to reduce the risk of climatic damage, the costs involved in achieving the standards are accepted as a quasi-insurance premium. While most industrial countries have so far merely formulated the stabilization of climate-relevant emissions as a climate protection goal, the Federal Government is pursuing a relatively ambitious goal with an absolute reduction of 25% by the year 2005 compared with the 1990 level.

Now the question is whether this national reduction goal can be attained with „no regrets" measures - and one can hardly expect more on the basis of voluntary agreements. An optimistic answer to this question is fed by estimates in the latest report submitted by the Intergovernmental Panel on Climate Change (IPCC 1995b:20), according to which „no regrets" measures have a reduction potential of 10 to 30 per cent in the next 20 to 30 years. The enormous potential of „no regrets" measures, which microeconomic studies have also identified in the OECD-countries including Germany, is attributed to the fact that profitable investments in improved energy efficiency have so far never materialized due to substantive market failure and to shortcomings in the coordination between institutions. Examples of these impediments are a lack of information and of economic incentives to conserve energy in the case of public utilities. The possibilities to overcome the impediments that are mentioned include deregulation measures, provision of advice and training and upgraded financial support programmes. In addition, cooperative solutions, for instance in the form of voluntary environmental protection measures, are seen as a way of overcoming institutional impediments. Seen against this background, voluntary agreements on a reduction in carbon dioxide emissions definitely have a potential.

In contrast, initial experience gained in the US with the Climate Change Action Plan (CCAP) of 1993, which essentially contains voluntary measures on the part of public utilities and companies, is not promising. It seems as if current measures are not sufficient to attain the US stabilization goal by the year 2000. Only individual states - such as New York, which is able to replace decommissioned oil- and coal-fired power plants with gas-fired ones in a cost-effective way - are expected to achieve the stabilization goal. Yet, wherever costly adjustments would be necessary that go beyond „no regrets“ measures, the voluntary agreements do not appear to be working (SANGHI 1995).

Once the goal in climate policy is in danger of being missed, the real Achilles heel of voluntary agreements is revealed. For now it becomes evident whether the tough stance in environmental policy that was threatened in this case will really be taken up. The Federal Government knows full well that the success of negotiated agreements greatly depends on how credible the threat looming in the background is. That is why it explicitly emphasized (BMU 1995:3) it would „ remain in control“ and not hesitate to „use regulatory and fiscal instruments once it emerges that the pledge given by the business community amounts to little more than ´business as usual´ or fails to be complied with.“

6. Evaluating the instrument

6.1 GOAL CONFORMITY

6.1.2 CFC-negotiated agreement: in certain applications in conformity with the goals

So far voluntary agreements to reduce ozone-depleting substances have definitely been successful in ecological terms. The goals were exceeded, although there were some critics who argued the phase-out could have been achieved even faster (KOHLHAAS/PRAETORIUS 1994:89). As a rule the monitoring of the negotiated agreements is done by someone who is neutral. Negotiated agreements governing the reduction of CFCs enjoyed special advantages (compared with the case study on the reduction of carbon dioxide), since substitutes had already been discovered which when applied did not lead to costs soaring and since additional pressure was exerted by the demand side (slump in sales of sprays containing CFCs). Thus, in an entrepreneur's cost-benefit calculations, opting for a reduction in CFCs involved little risk. On the other hand, continuing to manufacture products containing CFCs would have been much riskier from a managerial point of view. This is also borne out by the fact that although CFCs were speedily replaced in products, CFC-substitution in production processes was slow to materialize, because the switch was costlier and there was less pressure by the demand side.

6.1.2. No impetus for absolute CO_2-emission reductions

The business community's voluntary agreements on a reduction in carbon dioxide of 1991 and 1995 came in for a lot of criticism, because they did not contain any noticeable initiatives that clearly went beyond „business as usual". Criticism was levelled in particular at the fact that it was hard to check whether the goal was being achieved. The following minimum requirements for information (listed by the German Federal Environmental Protection Agency) that should be included in a negotiated agreement on carbon dioxide reduction (ÖKOLOGISCHE BRIEFE 1996 (2)) were not met:

- reference and target year as well as a reduction path in the form of a timetable with detailed information on partial goals,
- exact fixing of emission or energy conservation goals,
- absolute energy consumption listed according to fuel,
- development of primary consumption,
- development of specific consumption per technical unit,
- reductions achieved,
- in-depth comment on and analysis of the figures provided (e.g. information on whether reductions are attributable to additional climate protection activities, an economic slowdown or to modernization investments that would have been made anyhow) as well as
- detailed list of the „special efforts" promised.

The updated version of the negotiated agreement of March 1996 meets these minimum requirements. The plan on carbon dioxide monitoring presented by the BDI (Federation of German Industries) provides for associations to record their reductions in a total of eight tables (BDI 1996). Total fossil fuel input, net power supplied externally, energy input as well as specific carbon dioxide emissions calculated from this and specific energy input are to be stated for the base year, the previous year and the year under review. The demanded comment on and analysis of the figures and the list of special efforts are also included. The reports are checked by a neutral expert.

What is particularly striking is that no fewer than 12 out of 19 associations pledge to reduce absolute carbon dioxide emissions. According to the Federal Government (BUNDESREGIERUNG 1996a:3), the pledges correspond to a 20% reduction in emissions in these sectors. However, when one takes a closer look it is precisely the achievement of absolute reduction goals that turns out to be the agreement's real Achilles heel, something the statement made by the Vereinigung Deutscher Elektrizitätswerke (VDEW, Association of German Electric Power Stations) (VDEW 1996) illustrates.

According to the VDEW, the absolute reduction potential in the electricity industry until the year 2015 amounts to 25% compared with the base year of 1987. However, compared with the base year of 1990, which the Federal Government is now taking as a basis, estimates merely put the potential at 12%. The figures are even more off target if the target year of the Federal Government is taken as a basis: The potential for the period up until the year 2005 is, compared with the reference year of 1990, down to 8 to 10%. Compliance with these pledges is even conditional on ambitious prerequisites such as:

- a consensus in society on the exploitation of nuclear energy on the basis of existing law,
- an increase in the service life and capacity of existing nuclear power stations,
- the Mülheim-Kärlich nuclear power station going into operation,
- the undisturbed operation of existing nuclear power stations and
- unrestricted choice of fuels for the power stations on the part of the companies.

On the whole with regard to goal conformity, one can stress that shortcomings of the instrument of voluntary agreements have been identified and in part abolished. Still, when a comparison with the Federal Government's climate protection goals is made, one has to question the goal conformity of this instrument, in particular if all other measures are discontinued due to the agreement that was made. To date the development of absolute carbon dioxide emissions is by no means following a path that makes this goal seem feasible (KOHLHAAS/PRAETORIUS 1995:278). Nor does the updated declaration issued by the German business sector indicate a new trend concerning this path. Additional need for action can be deduced, in particular for the German states that made up the Federal Republic of Germany prior to unification. If the VDEW puts the absolute carbon dioxide reduction potential of German electric power stations, which are responsible for about a third of all German carbon dioxide emissions, at a mere 8 to 10% compared with the target year of 2005 (VDEW 1996:5), the question arises as to who is to contribute the above-average reductions needed to offset the expected increases in carbon dioxide emissions in areas like transport.

6.2 SYSTEM CONFORMITY

6.2.1 Non binding agreements: no sanctions against unfair players

Voluntary agreements can constitute solutions in keeping with the system, if they establish binding standards for the parties involved and if free-rider behaviour can be prevented. However, since associations normally do not possess any effective mechanisms to punish their members, and hence no binding rules of the game can be agreed, the approach to problem-solving shifts to the moves of the game, that is to say to the wrong level in the system. As KREUZBERG (1993:308) writes, voluntary agreements hinge on „a disproportionately high participation of 'honest companies' in environmental protection activities and therefore lead to a redistribution of burdens to the detriment of 'honest' players."

6.2.2. CFC-phase out: „soft" instruments appropriate for preventive strategies

Prohibitions are definitely measures to avert concrete and acute damage to the environment that are in keeping with the system. In such situations the liberal principle of maximising freedom cannot be used as a yardstick for policy design on constitutional grounds (Rennings et al. 1996). That means that instruments with low intervention intensity and a high degree of freedom, such as negotiated agreements, can only take precedence as long as it is safeguarded that acute environmental damage will be averted. If this is not the case, „tough" instruments will be applied. This is also evident in the CFC-example: The realistic alternative to a voluntary reduction in the production of CFCs and ozone-depleting substitutes would have been a quick ban on these substances. In order to bring about a complete phase-out of CFC-application in products, finally the complementary CFC-halon-prohibition ordinance was issued in Germany. The aerosol industry did not agree to a voluntary phase-out desired by the Ministry for the Environment (KOHLHAAS/PRAETORIUS 1994:89).[3]

6.2.3. Climate policy: long-term danger of intervention spiral

A characteristic of voluntary environmental protection measures in the field of climate protection frequently cited as being particularly in keeping with the system, is that these measures, compared with energy and carbon taxes, cause

---

3 The way to protect the ozone layer that is more in keeping with the system than anything else would be to apply „tough" instruments presenting economic incentives, as was done in the US. There politicians successfully opted for the application of a mix of instruments comprising levies and permits (COOK 1996:4).

fewer side-effects on goals relating to stability and distribution policies. The „double dividend“ of an ecological tax reform is often overrated, the argument goes, i.e. the simultaneous achievement of ecological (reduction in greenhouse gases) and economic goals (e.g. job creation) due to the revenue-neutral compensation of ecological taxes (KOSCHEL/WEINREICH: 1995).

However, if emitters are charged the adjustment costs caused by a climatically-sound restructuring of capital assets, this has to be regarded as being in keeping with the market, irrespective of the compensation question. A „policy of little steps“, which the German Council of Environmental Advisers advocates, would primarily send the necessary price signals to consumers for a more economical use of energy. This would be the crucial advantage over voluntary agreements. Secondly, a policy of little steps would minimise the side-effects on goals in the fields of stability and distribution policies. For voluntary agreements are primarily not in keeping with the market and the system, because such price signals usually are not included in the instruments they are provided with. A climate policy based on voluntary agreements is thus always associated with the danger that even a high „no regrets“ potential that could be siphoned off is overcompensated by increasing energy consumption - not only in the sphere of transport.

6.3. ECONOMIC EFFICIENCY

6.3.1 CO_2-agreement is not cost-efficient

Solutions to the problem of reducing emissions are economically cost-effective if each emitter fixes his contribution to the reduction in such a manner that the overall economic avoidance costs are minimised. Individual emitters neither know the avoidance costs nor are they interested in including them in their decision-making, hence, these signals have to be sent through environmental policy. Levies and permits in particular are classic textbook instruments meeting the economic efficiency criteria, because they send out these signals in the form of a correction of relative prices. On the other hand, a reduction in emissions with minimal costs for the overall economy is unlikely to be achieved via a voluntary agreement on carbon dioxide or eco-efficient cars, for instance, due to the restriction to certain sectors and the free-rider behaviour to be expected on the part of the members of the associations.

Moreover, voluntary agreements are not convincing when it comes to their dynamic efficiency, i.e. their effect on technological progress. Here it is possible to compare voluntary agreements to regulations. Once the goal has been achieved, there are no further incentives to reduce emissions. Compared with regulations, there does not even inevitably have to be an incentive to maintain the standard once it has been attained (KREUZBERG 1993:309).

6.3.2 *Cost-efficient CFC phase-out*

When assessing the economic efficiency of German agreements on reducing ozone-damaging substances, consideration has to be given to three unusual features in particular that accelerated the substitution on managerial grounds alone:

- the ban on CFCs as propellants for most aerosol products in the US as early as 1978,
- the declining demand for products containing CFCs (e.g. aerosols) and
- the availability of low-cost substitutes.

The economic cost efficiency of negotiated agreements on CFCs has to be regarded as positive. Each agreement covered the main polluters, which were enabled to implement the reduction at the lowest possible cost.

Concerning dynamic efficiency, here, too, one has to repeat: Once the goal has been achieved, there are no further incentives for progress in terms of environmental technology. However, with respect to the development of environmentally-friendly CFC substitutes, the bottom line is that owing to the aforementioned unusual features there had already been sufficient incentives for technological progress.[4]

6.4 INSTITUTIONAL CONTROLLABILITY

6.4.1 *Enforceability depends on induced costs and external pressure*

In principle the enforceability of voluntary agreements in associations depends on three factors:

- the costs of the agreement made,
- the costs of the government ordinance looming in the case of non-performance of the agreement and
- the effectiveness of possible punishment meted out by the associations themselves (e.g. expulsion from the association).

---

[4] What has turned out to be a particularly efficient instrument in the US to protect the ozone layer is the simultaneous application of permits and a tax for the protection of the ozone layer. The American environmental agency EPA estimates that the administrative costs amounted to only 10 per cent of the administrative work a regulation would have entailed. Furthermore, it was possible to quickly adjust the licenses issued to the modifications of the Montreal Protocol. Whereas in 1988 the cost of halving CFC consumption was put at $ 3.50 per kilogramm, only two years later it was possible to lower these estimates to $ 2.20 per kilogramm (COOK 1996:4f.).

Whereas „no regrets“ measures are unlikely to meet with much opposition in an association, pledges going beyond them involve the danger of an internal allocation struggle, which can let a withdrawal from the commitment seem worthwhile. In such cases certain passages in the declaration providing scope for interpretation, or preconditions for the agreement that were never met can provide a welcome opportunity to terminate cooperation.

Problems relating to the enforcement of voluntary agreements on carbon dioxide reduction in associations have been revealed, e.g. in talks conducted by the Federal Environmental Protection Agency (UBA) with representatives from the associations in connection with the climate protection initiative launched by the German business community. For example, UBA's annual report reads (1993:167, translated by the authors): „The talks revealed fundamental difficulties whenever negotiated agreements made by associations are to contain binding requirements that are usually laid down in a statutory basis. Many individual businesses refuse to recognize declarations made by their associations as binding. It is virtually impossible to impose sanctions following non-compliance of the agreements“. Thus, the instrument of negotiated agreements reaches limits which result from its voluntary nature.

6.4.2 Redistribution at the expense of third parties

Since the instrument of voluntary agreements per se does not send out any price signals that lead to automatic adjustments on the part of the players, decisions have to be made on a case-by-case basis on how the overall reduction targeted is to be divided up between individual groups. This, some people say, involves the danger that groups that are inferior in terms of the way they are organized, like households, ultimately have to bear the largest adjustment burden. This has to be qualified by saying that this argument can also be used against economic instruments. In real life the political representation of interests also plays an important part with economic instruments, for instance when exceptional areas are stipulated that will be exempted from energy taxes. Even Denmark, a country usually regarded as exemplary when it comes to introducing ecological taxes, has special arrangements for particularly energy-intensive businesses, and the burden of levies is mainly borne by private households (MEZ 1995:109 and 126). As the example shows, the uniform price signals economic instruments want to bring about are frequently watered down in the political process.

6.4.3 Relinquishment of political scope

The updated declaration presented by the German business community on the prevention of damage to the climate once again clearly indicates the limits of voluntary agreements. If the environmental goal obviously runs counter to the individual economic interests of the associations, people - like some representatives of the VDEW recently - suddenly talk of the „fetish year 2005“ in connection with the Federal Government's timetable. Although, for example, the VDEW's agreement already is a scaled-down version (providing for a reduction of merely 8 to 10% by the year 2005), undesired political measures (e.g. in nuclear energy policy) may prompt termination of the agreement. In any case, a government has to expect being asked whether it is possible for a pledge with little substance to be worth so much that the government in return, for decades to come, puts up with being deprived of a great number of potential courses of action relating to climate and energy policies.

7. General conclusions and future perspectives

7.1 GENERAL CONLUSIONS

The examples analyzed here have offered weaknesses of voluntary agreements with regard to goal-conformity, system-conformity, cost-efficciency and institutional controllability. However, it can be argued that hardly any instrument will meet all these criteria. If it does theoretically, it may be watered down in the political process (or perhaps will not be implemented at all). Thus, it should be mentioned that there may be reasonable applications for voluntary agreements, but the instrument should be used very carefully. Against this background, we want to draw some general lessons from the German experience and give some recommendations for the international discussion of using voluntary agreements within the European Union.

Opportunities for the application of voluntary agreements that are welcome from an ordoliberal point of view especially exist if they are used within a mix of policy-instruments. If economic incentives are introduced on a voluntary basis or if agreements are used to accompany economic instruments (example: combination of carbon taxes and voluntary agreements), agreements would really deserve the attribute of being market-based. If economic instruments cannot be employed, for instance when substances have been completely banned, negotiated agreements as „soft“ instruments of environmental policy can certainly serve to accompany „tough“ regulatory measures or - provided it is not a matter of warding off acute dangers - to replace them (example: CFC-phase out). Except for these areas, however, from an ordoliberal point other instruments are preferable.

By combining negotiated agreements and economic instruments, it is possible to avoid a lack of incentives, which is one of the worst flaws of most voluntary agreements in terms of design. To date a lack of incentives has led to voluntary agreements either being undemanding with regard to the contents and being phrased accordingly, or, in the case of more ambitious pledges, to them being associated with a great amount of time and effort needed for enforcement and monitoring. It is also due to a lack of incentives that as a rule, only specific and no absolute reductions are achieved. Without any signals in favour of a way of using environmental resources that is generally less harmful, even with demanding specific reduction goals there is the risk that an overall increasing consumption of environmental resources will over-compensate for these goals.

The increasing significance of voluntary agreements in current environmental policies may have to do with them being labelled „market based instruments“ and the fact that their increased use is associated with hopes of thus strengthening the market system. However, a more in-depth analysis reveals that although in principle it is possible to design voluntary agreements in a manner that is in keeping with a market economy, in most cases people avoid doing it. As the examples that were studied showed, such a free market-oriented design does not come about voluntarily and spontaneously, rather it requires standards and a framework set by government. However, once such a framework is set, agreements may be used as a tool for implementation within an environmental policy-mix.

For this reason one has to strongly advise the government against making a commitment to the effect that in return for voluntary declarations on environmental protection, the government will not make use of any other instruments. This imposes disproportionately severe restrictions on the politicians' latitude in how they act in the future and thus on their capacity for problem-solving. Such a policy tends to neglect other solutions that may be more appropriate to the problem, such as the application of a mixture of environmental policy instruments.

If a decision to give preference to voluntary solutions in general is made or if a decision in favour of such solutions is taken at an early stage, this too is counterproductive, because the substance of negotiated solutions - the governmental „potential for threats“ - is weakened and delays in the form of a stamina contest are provoked. The examples that were examined confirmed that without considerable governmental pressure, voluntary agreements do not yield any pledges that go beyond „business as usual“ or „no regrets“ measures. In order to make it absolutely clear that the government really „is in control“, the scope for design in environmental policy has to be kept unrestricted and flexible. On the other hand, giving priority to voluntary solutions on principle imposes disproportionately severe restrictions on government when it comes to quickly reverting to „tough“ environmental policy instruments following unsatisfactory negotiation results or delayed implementation.

7.2 EUROPEAN PERSPECTIVES

Although our case studies represent only German experiences, some lessons can be drawn for the use of voluntary agreements on a European level. As it seems, the European Commission is aware of several potential shortcomings of agreeements being stated here.

The European Commission (1997) issued a communication examining the use of environmental agreements as an instrument of EU environment policy. The (non-binding) paper deals with the instrument in general, presents guidelines on „environmental agreements“ and provides a survey over the use of voluntary (environmental) agreements in the Member States.

The guidelines given by the Commission point out seven topics that should be taken into account when public authorities design environmental agreements on national or local level:

- consultation,
- contractual form,
- quantified objectives,
- monitoring of results,
- public information and transparency,
- independent verification of results and
- additional guaranteees.

Before an environmental agreement is concluded, the paper says, all interested parties (companies, business associations, environment groups and public authorities) should have the opportunity to comment on the draft. Their opions should be taken into account.

The legal status of an agreement plays an important role for the sucess of an agreement. According to the Commission, binding agreements provide in general better safeguards in terms of achieving environmental objectives. Contracts (binding on both parties) offer a well-defined framework that may include sanctions for noncompliance and is enforcable through the courts.

The weakness and bad reputation of certain past agreements partly derive from the lack of quantified objectives, leaving room for the perception that agreements were used to avoid or delay effectice action, the Commisssion states. Objectives have to be quantified in figures as opposed to „best efforts“ clauses.Also intermediate objectives should be set to show the effectiveness of an agreement.

According to the Commission's findings, results of agreements have to be monitored, the agreement itself should define „how“.

The failure of agreements, the paper goes on, is often connected with a lack of public information and transparency. Transparency is crucial to assure third parties that non-regulatory obligations are kept. Agreements should be published in the national Official Journal or an equallly public document. Even a public register of agreements should be considered.

In some cases it may be appropriate to set up a committee or independent body to collect, evaluate, or verify results, This is particularly important in cases where the measuring methods differ or where the disclosure of business secrets has to be avoided.

As an additional guarantee that the agreement will be fulfilled, dissuasive sanctions such as fines and penalties could be foreseen for case of non-compliance, the Commission concludes.

Measuring the German way of using environmental agreements at the standard set by the Commission's guidelines, one will find that the national agreements fall behind in nearly every respect. Apart from the fact, that all the German agreements are non-binding, the procedure while negotiating an agreement and the structure of the final commitment show a whole lot of deficiencies.

Thus, fundamental flaws of the negotiations may be overcome when the guideline will be followed. From our point of view it is important to mention that, according to the communication of the Commission:

1. A cost-effective use of agreements will be made as a part of policy mix together with, for example, regulatory or economic instruments and
2. agreements may be under certain circumstances an efficient tool for implementing of environmental policy, but they are not appropriate for environmental target setting. General targets should be set through legislation.

These basic conclusions go along with the findings of our study. The politician's „art of the possible“ may now be to look carefully for reasonable applications of agreements without falling back to traditional command-and-control measures. As experiences in Belgium (Seyad et al.:1996) and the United Kingdom (Eden:1996) show, neither a strategy of strong regulation of voluntary agreements (which makes the agreements obsolete) nor a strategy of far-reaching deregulation (or self-regulation) of environmental target-setting seems to be promising. If not driven to one of these extremes, there may be several applications for more flexible ways of environmental policy.

References

BDI - Bundesverband der Deutschen Industrie e.V. (1996) Aktualisierte Erklärung der deutschen Wirtschaft zur Klimavorsorge. Köln.

BGW - Bundesverband der deutschen Gas- und Wasserwirtschaft (1996) 1. Zwischenbericht zur Klimaschutzerklärung der deutschen Gaswirtschaft vom März 1995 bis zum Jahr 2005. In: BDI - Bundesverband der Deutschen Industrie e.V.: Aktualisierte Erklärung der deut-schen Wirtschaft zur Klimavorsorge. Köln.

BMU - Bundesministerium für Umwelt, Naturschutz und Reaktorsicherheit (1995) Staatssekretär Jauck fordert mehr Kooperation von Staat und Wirtschaft im Umweltschutz. BMU-Pressemitteilung 117/95, Bonn.

Brohm, W. (1992) Rechtsgrundsätze für normersetzende Absprachen - Zur Situation von Rechtsverordnungen, Satzungen und Gesetzen durch kooperatives Verwaltungshandeln, DÖV, pp. 1025-1035.

Bundesregierung (1996a) Erklärung der deutschen Wirtschaft zur Klimavorsorge. Pressemittei lung Nr. 118/96, Bonn.

Cook, E. (1996) Marking a Milestone in Ozone Protection: Learning from the CFC Phase-out. In: World Resources Institute. Issues and Ideas, January 1996. Washington.

Eden, S. (1996) Problems of Environmental Consensus: Packaging Regulation and Self-Regulation in the UK. Paper presented on the Sixth Annual Greening of Industry Network Conference, Heidelberg, November 1996.

Grossekettler, H. (1991) Zur theoretischen Integration der Finanz- und Wettbewerbspolitik in die Konzeption des ökonomischen Liberalismus. In: Erik Boettcher et al. (Hrsg.) *Jahrbuch für Neue Politische Ökonomie* **10**, Tübingen, pp. 103-144.

Hoffmann-Riem, W. (1990) Reform des allgemeinen Verwaltungsrechts als Aufgabe - Ansätze am Beispiel des Umweltschutzes, *AöR*, 400-447.

Holzhey, M. and Tegner, H. (1996) Selbstverpflichtungen - ein Ausweg aus der umweltpolitischen Sackgasse? *Wirtschaftsdienst* **8**, 425-430.

IPCC - Intergovernmental Panel on Climate Change (1995a) Second Assessment Synthesis of Scientific-technical Information Relevant to Interpreting Article 2 of the UN Framework Convention on Climate Change. Geneva.

Kohlhaas, M. and Praetorius, B. (1994) *Selbstverpflichtungen der Industrie zur CO_2-Reduktion.* Berlin.

Koschel, H. and Weinreich, S (1995): Ökologische Steuerreform auf dem Prüfstand - Ist die Zeit reif zum Handeln? in O. Hohmeyer (Hrsg.), *Ökologische Steuerreform.* ZEW-Wirtschaftsanalysen Vol. 1, Baden-Baden.

Krause, F., Koomey, J. and Olivier, D. (1994) Incorporating global warming externalities through environmental least cost plannning: a case study of Western Europe, in O. Hohmeyer and R. L. Ottinger (eds.), *Social Costs of Energy*. Springer Verlag, Berlin, Heidelberg, New York, pp. 287-312.

Kreuzberg, P. (1993) Zur ökonomischen Rationalität „freiwilliger Kooperationslösungen" für das Klimaproblem, *Zeitschrift für Energiewirtschaft (ZfE)* **4**, 304-309.

Maier-Rigaud, G. (1995) Für eine ökologische Wirtschaftsordnung, in Jahrbuch Ökologie 1996. Munich.

Mez, L. (1995) Erfahrungen mit der ökologischen Steuerreform in Dänemark, in O. Hohmeyer (Hrsg.), *Ökologische Steuerreform*. ZEW-Wirtschaftsanalysen Vol. 1, Baden-Baden, pp. 109-128.

Muller, F. (1995) Energy Taxes, the Climate Change Convention and Economic Competitiveness. Paper Presented to the 3rd International Conference on Social Costs, Ladenburg, Germany, 27-30 May.

Murswiek, D. (1988) Freiheit und Freiwilligkeit im Umweltrecht, *Juristen Zeitung* **43**, 985-993.

Ökologische Briefe (1996) „Denkhilfe“ für die Wirtschaft aus dem Umweltbundesamt.

Rennings, K. (1994) *Indikatoren für eine dauerhaft-umweltgerechte Entwicklung*, Stuttgart.

Rennings, K., Brockmann, K.-L., Koschel, H. Bergmann, H. and Kühn, I. (1996) *Nachhaltigkeit, Ordnungspolitik und freiwillige Selbstverpflichtung*. Heidelberg, 1996.

Sanghi, A. K. (1995) Climate for Climate Change Actions in the U.S.: The New York Experience. Paper Presented to the 3rd International Conference on Social Costs, Ladenburg, Germany, 27-30 May.

Seyad, Akim, Marc de Clerq, Filip Senesael (1996) The Use of Voluntary Agreements as Instruments of Environmental Policy in the Energy Sector - the Belgian Electricity Voluntary Agreement Case. Paper presented on the Sixth Annual Greening of Industry Network Conference, Heidelberg, November 1996.

UBA - Umweltbundesamt (1993) *Jahresbericht 1993*. Berlin.

VDEW - Verband Deutscher Elektrizitätswerke (1996) Bericht 1996 zur „Erklärung der VDEW zum Klimaschutz“. In: BDI - Bundesverband der Deutschen Industrie e.V.: Aktualisierte Erklärung der deutschen Wirtschaft zur Klimavorsorge. Köln.

Weiland, R. (1995) Rücknahme- und Entsorgungspflichten in der Abfallwirtschaft - Eine institutionenökonomische Analyse der Automobilbranche. Wiesbaden.

SECTION III

POLITICAL AND ECONOMICAL INSTRUMENTS - AN OVERVIEW

TAXES

CARBON AND ENERGY TAXES IN OECD COUNTRIES[1]

RICHARD BARON*
Administrator, Energy and Environment Division
International Energy Agency
9, rue de la Federation, F-75739 Paris Cedex 15, France

* The views expressed here are those of the author and do not necessarily reflect the IEA Secretariat's views or that of IEA member countries.

1. Introduction

Taxation of externalities, together with tradeable permits, has long been advocated as the most economically efficient way to internalise externalities into society's choices; knowledge of marginal externality cost and marginal abatement cost curves indicates at which level the tax should be set to maximise social welfare through proper internalisation of externalities. In the case greenhouse gas emissions, however, there is still little knowledge about potential climate change damages, which prevents a straightforward application of economics findings. Nevertheless, under perfect market conditions, taxation of greenhouse gas emissions at a uniform level across all sources would still achieve emissions reduction at the lowest overall economic cost.

Political reality[2], however, has so far made it difficult for any country to apply uniform taxation of greenhouse gas emissions, or to the most significant subset of these, energy-related CO_2 emissions. This paper provides up-to-date information about the current experience with taxation as an instrument to limit greenhouse gas emissions in OECD countries, offers a discussion on the key features of that experience, and practical considerations about the possible use of taxation for

[1] This paper is partly based on the working paper: "Taxation (i.e. carbon/energy)", under the OECD/IEA project on Policies and Measures for Common Action, for the Annex I Expert Group on the UN FCCC. This and orther working papers can be down-loaded from the following address: http://www.oecd.org/enf/cc/cc2.htm. Sandrine Duchesne provided statistical support.

[2] This dilemma is well summarised by the OECD: "If fixed at an appropriate level, environmental taxes minimise the overall cost of achieving a given pollution control target, even though some firms or industries may face higher costs with ecotaxes than with regulation" (OECD, 1996, p.11)

J. Hacker and A. Pelchen (eds.), Goals and Economic Instruments for the Achievement of Global Warming Mitigation in Europe, 207–229.

climate change mitigation policies. The paper does not cover at great length other energy taxation features in OECD countries, nor does it cover in any detail the European Commission carbon/energy tax proposals, which is the subject of another presentation in this conference.

2. Carbon/energy taxes: OECD experience

In most OECD countries, almost all forms of energy are taxed to varying degrees, not primarily for greenhouse gas purposes, but to raise revenues for government spending or to internalize other externalities. In fact, there is often an almost inverse relationship between fossil energy price levels, including taxes and subsidies, and their carbon content: fossil fuels with higher carbon content have lower end-use prices than those with lower carbon content. On a sectoral basis, industrial energy use is generally subject to a low level of taxation, whereas transportation fuels are usually heavily taxed, although with some disparity across countries. In fact, a previous study by the OECD have shown that a restructuring of existing energy tax regimes on the basis of the carbon content of fuels alone would achieve significant reductions without increasing the overall tax burden on energy (Hoeller and Coppel, 1992).

Section 3 reviews implemented carbon/energy taxes. Five countries (Denmark, Finland, the Netherlands, Norway and Sweden) have adopted carbon/energy taxes which generally include some rebates or exemptions for industry on competitiveness grounds, or alternative measures to achieve similar objectives.

Section 4 covers failed attempts at introducing carbon/energy taxation. At least three other countries have considered carbon/energy taxes (Australia, Switzerland and the US) but those proposals were not accepted. New Zealand has decided that a carbon tax would be introduced in 1997 if emissions are not on track to achieve existing targets, but is also examining alternative approaches such as tradeable permits. The European Union has considered several proposals for coordinated carbon/energy taxation, put forward by the European Commission. These proposals are not covered in this paper.

3. Five European countries have adopted carbon/energy taxation

Several carbon, energy and carbon/energy taxes have been introduced, or are being discussed in OECD countries, as instruments to reduce energy-related CO_2 emissions. For those countries which have introduced such taxes, the following general comments can be made:

- none of the implemented schemes covers all energy uses resulting in CO_2 emissions in a completely homogenous fashion, either on a per unit of energy or per ton of CO_2 basis;
- carbon/energy taxes have sometimes been introduced in place of other taxes on energy, so as to minimize the additional fiscal pressure, while providing a proper signal to reduce CO_2 emissions;
- carbon/energy taxes are often part of a more general fiscal reform to solve structural issues such as high "distortionary" taxes on employment and capital;
- taxes are usually phased-in, providing a period of adaptation and avoiding the negative effect of a "price shock". Tax rates can be adjusted for inflation over time to keep the signal constant in real terms. Carbon/energy taxes are subject to the value added tax (as high as 25% in Sweden).
- countries rely on carbon/energy taxes as *one* policy measure *in a package* of measures to achieve their emissions objectives, which accounts for differences among types of energy-users and prevailing end-use energy prices more than a uniform tax would;
- exemptions and exceptions have been granted to energy-intensive industries or to industries facing acute international competition; electricity, a highly traded commodity, is also granted special treatment.

This section provides an overview of carbon and energy taxes implemented in Denmark, Finland, the Netherlands, Norway and Sweden.

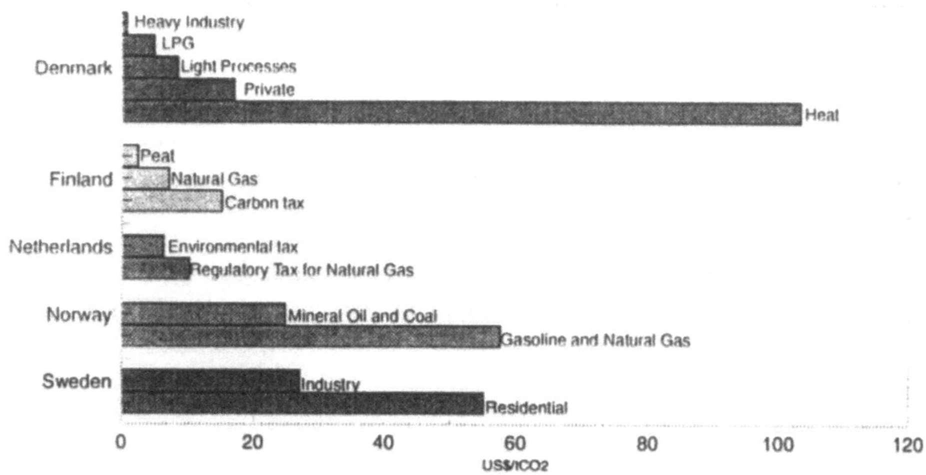

FIGURE 1. CO_2 or Related Taxes in OECD Member Countries
An Overview

Unfortunately, the taxation schemes for carbon and energy cannot be summarized in a single figure for each country; indeed, the complexity of each tax "package" makes it difficult to compare them in a straightforward fashion. TABLE 1 in appendix summarizes the information below and provide details on rates applied on a fuel-by-fuel basis[3].

3.1. WHAT IS TAXED? TAX BASE AND LEVEL

3.1.1. Denmark

A tax on energy use in the households sector has been in place in *Denmark* since 1977, and was significantly increased in 1986, offsetting the decrease in oil prices. This tax was broadly consistent with the energy content of fuels, with a lower rate on coal. Denmark introduced a CO_2 tax on energy consumption in 1993 as part of a package including various subsidy schemes for promoting means to produce electricity and heat from less carbon-intensive fuels and to increase energy efficiency. For the households sector, a *tranche* of the energy tax was converted into a carbon tax of DK 100 (16.3 US\$/tCO_2). Enterprises were also submitted to this tax, with a reimbursement of 50% generally available.

Starting 1996, a new, more comprehensive, tax scheme was introduced. It consists mainly of three rates, depending on the taxed activity:

1. Heavy industry (or industry facing competition): a tax growing from 5 to 25 DK (US\$ 4/tCO_2) over 1996-2000, with a possible reduction down to 3 DK (see '*who's taxed*');
2. Light processes (all not belonging to the above category) face a tax increasing from 50 DK to 90 DK (US\$14.7/tCO_2) over 1996-2000, with a possible reduction (68 DK in 2000).
3. Energy use for space heating is taxed at a much higher rate, growing from 200 DK to 600 DK (98 US\$/tCO_2) over the same period. This tax rate is equivalent to the sum of the carbon and energy taxes applied in the households sector in 1996.

Note that leaded and unleaded gasoline is not subject to the carbon tax, since it is already heavily taxed. The tax does apply, however, on diesel, but does not bring total taxes on diesel to the level of gasoline.

---

[3] For more detail on carbon/energy taxes in these countries, see OECD (1995): *Environmental Taxes in OECD Countries*, Paris. The author also relied on the document: *A comparison of Taxes on Energy in Eight European Countries*, provided by the Ministry of Housing, Spatial Planning and Environment of the Netherlands (1996), and on country submissions.

3.1.2. Finland

Finland introduced Europe's first "explicit" carbon tax, imposed on fossil fuels based on their carbon content, starting at a low level of Mk 24.5 per tonne of carbon (US$1.4 /t CO_2). The rate was doubled in 1993, to Mk 50, with a tax differentiation for diesel and gasoline. The effect of this new tax rate on fuel prices was:

- a 1-2 % rise in prices of electricity, light fuel oil and natural gas;
- a 5-8 % rise in prices of coal, gasoline and heavy fuel oil; and
- a 10 % rise in the price of diesel.

Until 1997, the tax was split into a "fiscal" component with tax differentiations for diesel and petrol, a carbon component and an energy component, replacing the pure carbon component (60 % of raised revenues through carbon, 40 % through the energy component). The energy tax is imposed on all primary energy sources except wood, wind power and waste used for energy production. In 1995, the tax rates were increased again to the following levels:

- Carbon component: Mk 38.3 t CO_2 (US$6.8, equivalent to US$25/tC);
- Energy component: Mk 3.5 per Mwh (US$0.62).

In 1997, Finland introduced two major changes to its carbon/energy taxation:

1. Taxes on fuels for heat generation have been restructured, from a carbon/energy basis to a carbon-only basis (Mk 70 t CO_2, about US$ 14). Natural gas has been granted a 50 % relief to the end of 1997. Peat is taxed at Mk 11 t CO_2.

2. The tax on fuel input to power generation was abolished, and replaced with a tax on all consumed electricity, with differentiated levels for households and industrial users (respectively Mk 0.031 and 0.01675 per kWh); in 1996 the total price for electricity was Mk 0.5 for households and 0.284 for industry, so the new tax, with VAT represents a 8 % price increase for households.

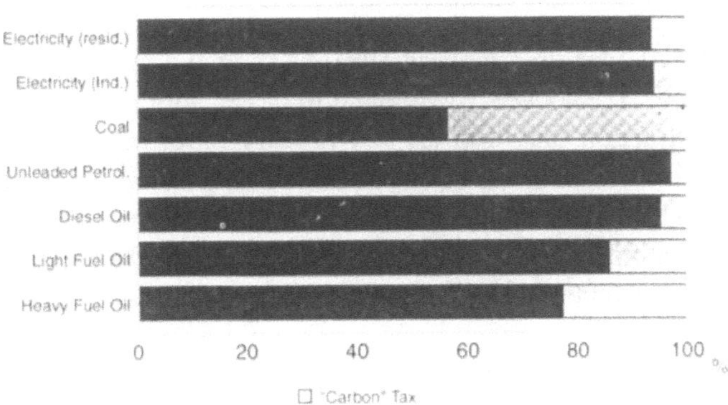

FIGURE 2. Proportion of 'Carbon' tax in total energy prices in Finnland (1997)

3.1.3. The Netherlands

The Netherlands have introduced two taxes on energy and carbon. The *environmental tax* introduced in 1988 and considerably increased in 1992, was initially intended to finance environmental policy expenditures. Since July 1992, the tax has been part of general revenue. It is set to achieve revenue requirements based on consumption in the previous year and, since 1992, on *energy and carbon* content (on a 50/50 basis). This tax also covers uranium used in power generation.

The level of the *environmental tax* on gasoline represents 1% of the total price; for light fuel oil used in industry, the *environmental* tax represents a 5% increase to the price for industry. The current tax rates per gigajoule and per tonne of CO_2 are roughly equal to a total of US$ 3 per barrel of oil equivalent, half on energy and half on carbon content.

A *regulatory energy tax* for small consumers was introduced to help achieve the country's CO_2 emission target. The tax, commenced on 1 January 1996 is designed to avoid the economic cost of an unilaterally imposed energy tax on industrial energy users competing internationally. The tax applies to:

- natural gas up to a consumption ceiling of 170 000 cubic meters per year,
- low amperage electricity to a ceiling of 50 000 kWh per year,
- mineral oil products which are substitutes for gas (home heating oil, light fuel oil, non-transport applications of liquefied petroleum gas, butane and propane) with ceilings comparable to the ceiling for gas, and restitution above these ceilings.

The *regulatory energy tax* applies to the gas and electricity use of households and companies; it covers about 40% of non-transport and non-feedstock energy use. The tax on gas will be introduced in three stages with the effect of raising gas prices by 20-25%; electricity prices will rise by about 15% in a single step.

3.1.4. Norway

Norway introduced a tax on 1 January 1991, starting at a rate of US$ 40 per tonne of CO_2 on gasoline. The tax was also applied to diesel, mineral oil, oil and gas used in North Sea extraction activities. As of 1 January 1996, the tax per tonne of CO_2 ranges from US$17 on petroleum coke to US$55.6 on gasoline and on gas use in the North Sea. In real terms, the tax has not changed in the recent years. The above rates should be compared with excise taxes which are equivalent to US$239.6 per tonne of CO_2 on gasoline, US$176 on diesel oil[4].

---

[4] An SO_2 tax is also applied to mineral oil which, if translated in CO_2 terms is equivalent to a $US 4.2 tax for auto-diesel and light fuel oil. The SO_2 tax increases with the sulphur content and reaches its maximum for heavy fuel oils at a rate equivalent to $US 35/tC.

For mineral oil (heavy fuel oil for industrial use, and light fuel oil for industry and households), the carbon tax has offset a decrease in the excise tax, resulting in a stabilization of prices for these oil products over 1991-1994. The Norwegian CO_2 tax covers about 60% of all CO_2 emissions.

3.1.5. Sweden

Sweden passed a bill in 1990 introducing a *carbon tax* and a value added tax on energy, and lowered the existing energy tax, as part of an overall fiscal reform. The original tax amounted to Skr.250/tCO_2 (US$ 35), levied on oil, coal, natural gas, LPG, gasoline, and fuel for domestic air transportation; fuel use for electricity production was exempted. Simultaneously with the introduction of the tax on carbon, general energy taxes of fossil fuels were cut by 25-50% but the net effect of increased carbon and reduced energy taxes was still to be an increase in tax revenues of Skr.3 billion.

In 1993, the carbon tax rate was increased to Skr. 370/tCO_2 (US$ 52) while the tax applied to industry was adjusted down to about 25% of the carbon tax paid by other energy users (Skr. 92.5 or US$13); in addition, the existing *energy* tax on industry (and horticulture) was abolished. At the end, the 1993 tax reform brought industrial total energy tax levels down to a much *lower* level than before the introduction of the carbon tax in 1991. A number of exemptions and ceilings on tax payments also apply (see section on exemptions).

On 1 July 1997, the Swedish government increased the CO_2 tax on industry to Skr.185 (now 50% of the tax paid by other energy users). The current CO_2 tax on services and households remains at Skr.370 (US$ 52/tCO_2). The tax is automatically adjusted for inflation, so as to provide a steady signal in real terms.

3.2. WHO'S (NOT) TAXED? EXEMPTIONS

Exemptions, tax rebates, tax ceilings and other mitigating measures have usually been applied to some activities or income groups in all countries.
The experience with these measures is as follows:

- Certain income groups would be more affected by the new tax than the rest of society. This effect can be offset through certain recycling systems, but in some cases a 'floor' under which energy is not taxed was introduced.
- Most importantly, certain activities are granted rebates or exemptions on competitiveness grounds. This has been the case where some governments wanted to introduce taxes without similar measures being taken by their trade partners (e.g., Sweden, in the context of the EC proposals). Certain countries have set programmes to improve the energy situation of their industry, so as to minimize the carbon/energy tax burden, or provided alternative approaches to reduce emissions in these activities (*de facto* exemptions).

- In some cases, the possibility of exemptions has led to tax evasion: firms could in some cases, isolate most energy-intensive operations and be granted exemptions on that basis for the rest of their operations. Criteria for exemptions were amended subsequently to solve this problem.
- It can be said that the need to accommodate for different situations across energy users, resulting in non-homogenous tax rates, sometimes necessitates additional adjustments in order to maintain the environmental effectiveness of the taxation system.
- In certain extreme cases, where emission leakages are more than likely to occur, total exemption has been granted, recognizing that the tax would not provide any environmental benefit.
- Last, certain fuels have been exempted from the tax (and sometimes subsidized) to account for their lower carbon content.

3.2.1 Denmark

In Denmark, an automatic exemption was granted to industry on competitiveness grounds, in the context of a disagreement about the adoption of a common carbon/energy tax in the European Union or OECD. At first, a set of criteria were applied for tax rebates, based on an energy-intensity indicator, roughly equal to the CO_2 tax payments over value added. A progressive refund was obtained automatically, if the carbon levy was more than 1% of value added: between 1 and 2%, 50% of the amount above 1% would be refunded; between 2 and 3%, the refund is 90% of the amount above 3%.

Subsidies for energy audits were also available, and if the recommendations for energy efficiency improvements were followed, a total tax refund was granted (so-called CO_2 subsidy). On the whole, before the new green taxes were introduced, the average carbon tax rate on energy use by industry was DK 35 per tonne of CO_2 (US$5). It became clear for the Government that the criterion for rebating the tax could easily be manipulated. This contributed to prompting the new carbon/energy tax package. Under the new tax package, lower rates are granted to "heavy industry", in fact to a list of companies either energy-intensive or operating on competitive markets; this was negotiated rather than determined with a single criterion. Firms on this list, and other firms with a ratio of CO_2 tax to value added above 3% are eligible for voluntary agreements, which implies tax rebates (from DK 25 to DK 3) *if successfully implemented.* The same option is available for light processes, with a tax reduction from DK 90 to DK 68.

For trade reasons, electricity is exempt from the CO_2 tax in Denmark; a subsidy to encourage the use of natural gas was introduced to compensate the non-taxation of coal as an input to electricity.

3.2.2. Finland

In Finland, industries are not granted reduced or zero tax rates. Exceptions include products used as raw material in industrial production, or used for air travel and some sea transportation as well. It can be assumed that the low level of the current carbon/energy tax may make it easier for the government not to exempt. Finland however removed its tax on fuels used in electricity due to increased competition in that sector following the opening of the Scandinavian electricity pool.

3.2.3 The Netherlands

By design, numerous exemptions from the two taxes apply to industry and households in *the Netherlands*, but these are not defined by branch or industry, but based on actual level of energy use. In the case of large-scale natural gas users, natural gas consumption in excess of 10 million cubic meters is taxed at less than half the rate applied otherwise. Further tax relief was granted to producers of residual fuels when those fuels are used where they are produced: the tax on these fuels will be zero until 1 January 1999. It was stressed by the Parliament that the financial margin created by this temporary tax relief would promote investments in energy conservation by the companies concerned.

The *regulatory energy tax* includes a number of exemptions. First, exemptions aimed to increase the environmental efficiency of the tax: district heating is not taxed, natural gas used for power generation is also exempted, so as to foster the penetration of combined heat and power generation.

Second, by definition this tax on small scale energy use does not apply to energy use above the defined ceilings (about 60% of non-transport, non-feedstock energy use in the Netherlands). The main argument for these exemptions was that Dutch companies which operate on the international market don't have the opportunity to pass on price increases to their international consumers, Dutch firms being price-takers rather than price-makers.

A tax-free energy "floor" has been defined for metered energy, set at 800 cubic meters of gas and 800 kWh of electricity use; about 6% of gas consumers and 5 to 10% of electricity consumers are thus exempted from the tax. The "floors" encourage energy efficiency and correct for the relatively high share of energy expenditures in total income for lower income groups.

3.2.4 Norway

In Norway, the CO_2 tax only covers 60% of energy-related emissions. Onshore use of natural gas is not taxed. The pulp and paper, and fish meal industries pay half the rate on mineral oil. Coal and coke, when used as reducing agent and feedstock, and in the cement and leca industries are not taxed. Air service, shipping coastal freight transportation, coastal fishing in distant water and the supply fleet in the North Sea are also exempted. Concern for competitiveness has been a major factor in determining what exemptions should be granted. In the particular case of Norway, a serious concern for CO_2 leakages has led to some exemptions: the taxation of fuels for fishing in distant waters, aviation and shipping would indeed have very little effect on overall energy use since it is easy to tank in other countries.

3.2.5. Sweden

Sweden granted exemptions from the carbon tax to industrial users, in addition to the lower rate (25% of the set level until 1 July 1997, 50% since then). For any single company, there would be a ceiling on the total amount of *energy* taxes to be paid. During 1991, this ceiling was 1.7% of the value of manufactured products; over 1992-94, it was brought down to 1.2%. It should have been eliminated by January 1995. This exemption rule ran into the same evasion problem as in Denmark, with firms isolating their most energy-intensive activities and being granted exemption on that basis. Additional to the tax evasion problem, the exemption was granted on a company-by-company (not on an industry branch basis), and had to be renewed every year by the government, an impractical and expensive scheme.

The electric power sector is exempted from the carbon tax, for reasons related to the open electricity market now in place in Norway, Sweden and Finland.

Biomass use in power generation, as a renewable energy source, is not subject to carbon and energy taxes, and a limited subsidy was introduced for investments in plants using biofuels.

3.3. TAX REVENUE RECYCLING: FROM GRANTS TO FISCAL REFORMS

What is the gross collected amount of these taxes, before any recycling takes place? This isn't a straightforward question, since several taxes are applied to the same energy sources in these (and other) countries: other environmental taxes (e.g., on SO_2), excises, taxes to finance maintenance strategic reserves, value added tax, to name a few.

Revenues of carbon/energy taxes have, roughly speaking, been 'recycled' through the following means:

- carbon/energy taxes have been introduced as part of fiscal reform packages, and helped reduce most distortionary taxes; or
- they have been allocated to the general budget, without explicit recycling (in *Finland*); or
- they were used to alleviate the economic cost and distribution impacts of the tax through targeted grants, subsidies, etc.

3.3.1. Denmark

In Denmark, carbon tax revenues from industry are to be entirely recycled in that sector, through:

1. Lower employers' social security contributions, to become the largest recycling item after 3 years;
2. Investment grants for energy efficiency improvements, available for four years;
3. A fund for small businesses which receive only a limited share of recycling via reduced social security contributions.

Energy efficiency grants, up to 30% of the initial outlay, are available on a project by project basis, subject to financial criteria to assure that the project is really driven by energy savings considerations, and that its profitability is not such that it would have been implemented anyway (i.e., it is not a "business-as-usual" investment opportunity).

3.3.2. The Netherlands

The *environmental tax* of *the Netherlands* goes to the general budget. The new *energy regulatory tax* is used to reduce direct taxes paid by households and businesses. The government has decided to recycle revenues back to households through three changes to the personal income tax: a decrease in the rate for the first income bracket (67% of revenue recycled to households), a raise of Dfl. 80 in the tax free allowance (32% of recycled revenue), and a Dfl. 100 raise in the standard deduction for senior citizen (ca. 1% of recycled revenues). Businesses are compensated through a 0.19 percentage points reduction in the rate of the employer's paid social security contribution (ca. 57% of revenues recycled to businesses), a raise of Dfl. 1300 in the standard deduction for small businesses (25% of recycled revenue), and a 3 percentage points reduction in the corporate tax rate over the first 100 000 Guilders of profit (the remaining 18%).

Together, these two taxes represent 2.5% of total tax revenue in the Netherlands (in 1995, the first tax raised Dfl. 1,350 Million; after 1998, the *energy regulatory tax* is expected to raise Dfl. 2,100 Million, including the increase in value added tax revenues).

3.3.3. Norway

Norway uses carbon tax revenues for its general budget although carbon taxation was introduced as part of an overall fiscal reform. Revenues from the CO_2 tax amounted to more than Nkr. 5 billion in 1993 (about 0.68 percent of Norway's total tax revenues).

3.3.4. Sweden

Sweden also introduced the carbon tax as part of an overall fiscal reform which shifted tax revenues away from personal income tax (the highest marginal income tax rate being 72%). As a result of the *whole* reform (including increase in VAT revenues), direct taxes on households and corporations amounted to 21.3% of GDP in 1991, compared with 25.3% in 1989[5]. Total tax revenues from the carbon tax alone amounted to Skr.10.3 billion in 1993, i.e. 2.5% of total central government revenues; other energy taxes contributed Skr.36 billion.

3.4. ACHIEVEMENTS IN TERMS OF EMISSION REDUCTIONS

The complexity of the above schemes, with numerous exemptions affecting the price signal of taxes, makes it difficult to assess what their effect would be on energy-related CO_2 emissions and the economy. An additional difficulty comes from the fact that carbon/energy taxes never cover 100% of fossil energy use; so that the effect of the tax on overall energy-related CO_2 emissions is uncertain. Still, some analyses of expected impacts are available and summarized here.

3.4.1. Denmark

Denmark has estimated the effect of its new carbon/energy tax scheme, taking into account the recycling of tax revenues through lower social security contributions and investment grants to promote energy efficiency[6]. The tax, along with other measures such as subsidies for energy efficiency improvements, would provide a 4.7% reduction in CO_2 emissions from 1988 levels in the year 2000. The relative high tax on space heating would have a modest effect, whereas subsidies have a key role in providing emission reductions.

The results on the macro-economy underline the difficulty of assessing the economic impacts of such a complex system. With respect to costs for industry *before recycling*, the tax burden accounts for less than 0.5% of total labor cost; over 5 years, such a fluctuation is modest compared with what variations in exchange rates, prices and wage rises may bring about. Recycling basically reduces the overall cost to zero, with lower labor cost and investment grants bringing about a 0.5% reduction in cost. Investment grants for energy efficiency improvements mean increased administrative costs in industry, but lower energy costs, and a potential improvement in competitiveness. The net effect on employment would be modest but positive (+2000), as a result of:

- a negative effect of the tax in some sectors,
- a positive effect of recycling through lower social security contributions,
- a positive effect from the investment grants,
- a positive effect through the substitution of labor and capital for energy (uncertain).

---

[5] See OECD (1994).

[6] Ministry of Finance, Denmark (1995).

3.4.2. *The Netherlands*

The Netherlands expect that their new *regulatory energy tax* will contribute a total CO_2 emission reduction on the order of 1.7 to 2.7 million tonnes per year in 2000 or 1.5% of the country's total emissions; CO_2 emissions from the groups targeted by the tax are projected to decline by about 5%[7], although the tax is not the only tool applied to reduce CO_2 emissions from these sectors. No information was available on what might have been the effects of the *environmental tax*, in the last 3 years. Note that by design, the level of this tax cannot be predicted, since it is based on revenue requirements for the overall budget.

3.4.3. *Norway*

Norway has tried to estimate the effects of its CO_2 tax for the first three years of its implementation (1991-1993). The results apply to only parts of total CO_2 emissions covered by the tax. They are based on a detailed sector-by-sector analysis which highlights the difficulty of assessing the relative contribution of prices, trade conditions, general technical improvements, regulations, and changes in industry structure in the observed changes in CO_2 emissions. One conclusion is that the CO_2 tax has probably had some effect on emissions form mobile sources in the households sector and from stationary sources. The total effect varies between three and four percent over the period; the price of fuel oil and gasoline increased by 11-17% and 9-11% respectively, as a result of the CO_2 tax. The implied price elasticity of oil demand ranges from 0.17 to 0.44, in line with conventional numbers, although maybe on the high side for such short term responses.

3.4.4. *Sweden*

Given the combination of the carbon and energy taxes, and their evolution over the past years, the analysis of their impacts on overall energy consumption and resulting carbon dioxide emissions proves difficult. A few observations can be made[8]. First, the *removal* of the energy tax on industry in 1993 was followed by a 20% increase in heavy fuel oils consumption over the first six months of the year, compared to the same period the previous year (this is however partly explained by a surge in economic growth); the sales of light fuel oil decreased by about 8%. In the paper and pulp industry (which represents about 40% of final energy consumption by industry in Sweden), the increase in HFO demand was 30%, with a corresponding decrease in the use of wood for energy. This demonstrates, if needed, that industry fuel choices are influenced by prices, and suggests that taxes have an impact on energy consumption.

---

[7] Ministry of Housing, Spatial Planning and Environment of the Netherlands (1995.a)

[8] Ministry of the Environment and Natural Resources of Sweden (1994).

In the district heating sector, where the carbon tax was raised in 1993, demand for carbon tax-free wood fuels increased by 30%. For other sectors, the effect of price changes introduced by the carbon and energy taxes on energy demand and CO_2 emissions are less clear.

4. Failed attempts at carbon/energy taxation

Several other OECD countries have considered carbon/energy taxes to help reduce their CO_2 emissions. These countries made proposals that were dropped after discussion, or considerably down-sized. Others, like the carbon/energy tax proposal made by the European Commission (not discussed here), have undergone several revisions. Other countries, inside the European Union are trying to introduce carbon/energy taxes with an environmental focus (Austria and Belgium). It is difficult to draw any firm conclusion from proposals under discussion, although they indicate what aspects of such taxes make them more or less acceptable in different countries. What follows is a brief presentation of the central features of the proposed taxes, as well as bones of contention that appeared in the discussion towards implementation.

4.1. AUSTRALIA'S "GREENHOUSE LEVY"

Australia discussed a "greenhouse levy" of approximately 3.5 US dollars per ton of carbon[9]. The tax was proposed as one of a possible set of measures to enhance Australia's greenhouse response. Complementary proposals were also discussed, some of which would have involved spending some of the levy revenues on other reduction measures, such as a Sustainable Energy Agency. The levy was set at a modest level, and was discussed as a possible signal for action to increase efficiency and consider less carbon-intensive fuels.

The proposal encountered strong opposition from industry. It was claimed that the levy would reduce profitability for energy intensive industries and might cause some industry to alter investment plans in Australia. The government decided not to proceed with the levy following industry commitments to achieve more significant emission reductions through cooperative approaches.

---

[9] In 1989, the price of steam coal was 63 Australian dollars in 1989 (industry price for one ton of oil equivalent). At 1996 exchange rate, without accounting for inflation, the greenhouse levy represents a 4.3 per cent increase for coal prices to industry.

4.2. THE US "BTU-TAX"

The BTU[10] tax proposed by the Administration in 1993 met the same fate. There are a rather wide range of energy taxes and tax expenditure policies in the U.S. at the state and federal levels of government that fall directly or indirectly on various forms of energy. Nevertheless, most energy sources in the U.S. remain lightly taxed relative to other OECD countries and, like many other countries, those energy taxes that do exist fall most often on end users or consumers of energy. In early 1993, the Clinton Administration proposed a new form of energy tax at the federal level that would have changed this feature of US energy prices. The proposed BTU tax proposal would have taxed virtually all forms of fossil fuel energy in the U.S. and increased energy prices for all end users (consumers and producers). While the total energy tax burden in the U.S. would have remained below those in Europe (primarily because of the gasoline tax differentials), taxes on some other forms of energy would have been higher. With the exception of a few European countries, the BTU tax would have been one of the only broad-based, inclusive energy taxes in the OECD. Although the BTU tax proposal ultimately never became law, the analytical and political debate accompanying the tax initiative provides important insights to the development of national and international tax policy. We try to summarize these insights below.

The BTU-tax did not originate in a debate on environmental or energy policy, but as part of a broader set of concerns involving the budget deficit and macroeconomic performance. For instance, its rate was not fixed based on the expected environmental outcome, but rather on the basis of how much revenues it would bring. Of course, the other benefits of the BTU tax were highlighted (environmental, reducing energy supply risks, etc.). The BTU tax was selected among alternatives because it offered a balanced coverage of all these issues, without affecting one fuel or energy source too specifically. Although a number of exemp-tions were included in the original scheme, the tax would still achieve environmental goals if implemented.

A central tension in the design of the BTU tax was where to place the burden for paying the tax in the marketing chain from production to ultimate end user. While the revenue and macroeconomic implications of one collection point over another are fairly minimal, it makes a big difference in the incentives for fuel switching, and ultimately the environmental impact of the tax, as well as the costs of collection and enforcement. The initial BTU proposal imposed the tax at the minemouth for coal, the refinery gate for oil, the pipeline for gas, and at the electric utility for hydro and nuclear generated electricity. By placing the tax "upstream", the intention was to send the price signal as early on in the production chain as possible thereby creating as broad a set of market responses as possible. It could have gone further - very

[10] BTU: *British Thermal Unit*

early versions of the tax placed the burden of the tax for oil and gas at the wellhead. As the debate went on, the tax was shifted further downstream. One important argument was made by the natural gas producers who stressed that they would not be able to pass on the tax to final consumers under their current fixed-price contracts. Also, the percentage price increase resulting from the tax in the price of coal, for instance, would be greater if the tax is placed at the minemouth before transportation costs have been incurred as compared to placing the tax at the electric utility gate.

At the end, the issue of who pays the tax, which determines the regional and energy-source impacts of the tax, probably dominated the debate much more than the overall cost of the tax to the US economy or competitiveness issues (the low level of the tax would not have raised US energy prices to levels observed in the rest of the OECD countries). This national example shows that the *perceived* distributional effects[11] or the "visibility" of the tax, determined by the tax design (energy versus carbon, upstream or downstream...) play a central role in the acceptability of a tax.

4.3. NEW ZEALAND'S "LOW LEVEL CARBON CHARGE"

In 1994, New Zealand proposed a package policy of policy measures to address climate change. One of the proposed measures was a carbon charge on energy-related CO_2 emissions. The charge would have been applied at the point of extraction or importation for fossil fuels, with no exemptions. The charge was intended to be revenue neutral and various revenue recycling options including income and corporate marginal tax rate reductions, debt reduction, and funding high priority expenditure such as energy efficiency programmes. The proposed charge was set at a low level ($NZ10-30/tC, or $US 4-12/t$CO_2$) in order to stimulate a change in investments toward lower carbon-intensive options while avoiding excessive short-term adjustment costs.

The carbon charge proposal met strong opposition from both within government and from industry. At that time, forest absorption was also projected to increase significantly over the decade to 2000 and net emissions were projected to decrease by 54% even though energy and industrial process emissions were projected to increase by 18-22% by 2000. In the face of industry opposition, the government decided to pursue a strategy of voluntary agreements with industry to reduce emissions. However, the possibility of a low-level carbon charge remains.

---

[11] It should be noted that in theory the point of collection for the tax does not alter its incidence on emissions, which is only dependent on supply and demand elasticities.

4.4. SWITZERLAND: EARLIER AND ONGOING PROPOSALS

In 1994, Switzerland considered the introduction of a CO_2 tax to help achieve the Rio target; the discussion was also triggered by debates at the EU level on the adoption of a harmonized carbon/energy tax. The CO_2 tax envisioned by Switzerland, reaching Sfr 36/tCO_2 (US$ 26) would have resulted in domestic energy price increases as high as 100% (in the case of coal). The tax was considered as beneficial to both the global and local environment, to Switzerland's energy security and, over the long run, to the Swiss economy as a whole. Tax rebates for energy intensive industries were included in the proposal. The tax would have been partly recycled through lower labour taxes and through a per capita amount.

An interesting feature of the proposal was a structure of tax rates for energy-intensive industries that would still encourage lower CO_2 emissions, avoiding to sacrifice the environmental effectiveness of the tax instrument in face of its economic impact. The tax rate was to decrease for companies whose consumed fossil energy represented more than 3% of the gross output value. The rates were lower for energy-intensive consumers, but set so that a shift to the next tax category through a reduction in energy-related emissions would still mean overall financial savings.

In its current discussion on how to achieve its 10% reduction goal in the year 2010, Switzerland has re-initiated a discussion on carbon/energy taxation. The introduction of the tax on carbon emissions would hinge on the level of emissions in the year 2004, at which time the government would decide whether the country is on track to meet its 2010 emission goal, or whether a price signal is required. The current law, still in discussion, indicates a maximum level of SF 210/tCO_2 (US$ 150).

The law on CO_2 emissions reduction also includes the possibility for economic activities to be exempted from an eventual tax, provided they formally agree to reduce their emissions. If a given sector/activity does not meet the previously-agreed target, the exemption is cancelled and CO_2 taxes have to be paid *retroactively* (with interest). This innovative approach combines a strong price incentive with the assurance for the regulator that certain emission levels will be met, since non-compliance would imply a very high "penalty" for excess emissions.

5. Summary: from the 'textbook' environmental tax to unilateral taxation decisions

A number of political, social, and technical considerations explain the departure from the economic textbook version of environmental taxation, i.e. a single tax rate applied to all emitters. Taxes have originally been introduced at modest levels, and sometimes substituted to other existing taxes, to account for already relatively high energy prices in these countries. Moreover, industrial activities have often been granted reduced rates, exemptions, or funding to alleviate the tax burden through energy saving investments. These exceptions were motivated by concerns for international competitiveness, since EU or OECD trade partners had not implemented similar carbon/energy taxes. The electric power sector has generally been exempted from paying carbon/energy taxes on its fuel inputs, so as to maintain competitiveness with imported electricity which could not practically be taxed on its embodied carbon. In Finland and Denmark, a tax has been applied to final electricity consumption to compensate for this exemption.

Measures aiming at lowering the burden on industry need not be counterproductive in terms of environmental outcome; Switzerland's rebate system (1994 proposal) was designed so as to still provide an incentive for less emissions. The tax revenues can also be recycled to finance more efficient energy use, although this adds to the otherwise straightforward administration of such taxes in governments. In other instances, however, exemption clauses have been manipulated and led to tax evasion, with detrimental effects for the environmental objective.

In all, exemptions, differentiated tax rates, etc. appear to be prerequisite for political acceptance of new carbon/energy taxes, at least in the context of unilateral decisions to use such instruments. The problem is likely to remain as long as not all trade partners take similar measures. Phasing in taxes, at a pace that would be in line with the capital stock turnover rate of the taxed sectors, may be a way to accommodate concerns about the economic effects of the tax.

Where economic theory would dictate to apply taxes (or tradeable permits) as the preferred approach, governments have generally considered taxes as one tool in their policy packages to deal with climate change. The role of taxes in the wider sets of policies and measures ranges from the straightforward price signal to a penalty for non-compliance, or a 'threat' helping the introduction of alternative mitigation policies. Until recently, energy taxes have mostly been used as a source of government revenues, and were consequently applied to sectors where short-term demand for energy was perceived as price-inelastic (private transportation), so as to minimise the overall welfare loss to society. In other words, these taxes were motivated by *fiscal* efficiency, and not by *environmental* efficiency (interestingly, recent changes in the taxation of transportation fuels across OECD countries are equivalent to surprisingly high taxes when expressed on a CO_2 basis). Carbon/energy taxes, to the extent that should reduce their own base and are applied more widely, depart from that approach. Yet, they also bring non-negligible tax revenues, which explains their introduction as elements in more wide-ranging fiscal reforms.

OECD and other Annex I countries are faced with a new challenge in their energy policies: energy markets are being de-regulated with, as a probable effect, a reduction in end-use energy prices, while the threat of climate change requires action now to 'de-carbonise' economic growth. Over the past decade, end-use energy prices have already declined in real terms, providing less of an incentive to invest in energy efficiency as a cost-saving measure, or to develop new, less energy-intensive processes and technologies. Declining fossil energy prices, as well as energy subsidies, hamper cost-effective market response to climate change concerns. The purpose of carbon/energy taxation is to restore incentives to save energy, and to provide a long-term signal for research, development and deployment of climate-friendly processes. Carbon/energy taxes introduced in five OECD countries have started to set such signals, while taking into account for the particular economic and social circumstances of these countries.

References

Baron R. et al. (1996) "Taxation (i.e. carbon/energy)". Working Paper 4, Policies and Measures for Common Action. Annex I Expert Group on the UNFCCC, OECD/IEA, Paris.

Hoeller, P. and Coppel, J. (1992) "Energy taxation and price distortions in fossil-fuel markets: some implications for climate change policy", Chapter 12 in OECD (1992)

Ministry of Housing, Spatial Planning and Environment of the Netherlands (1996) A Comparison of Taxes on Energy in Eight European Countries. Background paper prepared for the Ministerial Meeting on CO_2/Energy Tax Policies, The Hague, 31 January 1996. Directorate-General for Environmental Protection.

Ministry of Housing, Spatial Planning and Environment of the Netherlands (1995) The Netherlands' Proposed Regulatory Tax on Energy - Questions and Answers, October.

Ministry of the Environment and Natural Resources of Sweden (1994) The Swedish experience - taxes and charges in environmental policy.

OECD (1996) Implementation strategies for environmental taxes. Paris, France.

OECD (1994) Environment and taxation: the cases of the Netherlands, Sweden and the United States, OECD Documents, Paris.

OECD (1992) Climate change - Designing a practical tax system. OECD Documents, Paris.

TABLE 1. Carbon, CO_2, or Related Taxes in OECD Member Countries effective on 1 July 1997

Country	Tax in Original Units for Main Products	Tax in Original Unit/CO_2	Tax in US$/t$CO_2$	Effective Date	Exceptions	Comments
Denmark	Private Coal: DKr 242/t Electricity: DKr 0.10/kWh Fuel oil: DKr 320/t Heating oil: DKr 0.27/l Diesel oil: DKr 0.27/l LPG: Dkr 86/tonne	Private DKr 100	Private 16.35	Private 1/1/96		CO_2 taxes shown are part of an integrated CO_2 and energy tax system. The introduction of the CO_2 tax in 1992 was accompanied by a small reduction in energy taxes. The industrial and commercial sectors are exempted from ordinary energy taxes.
	Heavy Industry (or facing competition): Dkr 5 (US$0.8/t$CO_2$) to 25 Dkr/t$CO_2$ (over 1996-2000) Light processes: Dkr 50 (US$8/t$CO_2$) to 90 /t$CO_2$ (over 1996-2000) Energy use for space heating: Dkr 600 /tCO_2 (US$ 98)			1/1/96	Heavy industry: if successful voluntary agreements, rate decreased to 3 Dkr; Light processes: if successful voluntary agreements, rate decreased to 68 DKr in 2000. Electricity is exempted from CO_2 tax on its input for trade reasons.	A subsidy to encourage the use of natural gas was introduced to compensate the exemption of coal for electricity.

TABLE 1. Carbon, CO_2, or Related Taxes in OECD Member Countries effective on 1 July 1997 (continued 1)

Country	Tax in Original Units for Main Products	Tax in Original Unit/CO_2	Tax in US$/t$CO_2$	Effective Date	Exceptions	Comments
Finland	Carbon Tax From 1 January 1997: Coal: 169 Mk/t Light fuel oil: 0.186 Mk/l Gasoline (leaded): 0.164 Mk/l Diesel oil: 0.186 Mk/l Natural Gas: 0.071 Mk/m^3 (Note: to be raised to 0.142 Mk/m^3 in 1998) Peat: 11 Mk/tCO_2 (4.2 Mk/Mwh)	70 Mk/tCO_2	14.7	1/1/97	Products used as raw materials in industrial production. Fuels in overseas planes and vessels.	
	Electricity Consumption Industry: 0.0145 Mk/kWh Other: 0.033 Mk/kWh			1/4/97		These electricity consumption taxes went into effect on 1 April 1997. Between 1/1/97 and 31/3/97 a single tax of 0.024 Mk/kWh was applied to all electricity consumers.
Nether-lands	Environmental tax Gasoline: Gld 2.51/100 l Light fuel oil: Gld 2.75/100 l Gas oil; diesel: Gld 2.77/100 l Heavy fuel oil and other mineral oils: Gld 32.33/tonne	Gld 5.16 (+ Gld 0.39/GJ)	2.8 (to 0.22 US$/GJ)	1/7/1992	Non-energy uses and international sea/air traffic.	Previous general environmental tax restructured to 50% CO_2 and 50% energy-based.

TABLE 1. Carbon, CO_2, or Related Taxes in OECD Member Countries effective on 1 July 1997 (continued 2)

Country	Tax in Original Units for Main Products	Tax in Original Unit/CO_2	Tax in US$/t$CO_2$	Effective Date	Exceptions	Comments
Nether-lands	LPG: Gld 33.08/tonne Coal: Gld 23.38/tonne Natural gas: 0-10 mln m^3 - Gld 21.55/1 000 m^3 > 10 mln m^3 - Gld 14.10/1 000 m^3					
	Regulatory Tax Light fuel oil: Gld 2.82/100 l LPG: Gld 33.6/tonne Gas oil (non-propellent): Gld 2.84/100 l Natural gas: Gld 0.032/m^3 (up to a ceiling of 170 000 m^3) Electricity: Gld 0.0295/kWh (low amperage, for consumption between 800 and 50 000 kWh per year)			1/1/96	Note ceilings indicated. In addition, the greenhouse horticulture sector is exempted from the Regulatory Tax on natural gas.	

TABLE 1. Carbon, CO_2, or Related Taxes in OECD Member Countries effective on 1 July 1997 (continued 3)

Country	Tax in Original Units for Main Products	Tax in Original Unit/CO_2	Tax in US$/t$CO_2$	Effective Date	Exceptions	Comments
Norway	From 1 January 1996: Gasoline: NKr 0.85/l Mineral oils: NKr 0.425/l Natural gas: NKr 0.85/m³ Coal: NKr 0.425/k	Gasoline: 362 Mineral oils: 162 Natural gas: 362	Gasoline: 56.1 Mineral oils: 25.1 Natural gas: 56.1	1/1/91 (coal 1/7/92), revised several times	• Fuels in all air and sea transport. • Coal used as input to industrial processes	Increases in the CO_2 tax rate on mineral oils in January 1993 were offset by an elimination of the basic tax rate for mineral oil.
Sweden	Residential *Mineral oils: SKr 1058/m³ *Coal: SKr 920/t Natural gas: SKr 0.788/m³ LPG (non-propellent): 1.105/kg	Residential SKr 370	Residential 52.4	1/7/1997	Cap on total energy-intensive industrial CO_2 and energy taxes paid: • Electricity sector; • International sea and air traffic; • Biofuels; • Ethanol.	There also is a CO_2 tax of SKr 1.00/kg of fuel used in domestic air traffic. Tax automatically adjusted for inflation.
	Industry ^ Mineral oil: SKr 529 /m³ ^ Coal: SKr 460/t ^ Natural gas: SKr 0.394/m³ ^ LPG (non-propellent): 0.552/kg Gasoline: 0.86/l * Auto-diesel oil SKr 1112/m³	Industry SKr 185	Industry 26.2			*Rates as of 1 July 1997.* ^ *Estimated new rates as of 1 July 1997*

Source: Countries submissions.

EUROPEAN COMMISSION'S EXPERIENCE IN DESIGNING ENVIRONMENTAL TAXATION FOR ENERGY PRODUCTS

STEPHEN BILL
European Commission
Rue de la Loi 200, B-1049 Bruxelles, Belgium

1. Introduction

Two major factors have influenced the use in European Community Member States of tax instruments as a tool to achieve environmental policy goals. The first is the role which the taxation of energy products has traditionally played in national fiscal policies, and the second is the fact that, since the establishment of a European Single Market on 1.1.1993, there has been a substantial degree of harmonisation of the rules governing indirect taxation in the Community.

Excise duties on mineral oil products used as motor fuels and heating fuels have been, for several decades, an important source of Government revenue in all EU Member States. They provide a stable fiscal income because of the relatively inelastic demand for, in particular, motor fuels. Revenue from this source has increased over the last 15 years from 2.1% to 2.6% of GDP for the European Union as a whole.

Since 1 January 1993 the basic rules concerning the scope, collection and control of mineral oil excise duties have been harmonised in the Community. This was done to enable goods to be moved freely whilst permitting Member States to continue to collect their national revenues following the removal of all frontier controls within the newly created European Internal Market.

Part of this process of harmonisation has been the adoption of minimum rates of duty which Member States are obliged to respect. These minimum rates are considerably higher than, for example, the comparable tax rates applied in Canada and the U.S. and the actual rates applied by the different Member States are even higher (see TABLE 1).

This relatively high level of taxation has been justified increasingly over the last decade by the principle of internalisation of external costs i.e. the price of energy should include all its costs to society not just the production costs. This explains, in particular why motor fuels are taxed so high.

J. Hacker and A. Pelchen (eds.), Goals and Economic Instruments for the Achievement of Global Warming Mitigation in Europe, 231–237.

TABLE 1. Total Level of Indirect Taxation on Mineral Oils

	Leaded Petrol	Diesel	Heating Oil
EC minimum rate	337	245	18
USA	75	87	-
Canada	148	121	-

2. First proposal for a CO_2/ Energy Tax

At the same time as the Community was reaching agreement on the harmonised excise system for mineral oils, the Commission launched a new initiative in the form of a separate proposal for a carbon/energy tax. This was part of a package of measures designed to stabilise CO_2 emission with the principles laid down in the Climate Change Convention.

This proposed instrument had the following features:

1) it was intended to be a broad-based tax on all energy products except renewables. (The tax base would be effectively twice as large as the mineral oil excise duty since mineral oils account for only about 47% of total energy consumption in the Community);
2) the prime aim of the tax was two-fold - to improve energy efficiency and to reduced the level of emissions by encouraging the use of fuels which emit little or no CO_2 emissions. The tax was therefore calculated on a 50/50 basis according to the energy content and the carbon content of the various energy products (see TABLE 2);
3) the tax was to be totally harmonised (rates, scope etc.) at the European level and would be in addition to all other existing taxation. The revenue would accrue to the individual Member States (estimates were that the tax yield would be around 1% of the total EU GDP);
4) It was intended that the tax could be revenue-neutral, however, with Member States using the increased revenue to offset reductions in direct taxes and charges on labour, with the aim of making the costs of employing labour cheaper and thus creating new job opportunities (the so-called "Double Dividend);
5) the tax was to be phased in over a 7 year period. Initial rates would start at a level equivalent to $3 a barrel with annual increases taking it to $ 10 a barrel. It was to be applied in addition to other taxes, notably the mineral oil excise duty but using the same tax mechanism;
6) several specific provisions were incorporated to mitigate the impact on industrial competitiveness (including special treatment for energy intensive industries and incentives for energy saving investments). Most crucially, however, implementation of the tax was made conditional upon the introduction of similar measures in the other main OECD countries.

TABLE 2. A few examples of taxes on products in ECU on the basis of $3/barrel

Product	Unit	CO_2 tax[1,5]	Energy tax	Total tax[5]
1. Petrol[2]	1000 l (=32,7 Gj)	6,59 (0,202)	6,87	13,46 (0,412)
2. Kerosene/aviation spirit[3]	1000 l (= 35 Gj)	7,05 (0,201)	7,35	14,40 (0,411)
3. Diesel/fuel oil[4]	1000 l (= 37 Gj)	7,66 (0,207)	7,76	15,42 (0,417)
4. Heavy fuel oil	1000 kg (=40,2 Gj)	8,77 (0,218)	8,44	17,21 (0,428)
5. Petroleum coke	1000 kg (=31,5 Gj)	8,74 (0,277)	6,62	15,36 (0,488)
6. LPG	1000 kg (=48,1 Gj)	8,66 (0,18)	10,10	18,76 (0,390)
7. Natural gas	1000 m³ (=34,6 Gj)	5,44 (0,16)	7,14	12,58 (0,370)
8. (a) Hard coal	1000 kg (= 25 Gj)	6,5 (0,26)	5,25	11,75 (0,48)
(b) Low-grade anthracite	1000 kg (= 19 Gj)	4,94 (0,26)	3,99	8,93 (0,48)
(c) Sub-bituminous coal	1000 kg (=12,6 Gj)	3,28 (0,26)	2,64	5,92 (0,48)
9. Coke	1000 kg (= 26 Gj)	7,8 (0,30)	5,46	13,26 (0,51)
10. (a) Opencast Lignite	1000 kg (= 7,5 Gj)	2,25 (0,30)	1,58	3,83 (0,51)
(b) Lignite briquettes	1000 kg (= 20 Gj)	5,6 (0,28)	4,2	9,8 (0,49)
(c) Deep-mined Lignite	1000 kg (= 17 Gj)	4,76 (0,28)	3,57	8,33 (0,49)
11. Peat	1000 kg (= 10 Gj)	3,0 (0,30)	2,1	5,1 (0,51)

[1] CO_2 and thermal conversion rates: source Eurostat.
[2] Specific gravity: 0.74 t per 1 000 l.
[3] Specific gravity: 0.81 t per 1 000 l.
[4] Specific gravity: 0.87 t per 1 000 l.
[5] The values in brackets represent the tax per gigajoule; the amount of the energy tax is ECU 0.21 per gigajoule in all cases.

3. Discussions on proposal

The proposal was put forward in 1992 and was the subject of intensive discussions in the EU Council over a two year period. It met with considerable opposition for a variety of reasons:

1) some (lesser developed) Member States opposed the concept of harmonised levels of taxation arguing that a greater effort should be made by the heavily industrialised countries of Northern Europe (burden-sharing);
2) some Member States were opposed to an increase in the fiscal competence of the Community and thus opposed, on grounds of principle, the introduction at a European level of a new tax (fiscal sovereignty);
3) at least one Member State opposed the concept of a mixed (energy and CO_2) tax and argued for a pure CO_2 tax;
4) several Member States were concerned about the effects of the new tax on the international competitivity of energy intensive industry (despite the conditionality clause);
5) not all Member States were convinced that a new tax was necessary to enable Europe to meet the targets;
6) finally, with a mixed system of taxation (CO_2 plus energy) there were practical problems designing a workable system of taxation for electricity.

4. Revised proposal for a CO_2 Energy tax

In order to overcome these problems a revised proposal was put forward in May 1995 which introduced the concept of a transitional stage whereby Member States would agree the scope and structure of a common tax but would be free to decide, product by product, whether to introduce a positive rate of tax and the level of tax which they wished to apply. The rates set out in the original proposal were to become "target" rates which Member States would aim to reach by the year 2000.

This approach was certainly acceptable to those Member States who did not wish to introduce the proposed tax since they would not be obliged to change their taxation systems. It also had some attractions for those Member States in favour of a tax since it implied that the other Member States would at least co-operate in helping them to control the movement of energy products into their territory. However, discussions *on this revised proposal finally broke down on the question of what was to happen at the end of the transitional period.* A number of Member States wanted an assurance that there should be a harmonised tax introduced whilst a number of other Member States argued that the so-called "transitional regime" should be, in effect, regarded as a permanent arrangement.

5. Proposal for restructuring the Community framework for the taxation of Energy products

Finally, in March 1996, the Council of Ministers recognised that there was no hope of any further progress and requested the Commission to draw up a new approach which would expand the already harmonised excise duty on mineral oils into a new taxation scheme for all energy products.

The Commission proposal for a Directive restructuring the Community framework for the taxation of energy products responds to this request, but in addition fulfils the obligation which the Commission has, under Article 10 of Directive 92/82, to review the existing minimum rates of duty on mineral oils. The basic features of this new taxation framework are as follows:

1) it will incorporate and replace, rather than be additional to, the existing harmonised excise duty on mineral oils as well as existing national taxes on gas, electricity etc.;
2) it will have a tax base which incorporates nearly all energy sources (except for renewable and other "clean" sources of energy);
3) it will be based on the concept of "minimum level of taxation" rather than minimum tax rates. In other words, in order to meet Internal Market and revenue requirements (i.e. to avoid undue tax competition) Member States will be required to apply a certain level of tax to each product category;
4) the way in which the tax is calculated, however, will be left entirely up to Member States. This will leave them free to decide whether to base their tax systems purely on an old-style excise approach or to modulate their tax rates depending upon energy content, carbon content, SO_2, emissions or whatever. The only restraint will be that, for any given product, the sum total of all taxation in each Member State must reach the minimum level laid down in Community law;
5) above the minimum levels of taxation, Member States would be free to introduce tax differentials to encourage use of "cleaner" fuels (for example, low sulphur diesel). At the current time, tax differentials for mineral oils can be introduced but only if prior authorisation is obtained from the EU Council.

6. Benefits of new proposal

The principle objective of this proposal is to improve the functioning of the Internal Market by removing existing tax-induced distortions between competing fuels and between energy consuming industries in different Member States.

The second main objective is to prevent fiscal erosion caused by unhealthy tax com-petition. This arises because Member States lose complete freedom to shape their own tax policy because neighbouring countries tax the same products at lower rates.

Although these are the two main goals of the proposal, it is also hoped that it will serve to further the following policies which Member States may wish to enact at national level:

1) Employment

The proposal gives Member States a framework within which to reconstruct their national tax systems in a more employment-friendly fashion notably by using additional revenue raised through increased taxation of energy products to offset direct charges on labour. Macro-economic modelling has shown potential employment increases of between 155,000 to 457,000 jobs.

2) Energy

The proposal will permit Member States to avoid tax induced distortions for energy products, notably those used for heating fuels of which currently a large proportion is not obliged to be taxed under Community rules. In addition it contains provisions designed to reinforce the Union's security of supply through a process of diversification notably by encouraging the introduction of natural gas in new markets and the development of alternative renewable sources of energy.

3) Transport

The proposal offers Member States the possibility to reduce fuel taxes to compensate for the introduction of more efficient transport pricing instruments (such as electronic road pricing).

4) Environment

The proposal contains a number of provisions which will enable them to use the fiscal instrument to pursue environmental policies tailored to their own national circumstances. Firstly, those Member States who wish will be able to calculate their tax by reference to CO_2 emissions and/or energy content. Secondly, they will, subject to respect of the minimum rates, be able to differentiate the rates of taxation applicable to a particular product on the basis of environmental standards. Thirdly, a number of exemptions or reduced rates have been proposed on the basis of environmental considerations (e.g. for renewable energy sources, rail transport, heat in combined power installations etc.).

7. Conclusions

Indirect taxes on energy sources have come to be recognised, in Europe, not only as an essential tool for the traditional purpose of raising revenue but also as a useful means of promoting a more efficient use of a limited resource. The relatively high level of taxation of motor fuels in Europe has, for example, resulted in a higher overall fuel efficiency in the European car fleet compared to the American. Moreover, although this has not resulted in an absolute reduction of emissions related to car use, it has without doubt kept down their rate of increase.

As a result, the fiscal instrument is being looked to increasingly as a means of influencing consumer behaviour by differentiating the price of competing products in favour of "cleaner" products. Thus the compulsory tax differential which applies in the Community in favour of unleaded petrol has made it the predominant fuel on the European Market.

However, a harmonised approach to the use of fiscal instruments has proved impossible to achieve in a European Union of 15 Member States with

- different structures of energy supply
- different energy resources,
- different fiscal policies
- different degrees of economic development and industrial structures and
- different environmental priorities and problems.

Against this background we are now trying to achieve a tax framework which allows:

1. better competition between energy products in a situation where gas and electricity markets have been liberalised;

2. goods to continue to be traded freely throughout an Internal Market without border controls;

3. industries to compete on a relatively level playing field without too much tax induced distortion;

4. Member States to be able to maintain stable revenues; and

5. Member States to modulate their tax systems to pursue environmental, energy, transport and employment policies in the light of their own national circumstances and national problems.

LIMITS OF THE TAX APPROACH FOR MITIGATING GLOBAL WARMING*

Dr. TERRY BARKER
Department of Applied Economics
University of Cambridge and
Cambridge Econometrics
Sidgwick Avenue
Cambridge CB3 9DE, England

* This chapter draws on published work of the author and his co-authors as referenced.

1. Introduction

Carbon taxation, in the form of taxes on fossil fuels in relation to their carbon content is an obvious instrument for introducing the long-term costs of climate change into the price system. At first sight it has all the hallmarks of good taxation: it tackles an accepted economic problem, helping to bring the private costs of emitting CO_2 into line with social costs of global warming; its revenues can be expected to grow with income because energy demand tends to rise with income, and it is not easy to substitute away from fossil fuels in energy supply; it should be simple and cheap to administer through use of existing tax structures for excise duties; it might help to stimulate energy-saving and innovation and hence economic growth; and its side effects on equity are likely to be small enough for affected social groups to be compensated easily. The carbon tax could become one of the main pillars of fiscal systems in the 21 century, if necessary replacing income tax, VAT or import duties (or all three forms of tax) and improving social welfare.

However proposals to introduce carbon or energy taxes have met with substantial resistance and have not been agreed by member states of the European Union. Some of the objections amount to special pleading by the industries most affected by the tax, eg the coal industry in Europe, already extensively supported by subsidies, would face further loss in markets. Other objections have more force; this paper explores the limitations to carbon taxation and suggests how these may be ameliorated.

J. Hacker and A. Pelchen (eds.), Goals and Economic Instruments for the Achievement of Global Warming Mitigation in Europe, 239–251.

2. Carbon Taxation as an Instrument to achieve the Global Warming Mitigation

The objective of a carbon tax is to slow global warming, usually in the context of agreed targets for limiting or reducing GHG emissions, taking into account the scientific evidence, the risks of inaction and the low cost of a no-regrets strategy.

If such a tax is implemented throughout the European Union (EU) at a suitable rate then of itself the tax could achieve levels of greenhouse gas (GHG) and CO_2 emissions in the year 2010 in line with the European Commission's (EC) targets of 10% to 15% lower than the level estimated for 1990. The tax could take the form of a harmonised increase in excise duties on fossil fuels, in proportion to their carbon content along the lines of the European Commission's proposals for additional excise duties on fuels. However, as National Plans under the Rio Convention have shown, increases in excise duties on fuels are one component, and often a small one, in achieving reductions in GHG emissions. Additional regulatory measures and incentives proposed for example in the SAVE programme and mentioned by the EC as contributing to the target alongside the carbon/energy tax are also important. In addition, other policies and events have considerable effects on emissions eg the effects of the liberalisation of energy markets in member states and the development of a common energy market. This will help to reduce emissions if the electricity generators build up gas-fired generation and close down coal-fired generation; but overall energy demand may rise if liberalisation reduces energy prices to offset any tax effects.

Nevertheless, emissions are expected to be on a rising trend through to 2010 so that any tax would have to go on rising after the year 2000, probably for some years at an accelerating rate, in order to achieve long-term stabilisation and reduction. The 1997 additional excise duty proposal with stepped increases in 1998, 2000 and 2002 has three drawbacks. First, global warming is a long-term problem and the danger is that a policy which sets a limit on the increase in tax gives a message that when the limit is reached the problem is solved; this is not the case. Second, other GHGs should be taken into account in terms of CO_2 equivalence in global warming; this can be done to some extent by adjusting excise taxes by fuel and user, but gaps will remain especially in reducing emissions of methane. Third, the proposal is fixed in nominal terms for convenience; the effects will erode through inflation.

Fiscal Neutrality in Implementing a Carbon Tax

The assumed use of the revenues from the tax is critical in any assessment of the macroeconomic effects of the tax. If VAT is reduced then the inflationary consequences of the new tax (it is after all designed to raise energy prices) will probably be neutralised and inflation may even be reduced. If employment taxes are reduced, as recommended in the EC proposal, inflation may rise slightly, although the effects on employment may be larger. There is another more subtle argument in favour of reducing VAT in order to maintain inflation-neutrality. This is to do with the great uncertainty in forecasting the economy and the associated risks of making a mistake. The intention of the tax is to reallocate spending away from GHG emitting activities. It is not intended for macroeconomic management. It seems therefore appropriate to offset an increase in one indirect tax (the carbon tax) by a reduction in another (VAT), so maximising the relative price change and the effect on switching expenditures. The highly uncertain and controversial macroeconomic effects are then reduced to a minimum. Of course it may be the policy of the governments of the day to change the burden of taxation from direct to indirect taxes, in which case this intention is best stated explicitly.

Following from the importance of the use of revenues is the neutrality aspect of implementation. The proposal is that the tax should be implemented as tax-revenue-neutral, ie in each year the net receipts have been spent in reducing another tax. However, it is clear that the assumption of revenue-neutrality in an assessment of the tax is not appropriate, at least for the UK economy, because it does not simulate properly the long-term budgetary process. The macroeconomic balance of different taxes and expenditures is decided by the Chancellor or Minister of Finance from year to year on the basis of current concerns, and revenue-neutrality per se for one particular tax is of little importance. It seems preferable to assume that the change in the fiscal system is inflation-neutral and to leave aside questions of short-term macroeconomic demand-management or the merits of indirect versus direct taxation.

3. A Summary of the Limitations of the Tax Approach

This paper considers four sets of limitations in the use of carbon taxation:

- the use of the price mechanism
- the possible reduction in international competitiveness and associated carbon leakage
- effects on equity and
- dangers of over-taxation.

These are discusses one-at-a-time in the sections below and a final section considers how such taxation can be designed to be efficient and equitable in achieving a given objective.

4. Limitations of the Price/Tax Mechanism

4.1 EFFICIENT MARKETS

Carbon taxation is a market-based instrument and depends fundamentally on the efficient working of the market system for its success. This efficiency has many requirements and implications. First the legal and institutional structure should ensure that contracts are (1) available, (2) freely entered into by both or more parties and (3) enforceable under clear and widely accepted laws and rules; thus countries beset by bribery and corruption may not be able to use taxation because the taxes will be evaded or become an excuse for further corruption. A second requirement is that prices should reflect costs to some degree, so that the carbon tax will increase the price of carbon-intensive production; in some special circumstances, eg if the carbon-based energy is rationed, then the extra tax may have no effect on demand. Third, buyers and sellers should be well informed as to the costs and availability of alternatives and that future outcomes (even if not known) should at least be considered; in some cases, especially amongst some socially disadvantaged groups such as the elderly, there may be an unwillingness to consider alternatives, so extra taxation may have very inequitable effects.

This set of limitations is associated with the free working of markets and there is a close link between market failure and equity in some special cases. However, in comparison with many other countries, markets work well in the EU member states, and these limitations are not sufficient to rule out general carbon taxation.

4.2 UNCERTAINTY IN ACHIEVING TARGETS

A further important limitation in using the price mechanism is the fact that the outcome may be more uncertain than following direct regulation of the industries and others emitting GHG. Of course the outcome of a regulatory regime is also uncertain: the rules may not be clear; they may not be enforceable; and it may be impossible to monitor the effects of the regulation on the emissions. However, even assuming that the regulations are effective, the uncertainty of taxation is not much of a limitation for two reasons. First, the dichotomy between taxes and regulations is a false one; an effective schemes to abate GHG would combine both taxes and regulations, with each instrument supporting the other depending on the member states, sectors and institutions concerned. Second, since GHG abatement does not require a precise abatement by a definite date, the aim of policy must be to achieve significant reductions over a number of years, with policies adjusting to outcomes repeatedly over the years. In fact the risk is that fixed targets for individual sectors achieved by regulation will be highly inefficient in that the same target might be reached by means of taxation at much lower costs. The whole point of Joint

Implementation is to reduce the costs of abatement by shifting abatement to areas where the costs are lower.

4.3 CARBON TAXES MAY HAVE TO BE HIGH

A final limitation here is that in order to achieve the targets, the taxes may have to be very high, as suggested in some studies. It is argued that they may be unenforceable or they may distort the market. The first point to note is that proposed and implemented carbon taxes have started at very low levels with the intention of assessing the outcomes before tax rates are increased. The second is that the usual case will be that the tax signal is made clearer and indeed amplified by changes in regulations, advertising and energy-saving campaigns, so that the model results are likely to be over-estimates of the required increases in taxes to achieve a target.

The argument regarding distortion is completely misplaced. The whole point of the carbon tax is to improve the market signals so as to include the costs of GHG emissions leading to climate change; this is not a distortion but a correction. It is the market prices for carbon-based fuels without the taxes which are distorted; indeed, in terms of the neoclassical model, it is the whole allocation of resources which is distorted and inefficient if the global warming externality is excluded from prices at whatever taxation level required.

5. Issues of International Competitiveness

5.1 DEFINITION AND OVERVIEW

Competitiveness is taken to mean 'the ability to compete in international markets by industries or nations as depending on the prices and qualities of the goods and services they produce'. Changes in price competitiveness are distinguished from those in non-price competitiveness depending on whether the changes arise from changes in prices or in qualities. Sectoral competitiveness means that of the goods and services of a particular industrial sector, assuming that exchange rates are constant; national competitiveness is affected by the exchange rates. These definitions are similar to those used by Porter (1990) rather than those used by Pezzey (1992), which are more related to relative labour productivity performance. Competitiveness should be distinguished from comparative advantage, which refers to the relative competitiveness of different industries in an economy.

The most widely quoted example is the loss of price competitiveness in carbon-intensive goods and services which arises from the introduction of a carbon tax. It is argued that the tax would raise costs of burning fossil fuels and hence the costs of producing goods for export; these exports would then lose price competitiveness, and hence market share; industries in other countries which do not face the increase in costs will take that market share, increasing their use of fossil fuels leading to carbon leakage.

The argument is persuasive, but it is incomplete. The effects of unilateral action also depend on the use of the revenues from the tax. If they are returned to the economy by way of reductions in taxes which bear on sectoral competitiveness, such as taxes on employers, the initial loss of price competitiveness would be reduced depending on each industries use of fossil fuels and labour. Although energy-intensive industries may still lose out, there may be a *net increase* in price competitiveness of labour-intensive industries. If all the tax collected from energy-intensive industries were to be recycled to those industries, in the form of direct subsidy or support for energy-saving R&D and investment, then their competitiveness may increase in relation to other industries, although the extent of reductions in other taxes would be lower and the efficiency of the outcome would be in doubt.

Furthermore there may also be an increase in more general non-price competitiveness. Three factors could give rise to such a result: firstly, the encouragement given to domestic industries specialising in CO_2 abatement technologies may lead them to develop international markets in these technologies or improve their ability to compete in these markets; secondly, any associated improvements (eg reduced traffic congestion or reduced emissions of other pollutants) caused by the abatement policy could boost the attractiveness of the country; and thirdly, technological development and hence growth of exports may be stimulated by efforts to reduce CO_2 emissions. The scale of these improvements in non-competitiveness may be difficult to judge since it depends on the initial energy efficiency of the industries bearing the tax and the unexploited innovations available to them.

Therefore in macroeconomic terms, competitiveness (ie price and non-price competitiveness) is not necessarily reduced over the long-term by higher energy prices. The outcome depends on the use of the extra revenues by the government and the response of governments and other economic agents to the higher prices. If there is a reduction in labour or other costs of industries and/or a resurgence in energy-saving innovation and technology, then competitiveness may even increase. In addition any overall loss in competitiveness of the whole EU (in relation to the rest of the world) would be compensated in the long term by an adjustment of the exchange rates. For particular industries or for particular firms, however, the effects may be serious. The problem is already evident in the new competition from heavy energy-using industries in Eastern Europe, which are located in countries with low environmental standards. This is a sensitive issue because WTO rules do not allow the imposition of tariffs in these circumstances.

5.2 STUDIES ON COMPETITIVENESS EFFECTS OF CARBON TAXATION

Few studies have been designed to tackle the competitiveness and leakage questions directly, and most report the relevant results in a summary fashion if at all. Competitiveness effects have been seen as incidental to other questions such as: the rate of tax required to achieve the target reduction in emissions, the effect of the tax on economic growth and inflation, or a comparison of the tax with other taxes or other methods of reducing emissions. This aspect of the literature means that caution is required in interpreting the results; if these are incidental to the main arguments and largely the effect of simplifying assumptions, the results should be given limited weight. A second caveat relating to the results is worth mentioning. Although the models and applications are often given equal weight, for example by being listed in tables and described in the text, each has absorbed different levels of resources. Some are off-the-shelf applications of a standard model by one researcher; others, such as the OECD GREEN model, have involved a team of researchers over several years building and applying a much larger and more refined model (Oliveira-Martins, 1992). Extra resources do not necessarily imply that the results are more reliable, or more 'right' in any sense, but they do allow for more detailed modelling and more accurate representation of economic behaviour and technological change.

Some evidence of the price competitiveness effects comes from studies of the determinants of international trade. These studies are hard pressed to find significant relative price effects, and when they do so it is clear that changes in exchange rates, in labour costs and or even in raw material prices are more important than changes in taxes on energy. If national competitiveness of economies is considered, it is clear that high energy prices, brought about by high taxes or lack of domestic energy supplies, are no obstacle to industrialisation or rapid economic growth (witness the successes of Singapore, Hong Kong and Japan on a larger scale). Conversely the availability of ample reserves of low cost oil and gas, let alone coal, is not sufficient to provide a high degree of success in international trade (consider Nigeria, Iran, or indeed the UK). The root causes of competitiveness lie elsewhere.

More evidence comes from explicit studies of the effects of carbon taxation. Models considered in a recent review (Barker and Johnstone, 1998) can be divided into four broad general categories: static general equilibrium models (SGE), dynamic general equilibrium models (DGE), aggregated macroeconometric models (MACRO-A) and disaggregated macroeconometric models (MACRO-S), whether national or global. They differ significantly not only in terms of basic methodology and parameter assumptions but also by sectoral and regional classification. These differences fundamentally affect the results obtained and their interpretation.

The effects on the competitiveness of energy-intensive industrial sectors, such as chemicals or metal products, of unilateral action at the EU, NAFTA or OECD level appear to be very small in relation to the effects of exchange rate or wage rate fluctuations. The DRI studies estimate a change in the growth of EU exports of chemicals from 5.6% pa to 5.4% pa as a result of the EC tax. The reasons for the small response are transport costs and other barriers to trade, protection of domestic markets, especially in the case of iron and steel, and the relatively small increase in costs, especially when exchange rate changes, revenue recycling and carbon-cost-saving at all stages in the industrial system (especially in electricity generation) are taken into account.

5.3 THE EXTENT OF 'CARBON LEAKAGE'

Carbon leakage would occur if a country or trading block takes unilateral action to reduce CO_2 emissions, and this action results in higher emissions elsewhere in the world. Carbon leakages reflect the effects of unilateral (or regional but non-global) policies to reduce greenhouse gas emissions on emissions from countries which do not restrict their own emissions. The carbon leakage rate can be defined as the ratio of the difference between emissions from the rest of the world in the policy scenario and the base scenario over the difference between emissions in the regulating coalition in the policy scenario and the base scenario:

$$L_i = \frac{C_{w,p} - C_{w,b}}{C_{i,p} - C_{i,b}} * 100 \qquad (1)$$

where L = the leakage rate, expressed in %
C = carbon emissions
I = the country/trading bloc
w = the rest of the world
p = the policy scenario
b = the base scenario.

If the rate is greater than 100, then the effects of the policy are environmentally perverse, increasing global emissions. If the rate is less than zero then the effect of the policy is to reduce emissions elsewhere, reinforcing its beneficial environmental effects. And if the rate is between zero and 100 then some, but not all, of the effects of reduced emissions are negated by increased emissions elsewhere.

Leakages take place through two principal channels: the relocation of trade in manufactures and the substitution effects arising from a fall in oil prices. When a set of countries restricts emissions by imposing taxes (or instituting an emission permit regime) this increases the relative cost of carbon-intensive production, potentially shifting the comparative advantage of non-constrained economies toward the production of those goods which tend to be carbon-intensive, thus increasing emissions by changing the composition of world output *across* sectors. Similarly, the constraint will also reduce the demand for oil, potentially lowering the world pre-tax price depending on the response of suppliers. Depending on fuel substitutability in individual economies this may increase the carbon intensity of production processes *within* individual sectors. There may, therefore, be increased emissions due to both sectoral composition and sectoral production processes. To some extent, these two may be counter-balanced by income effects if the effect of the demand reduction implied by the carbon constraint reduces emissions from oil exporters, and if the coalition which imposes the tax experiences a significant fall in income and its imports from the rest of the world are relatively income elastic.

Intuitively, leakage rates for a given policy constraint should decrease with the size, expressed in terms of energy consumption, of the regulating coalition. There are two reasons for this. On the one hand, there are fewer 'free riders' able to exploit a given change in relative prices of fuels and manufactured goods. On the other hand, since the demand shock will be relatively more severe with a larger coalition, relative prices will tend to change more significantly. Leakages should also increase as the constraint is made more stringent, ie the higher tax rates implied by a shift from the Rio targets to the Toronto targets. This also reveals the importance of distinguishing between average and incremental leakage rates. For instance, in order to include leakage rates in the determination of effective policies by the regulating coalition, incremental leakage rates must be subtracted from incremental emission rates in the coalition country. Given these two points it is important to distinguish between the size of the coalition and the stringency of the constraint when discussing alternative estimates of leakage rates.

The high carbon leakage estimates of 80% for EU and 70% for OECD uni-lateral action which were given prominence following the DTI report on competitiveness (see Pezzey, 1992) and which have been quoted repeatedly in the literature, are not confirmed by the other studies reviewed. The estimates arose from the mechanical application of an unsuitable model. One of the few studies for Annex 1 signatories to the Rio Convention (Oliveira-Martins, 1995, using the OECD GREEN model) concludes that substantial cuts in emissions can be achieved with virtually no leakage. Indeed, far from there being leakage, it seems likely that if the carbon tax reduces economic activity as the OECD area introduces the tax, as some of the studies suggest, then emissions in the rest of the world will fall ('negative leakage').

6. Equity and the Carbon Taxation

The issue has been extensively explored and debated in the global warming literature especially with regard to energy taxation. A recent review of the literature and state of the art in research on equity effects of environmental policies in relation to climate change has been made by the OECD (1995). Two of the conclusions of this review are:

(1) 'With regard to income distribution effects, empirical studies suggest that a national carbon tax or trading programme would be at least mildly regressive ... in many OECD countries, although there is some evidence that such programmes might actually be progressive in developing countries' and
(2) 'Perhaps the major conclusion of the empirical studies is that distributional effects need to be considered very carefully as policy proposals move from concept, to design, and then to implementation.'

The literature is very sparse in this area, and much of it is focused on the effects of a carbon tax. For instance, it is frequently argued that the distribution effects of a carbon/energy tax - such as that proposed by the EC - will be largely regressive: the burden of the tax will fall disproportionately on lower income households. (See Poterba 1991, Dower and Zimmerman 1991, CEC 1992 and Johnson et al 1990 for discussions.) The basis upon which it is usually assumed that carbon/energy taxes are regressive is intuitively obvious: lower income households tend to spend a larger proportion of total household expenditures on fuel for domestic energy services (ie heating, hot water, cooking and lighting).

The most comprehensive study of the income distribution effects of carbon/energy taxation in Europe is that undertaken by the European Commission (Smith, 1992; CEC, 1992). The study mainly relied on EU household expenditure surveys for six countries usually for 1985 (Spain, Italy, Netherlands, France, Germany (1983) and Ireland (1987)). The analysis was mainly static, assuming (1) no indirect effects, (2) that any taxes do not change demand patterns, (3) that all the tax is passed on to prices, and (4) that there is no revenue recycling. Under these assumptions, the ratio of regressivity was calculated for 10 countries - defined as carbon tax payments as a percentage of household total expenditure of the poorest quartile over the corresponding percentage for the richest quartile. The conclusion was that carbon and energy taxes were weakly regressive for most countries, but more strongly regressive for the UK and Ireland. The analysis (CEC, 1992) goes further in considering indirect effects in some illustrative calculations for France and Germany, which confirm the static calculations (CEC, 1992, p.134). Smith was able to go further in considering the effects of recycling either as lump-sum payments or across-the-board reductions in tax rates, but only for the UK in 1988 (Smith, 1992, pp. 264-266), and again without considering indirect and dynamic effects.

The main problem with the EC study is that it concentrates on one side of fiscal reform, viz the effects of tax increases on expenditures; the use of revenues to reduce employers' taxes and increase employment is not treated specifically. Fiscal reform is intended to increase employment and reduce unemployment as well as improve environmental performance; it is likely that the extra employment will have favourable effects on the distribution of income, given that low-income households are likely to benefit more than other households from reductions in unemployment. In the large quantitative studies of the effects of the EC's proposed carbon/energy tax (DRI, 1991 and 1992) or of a substantial package of environmental policies (DRI, 1994), distributional questions have been incidental, rather than fundamental, components of the analysis. The reason is partly one of the lack of data and the poor quality of the data available. This remains a problem in all applications relating to the equity effects and any analysis has to be based on strong assumptions.

7. The Dangers of Over-taxation

It is argued that taxes on some energy products, particularly petrol and diesel, are already too high and that a new carbon tax would encourage governments to set rates at levels which amount to over-taxation. This is a controversial argument because it depends on valuations of the social costs associated with existing taxes and with the carbon tax. These costs (eg the external costs of the transport system - danger, congestion, damage to health and the environment) are already the reason why present taxes have been introduced. The additional carbon tax relates to global warming, but the costs of global warming are by no means agreed. In fact not even the methodology of a costing is agreed.

8. Overcoming the Problems: Designing Taxation to be Effective, Efficient and Equitable

8.1 THE USE OF EXEMPTIONS

The exclusion of certain industries from coverage of the tax on grounds of competitiveness increases the political acceptability of the tax but it must reduce its effectiveness in achieving the objective of reducing GHG emissions. Carbon taxation is intended to bear most heavily on goods and services which release most CO_2 in their production, yet some of the most energy-intensive industries are to be excluded. Clearly this will limit the impact of the tax on energy consumption and CO_2 emissions, but there are two additional problems. The first is that the industries which do not pay the tax will improve their competitive position in relation to those industries which do pay. There will therefore be some switching of demand towards the products of energy-intensive industries, precisely the reaction that such a tax should avoid. The other problem is that companies which find themselves paying the tax will try to be reclassified as exempt or eligible for rebates if at all possible, thereby limiting the impact. The reason for the proposed exemption is of course the risks to international trade competitiveness of introducing a unilateral carbon tax. The EC proposals are making allowance for the industries most exposed to this risk by excluding them in some way from paying the tax, at least until some more general OECD-wide tax is adopted. The exclusions reduce the efficiency of the tax (Oliveira-Martins, 1995), but they do mean that the European and UK industries most vulnerable to competition are protected in third markets.

8.2 RECYCLING REVENUES

The key to an effective and equitable introduction of carbon taxation is the use of the revenues to encourage energy saving and to compensate those groups which suffer the most inequitable effects. If the revenues are recycled via reductions in labour taxes, a switch in demand away from energy-intensive industries will tend to reduce the demand for capital equipment relative to that for labour. The levels of energy tax increases in most government proposals or actions in the EU are sufficient to generate small employment effects. For example the UK road fuel duty escalator (from 1997 a 6% increase in real road fuel duties every year for an indefinite period) with revenues recycled via reductions in employers' taxes, is estimated to increase employment above the base line by some 300,000 (about 1.2%) (Barker, 1997). The reduction in unemployment will undoubtedly be equitable; the equitable effects can be enhanced by using some of the revenues to insulate dwellings and improve fuel efficiency of domestic equipment.

8.3 SECONDARY BENEFITS

Finally carbon taxation has additional environmental benefits which are potentially so important that they may more than offset any estimated economic costs. These benefits come from the reduction in emissions and social costs associated with the burning of fossil fuels, apart from the GHG emissions, eg SO_x, small particles and noise. Taxation reduces these costs as a side-effect, so that expensive end-of-pipe abatement is not required. Such benefits should also be taken into account when designing the tax change.

References

Barker, T. (1993) The carbon tax: economic and policy issues, in C. Carraro and D. Siniscalco *The Carbon Tax: An Economic Assessment*, Kluwer Academic Publishers, Dordrecht.

Barker, T. (1997) Taxing pollution instead of employment, in T. O'Riordan (ed), *Ecotax Reform Taxation*, Earthscan.

Barker, T., Ekins, P. and Johnstone, N. (eds) (1995) *Global Warming and Energy Demand*, Routledge, London.

Barker, T. and Johnstone, N. (1998) Carbon taxation and international competitiveness in T. Barker and Köhler, J. (eds) *International Competitiveness and Environmental Policies*, Edward Elgar, forthcoming.

Cornwell, A. and J. Creedy (1996) Carbon Taxation, Prices and Inequality in Australia, *Fiscal Studies* **17** (3), 21-38.

Commission of the European Communities (1992) European Economy: The Climate Challenge - Economic Aspects of the Community's Strategy for Limiting CO2 Emissions in *European Economy*, **51**, Brussels.

Dower, R. C. and M. B. Zimmerman (1992) *The Right Climate For Carbon Taxes: Creating Economic Incentives to Protect the Atmosphere, August*, World Resources Institute, Washington.

DRI (1991) The Economic Impact of a Package of EC Measures to Control CO_2 Emissions, Final Report prepared for the CEC, November.

DRI (1992) Impact of a Package of EC Measures to Control CO_2 Emissions on European Industry

DRI and others (1994) Potential Benefits of Integration of Environmental and Economic Policies, Report prepared for the European Commission, DG XI, Graham and Trotman and Kluwer.

Johnson, P., McKay, S. and S. Smith (1990) The Distributional Consequences of Environmental Taxes, Commentary No. 23, IFS, London.

Mabey, N., Hall, S., Smith, C. and Gupta, S. (1997) *Argument in the Greenhouse, The International Economics of Controlling Global Warming*, Routledge, London.

Nordhaus, W.D, (1991) The cost of slowing climate change: a survey, in *The Energy Journal*, **12**, No. 1, 63.

OECD (1995) Climate Change, Economic Instruments and Income Distribution, OECD, Paris.

Oliveira-Martins, J., (1995), Unilateral emission control, energy-intensive industries and carbon leakages, Annex B pp. 107-124 in *Global Warming Economic Dimensions and Policy Responses*, OECD, Paris.

Oliveira-Martins, J, et al (1992) Trade and the Effectiveness of Unilateral CO_2 Abatement Policies: Evidence from Green in *OECD Economic Studies* **19**, 123-40, Winter.

Pezzey, J. (1992), Analysis of Unilateral CO_2 Control in the EU and OECD in *The Energy Journal*, **13**.

Porter, M.E. (1990), *The Competitive Advantage of Nations*, MacMillan, London.

Poterba, J. (1991) Tax Policy to Combat Global Warming: On Designing a Carbon Tax in R. Dornbusch and J. M. Poterba (eds.) *Global Warming: Economic Policy Responses*, MIT, Cambridge, MA, pp. 71-98.

Smith, S. (1992) The Distributional Consequences of Taxes on Energy and the Carbon Content of Fuels, *European Economy* Special Edition 1992 No.1 The Economics of Limiting CO_2 Emissions, 241-268.

Symons, E., J. Proops and P. Gay (1994) Carbon Taxes, Consumer demand and carbon Dioxide Emissions: a simulation analysis for the U.K'; *Fiscal Studies* **15** (2), 19-43.

SECTION III

POLITICAL AND ECONOMICAL INSTRUMENTS - AN OVERVIEW

JOINT IMPLEMENTATION

ACTIVITIES IMPLEMENTED JOINTLY (AIJ) AS AN INSTRUMENT FOR THE MITIGATION OF GLOBAL WARMING

Prof. CATRINUS J. JEPMA
University of Groningen/Amsterdam
P.O.Box 800, NL-9700 AV Groningen, The Netherlands

This contribution is based on the transcript of the tape recorded during the lecture of Prof. Jepma including the relevant parts of the discussion at the Advanced Study Course made by the editors. It has been revised and authorised by Prof. Jepma.

1. Introduction

The aim of this paper is to give an introduction to Joint Implementation (JI). Therefore the first two parts outline the scientific and economic background of this concept. This is followed by an explanation of the basic concept of JI with its advantages and problems. Afterwards JI is distinguished from and compared with emissions trading. A section on the actual status of JI in the UNFCCC process rounds off the paper.

2. Scientific background: future development of global climate

Since 1800 the concentrations of greenhouse gases in the atmosphere have increased. Additionally the scenarios of the IPCC indicate that they will increase further, if there is no policy action. TABLE 1 shows the so-called IS scenarios, which are considered rather authoritative, for the year 2100. These scenarios combine various assumptions with respect to a.o. economic growth, type and amount of energy supplies and size of the population in the period up to 2100. Depending on these assumptions different concentrations of greenhouse gases in the atmosphere are projected in TABLE 1 for the year 2100. Additionally the table shows the implications in terms of the projected global mean temperature and the sea level rise. In this manner the greenhouse problem, as it has been put forward by the scientists of the IPCC, has been outlined as simply as possible.

J. Hacker and A. Pelchen (eds.), Goals and Economic Instruments for the Achievement of Global Warming Mitigation in Europe, 255–271.

TABLE 1. Range of likely Climate Change Scenarios (without intervention)

Scenario	Basic Assumptions	Net anthropogenic Carbon Emissions (Gt/year)	Carbon Concentration (ppmv)	Temperature Rise (°C)	Sea Level Rise (cm)
1900	-	<0.5	315	0.3-0.6 below 1990	10-25 below 1990
1990	-	7	350 (rising)	-	-
Expected (in 2100)	*Yearly Economic Growth* 1990-2025: 2.9 % 1990-2100: 2.3 % *Energy Supplies* 12,000 EJ Oil 13,000 EJ Natural Gas *Population* 11.3 billion (2100)	20 (IS92a)	680 (rising)	2 (rising)	50 (rising)
Best (in 2100)	*Yearly Economic Growth* 1990-2025: 2.0 % 1990-2100: 1.2 % *Energy Supplies* 8,000 EJ Oil 7,300 EJ Natural Gas Nuclear costs decline 0.4 %/year *Population* 6.4 billion (2100)	5 (IS92c)	480 (stabilising)	1 (rising)	15 (rising)
Worst (in 2100)	*Yearly Economic Growth* 1990-2025: 3.5 % 1990-2100: 3.0 % *Energy Supplies* 18,400 EJ Oil 13,000 EJ Natural Gas Phase out nuclear costs by 2075 *Population* 11.3 billion (2100)	36 (IS92e)	900 (rising)	3.5 (rising)	95 (rising)

Source: Jepma and Munasinghe, 1998.

TABLE 2 summarises the development of carbon emissions in the different regions of the world between 1990 and 1995 and facilitates the comparison of the predicted scenarios with the actual emissions.

TABLE 2. CO_2 emissions (MtC) between 1990 and 1995

	1990	1995	Difference (%)
OECD	3035	3165	+4 %
Eastern Europe	1311	910	-31 %
Developing Countries	1774	2226	+26 %
World Total	6119	6301	+3 %

Although the OECD countries are committed by the United Nations Framework Convention on Climate Change (UNFCCC) to stabilise their carbon emissions on 1990 levels by the year 2000, their actual emissions have not been reduced but increased between 1990 and 1995. In opposition the Central and Eastern European countries show, because of their poor economic performance in the transition, a strong reduction of 31%. But in the developing countries there has been an increase of 26%. Global emissions increased by 3%. This indicates a move along scenario IS92a (TABLE 1) with the related implications for global temperature and sea level. Hence from the perspective of the scientifically founded scenarios there is a real problem.

3. Economic background: the costs of CO_2 abatement

Given the problem of global warming it is interesting to know about the economic costs related to its mitigation and possible ways of reducing these costs. FIGURE 1 presents various studies related to the costs of CO_2 abatement in the United States. According to the figure a reduction of the annual carbon emissions by 40 to 60% implies a cost for the US anywhere between 2 and 4% of its GNP. These significant costs have to be considered carefully and the question arises, how the investment in greenhouse policies can be carried out most cost-effectively. A possible solution is based on the optimal use of the various marginal abatement costs, that differ for individual reduction options as well as for individual countries. Take for example a forestry project: it is fairly obvious that the optimal use of the various marginal costs to sequester carbon by afforestation in an industrialised country are higher than the marginal costs of the same option in many of the developing countries. In this respect one can take advantage of the fact that the greenhouse policies or abatement strategies have the same effect on the concentration irrespective where

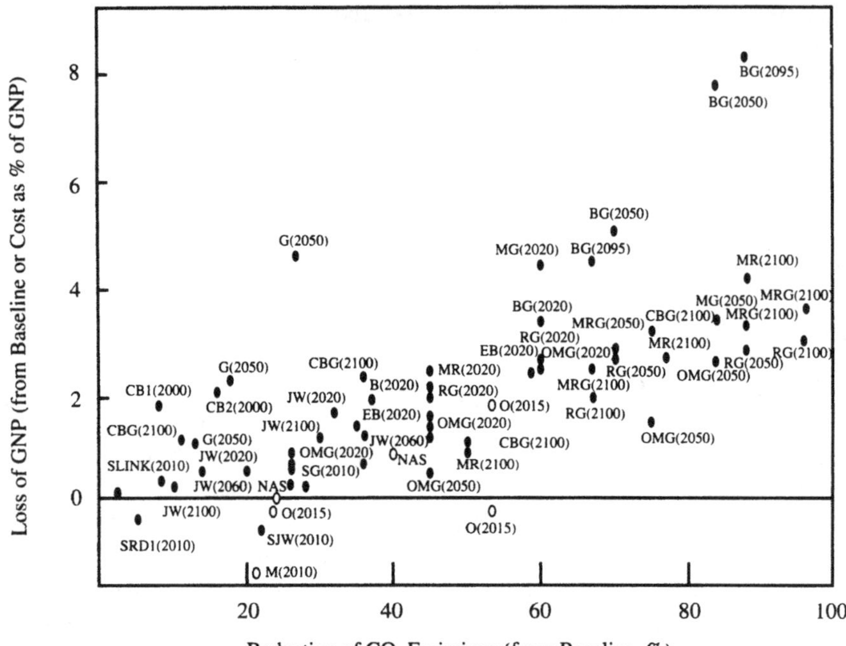

FIGURE 1. U.S. Studies on the cost of CO_2 abatement relative to baseline projection
Source: Grubb et al. (1993).

TABLE 3. Optimum mix of abatement options per region

Option	Level of emission reduction (MtC)			
	OECD	Eastern Europe	Rest of the World	Total
1. energy efficiency improvement	250	250	100	600
2. fuel switch	50	50	50	150
3. removal & disposal	100	50	0	150
4. nuclear energy	50	50	0	100
5. renewable energy	50	50	100	200
6. forestry	250	250	700	1200
Total	750	700	950	2400

Source: Jepma and Lee (1995).

the greenhouse action is taken, because CO_2 emissions have atmospheric hardly no local but only global consequences.

Therefore it is interesting to try to find the optimum mix of abatement options in the different regions of the world. A theoretical solution based on region- and option-specific cost functions, all characterized by increasing marginal abatement costs, is calculated by combining these cost functions and solving by linear programming. The results are presented in TABLE 3 both in terms of the composition of the options and in terms of the regions where the options should be applied.

In this specific study with a predetermined emission reduction target of 2,4 Mt carbon, about one third of the various options should be carried out in the OECD, in Eastern Europe and in the developing countries. Concerning the different options about half of the reductions are achieved by forestry projects and about a quarter by energy efficiency projects. For this particular solution the marginal abatement costs are equalised at around $ 50/tC. It represents cost savings of 60-80% compared to a solution without interregional compensation.

Similar studies to figure out what the best regional policy mix would be have been carried out by Richels et al. (1996). They assumed total freedom to carry out the greenhouse policies in those regions where they can be carried out most cost-effectively. Under those assumptions and based on four model exercises the overall costs of abatement could be reduced with some 70%. The results are summarised in FIGURE 2a and 2b.

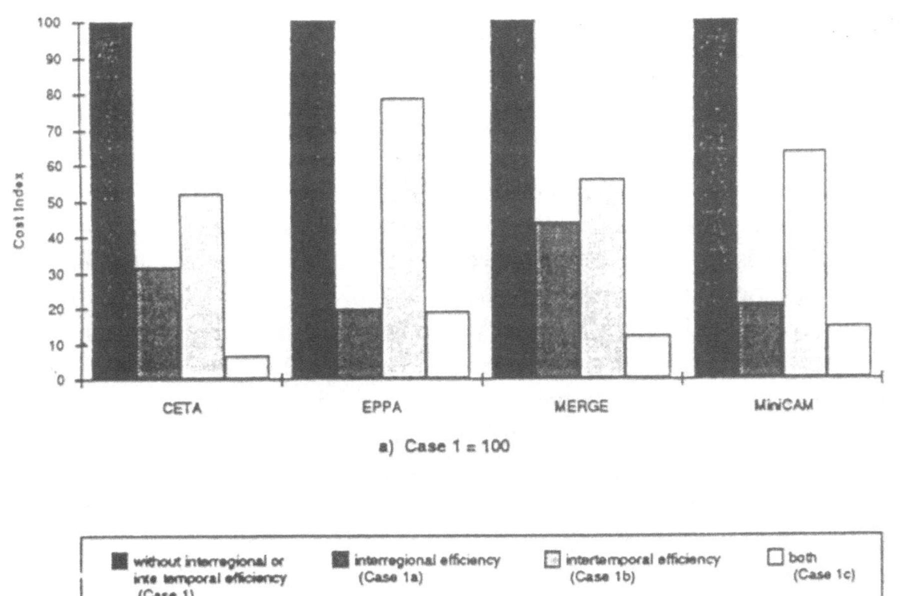

FIGURE 2a. Global costs of abatement under 4 alternative cases accord. to 4 individual models Source: Richels et al. (1996).

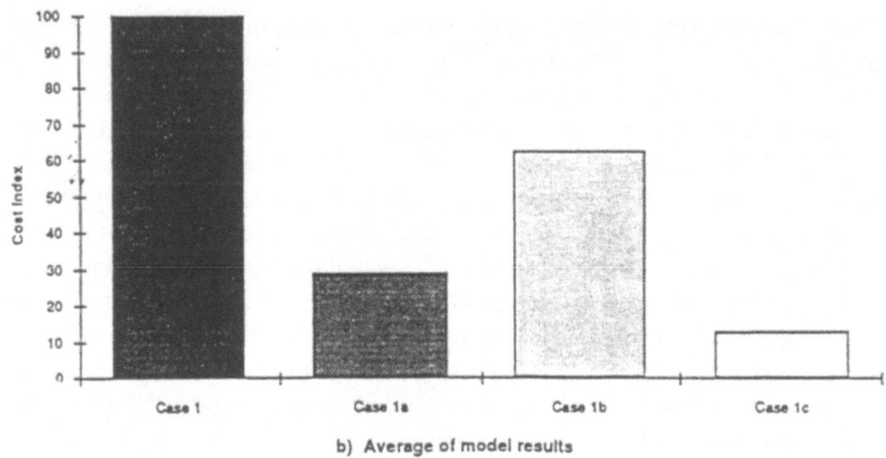

FIGURE 2b. Global costs of abatement under four alternative cases (Average of four models)
Source: Richels et al. (1996).

Summarising these results it can be stated that using the differences in marginal costs leads to interregional and intertemporal efficiency gains, and consequently to a significant reduction of the global costs of carbon abatement..

4. Joint Implementation

4.1 THE ECONOMIC PRINCIPLE

The principle idea behind JI is to make use of the above mentioned interregional cost differences and consequently enhance the cost effectiveness of the mitigation of global warming. This is visualised in a simplified example in FIGURE 3. Let us assume there are two countries with an emission reduction target X. Country A (origin in Y) has to achieve a bigger reduction YX, country B (origin in W) a smaller reduction WX, so jointly they have to fulfill an emission reduction of WY. Of course, both countries will start with the cheapest emission reduction option first following their respective marginal cost functions (starting from Y upward to the left, and from W to the right, respectively). If both countries carry out their obligations in their own territory, country A is facing total reduction cost equivalent to the two diagonally shaded areas under its cost curve between Y and X. Country B faces total reduction costs equivalent to the vertically shaded area under its cost curve between W and X. It is easy to see that between X and P the costs in country B are lower than in country A (both equivalent to the integral of marginal costs in the relevant interval). Therefore it makes sense from a global cost perspective, if both countries achieve emission reductions up to P, because

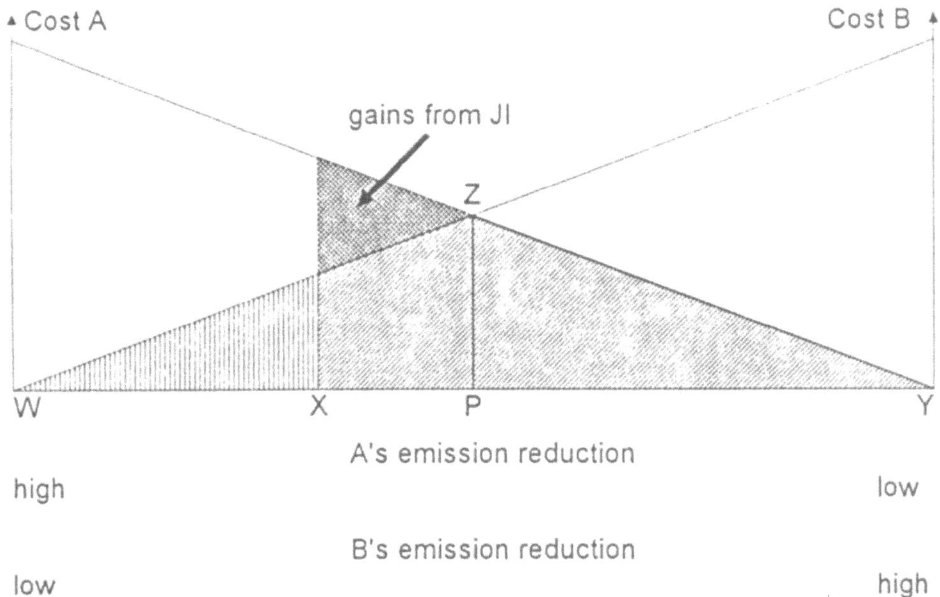

FIGURE 3. Theoretical gains from Joint Implementation

this reduces the overall reduction cost equivalent to the double shaded area between the two cost curves, WZY. Making the trade XP between country A and B and saving the small dark shaded triangle, is basically the principle behind JI. This just represents an application of the economic efficiency concept.

4.2 THE CONCEPT OF JOINT IMPLEMENTATION

FIGURE 4 summarises the concept of JI and its main arguments pro and con. Despite the clear economic advantages associated with JI there was, preceding to COP 1 in Berlin 1995, a strong debate going on between NGOs, developing countries' representatives, people from the industry and others about the issue, whether or not JI is a good concept. Of course, it is complex to apply it, but that's to a large extent an institutional and political issue. The bottom line, however, is that through JI all can benefit, because it is not only the investing countries that can benefit, by realizing their greenhouse policy targets cheaper, but also the recipient or host countries are going to benefit because of the net positive local effects of the projects. And finally the environment can benefit, insofar as more cost-effective abatement policies mean, that can better realize the overall environmental targets or even set more ambitious targets. Hence then JI is a win-win-win situation.

THE CONCEPT OF JOINT IMPLEMENTATION

TO PERMIT INDUSTRIALISED COUNTRIES TO FULFILL (PART OF) THEIR GHG COMMITMENTS UNDER THE FCCC IN ANOTHER COUNTRY FOR REASONS OF COST-EFFECTIVENESS

ARGUMENTS PRO:

- STIMULATE NEW AND ADDITIONAL FINANCIAL FLOWS FROM INDUSTRIALISED COUNTRIES TO THE REST OF THE WORLD
- STIMULATE TRANSFER OF ENVIRONMENTALLY-SOUND AND ENERGY-RELATED TECHNOLOGIES TO ECONOMIES IN TRANSITION AND DEVELOPING COUNTRIES
- CONTRIBUTE TO CONTINUOUS INTERNATIONAL DIALOGUE ON GREENHOUSE PROBLEM
- BRING DIFFERENT ENVIRONMENTAL PRIORITIES TOGETHER

ARGUMENTS CON:

- INDUSTRIALISED COUNTRIES BUY THEIR WAY OUT CHEAPLY
- BASELINE SCENARIO DIFFICULT TO DETERMINE
- PICKING THE LOW HANGING FRUITS FIRST
- ARE JI PROJECTS SUSTAINABLE?

FIGURE 4. The concept of Joint Implementation and arguments pro and con

The main element in the debate about JI was on the question if it is true that the host countries are actually going to gain: one argument in favour was that JI stimulates new financial flows from North to South. Another important argument was that new environmentally sound and energy related technologies will flow more rapidly from North to South. So, not only would the host countries be enabled to get their energy more cost-effectively, but also there was clear evidence that through this their local and regional environment would improve.

Take for example coal, which is the predominant energy carrier in India and China, the two countries with the largest coal reserve worldwide. The use of coal in power stations causes not only greenhouse problems, but also a lot of emissions of sulphur dioxide leading to local pollution and to less agricultural output. This loss is estimated to reach several billion dollar annually in these countries alone. In addition, in China for instance, there is a regional distribution problem emerging also, because the sulphur is coming from the rich areas, but it is the poor areas that are suffering from its impact. Consequently this can lead to a distribution issue within China between the rich and the poor regions. In other words, there are many aspects to the greenhouse issue which makes it attractive for China, but more generally for developing countries, to have access to modern energy efficient technologies. More generally, JI can, next to any technology or financial transfer, also reinforce the dialogue, the discussion, the contacts between North and South, and it can so bring different environmental priorities together.

Because of all of these advantages the Americans have demanded to include trading and JI in the Kyoto Protocol. But there is now a discussion going on between especially the Americans and European Union representatives on the three main aspects of the Protocol, namely: emission caps, JI and trading. The position of the Americans is that these three aspect cannot be disentangled. This is because how ambitious the caps are that may be, depends on the costs involved with the policy to achieve these caps, with the costs in turn depending on the opportunity of trading and JI. Up to now the Americans have refused to formulate their caps proposed to enter the Kyoto negotiations. In opposition the European Union has defined its cap proposal of minus 15% in 2010 vis-à-vis the 1990 GHG emissions level. They are prepared to talk about trading and JI once the Americans have formulated their own cap.

Another important argument of the opponents of JI is that it allows the industrialised countries to buy their way out in a cheap manner. They argue that JI is a way of avoiding measures at home, so that industrialised countries can continue with their economic process as they have done in the past, by starting forestry projects in Costa Rica, or energy efficiency improvements in China, and so on. In my personal opinion, the above argument is a moral one, however, which is irrelevant given the target of the Framework Convention on Climate Change: This Convention defines only the overall aim to globally reduce the greenhouse gas emissions and to make sure that the concentrations of greenhouse gases in the atmosphere are stabilised. Hence this and nothing else is the Convention's target, and the more cost-effective this can be achieved the better.

Another point in the discussion is the question of additionality. Let us take for example a highly efficient power plant in the Philippines designed by Royal Dutch (RD) and financed by it as a foreign direct investment, which is commercially feasible. Then the question arises, whether in a JI regime RD could get credits out of this foreign direct investment? The probable answer is: yes they can. Because the criterion whether or not there will be credits is not whether there is a financial benefit to the investor, but, according to the Framework Convention, whether or not there are net carbon implications. Hence, according to the FCCC text the way I read it, even a direct investment on a commercial basis can yield credits.

4.3 BASELINE DETERMINATION

One of the technical problems of JI is the formulation of the baseline. Since JI is a project oriented approach a specific project needs to be set up in a specific host country. The investing country is usually (co-)financing the project, that either sequesters carbon or reduces its emissions, and therefore will receive (part of) the credits. The question now is, how much credits will the investor get? The amount of credits received depend on the net carbon contribution of the project, which can only be figured out if a baseline is established. The baseline represents what would have happened without the project. To establish such a baseline is only possible by making assumptions about future developments, which in turn cannot only be technically difficult, but quite often also is arbitrary and therefore sometimes even debatable.

Let us take a forestry project in Cost Rica as an example: The project's aim is to protect a forest from felling. Then a possible assumption is, that within five years the forest would have totally disappeared, because the people in the area need the land for agriculture. By saving the forest there is a carbon sequestration activity, and it is possible to calculate what the carbon implications are, given the mentioned assumption. Obviously the assumption is debatable, because it could have been only two years or even twenty years before the forest would actually have been cut down, consequently creating less or more credits. The credits should actually amount to zero, if the specific forest would never be cut down, but instead a neighbouring forest is cut. Again, also about the scope of such 'leakage' effect, one can commonly only speculate. Hence this little example already shows, that a baseline is just a set of assumptions with many subjective, arbitrary elements.

Another example is an energy efficiency project in Russia: The baseline has to include assumptions for the economic development of the next 20 years, for example on the speed with which the Russian authorities are reducing the energy subsidies between 1998 and 2018. Because if they would rapidly reduce energy subsidies, this, of course, affects the baseline of the specific project. But no one knows what the Russian authorities are going to do on this in the next 20 years. Hence only best guesses can reasonably be made on this. Again, this is another way of saying that determining the baseline, which is the key to the credits that will be attached to a specific project always will be a matter of arbitrary estimates.

TABLE 4 shows an additional theoretical example for choosing a baseline for a certain project, where a power plant is built in India with the help of Western technology (hence the low level of emissions) and finance.

The question now is, what is the point of reference determining the project baseline: the emissions of a comparable average or marginal plant in the host country or region, or the average power plant in the investing region. Or: for that matter, is the criterion for crediting if the project is commercially viable, as discussed earlier?

TABLE 4. Possible Baselines for an JI gasification project in India

	Possible baseline emissions	Project emissions	Credits
	(tonnes GHG emissions per unit of energy)		
I South Asian average	700	400	700-400=300
II Indian average	600	400	600-400=200
III Most efficient Indian power plant	500	400	500-400=100
IV OECD average power plant	450	400	450-400=50
V Project commercially feasible	400	400	400-400=0

There are arguments in favour of each of these possible baselines with great consequences for the number of credits created by the project and the incentives for foreign investors to finance a project like this.

All the examples given above underline the technical problem of establishing a baseline. I personally think that a possible solution for this problem is, at least to a large extent, in, what I call "best professional judgement": Committees of experts who are reliable, who are respected, who are recognized, who set the baseline, which is an acceptable baseline to everybody, combined with a good monitoring and verification process.

4.4 MONITORING, VERIFICATION AND REPORTING

A possible framework for monitoring, verification and reporting is outlined in FIGURE 5. It shows a stepwise procedure starting with internal procedures within the project reported to the national institutions. These include the information in their national reports to the UNFCCC Secretariat. After a verification by independent experts the UNFCCC Secretariat approves the national reports and allocates the credits.

This system with several steps of monitoring and verification on the one hand by the involved project partners and on the other hand by independent experts on behalf of the UNFCCC will reduce the risk of corruption and cheating and at the same guarantees a reliable baseline.

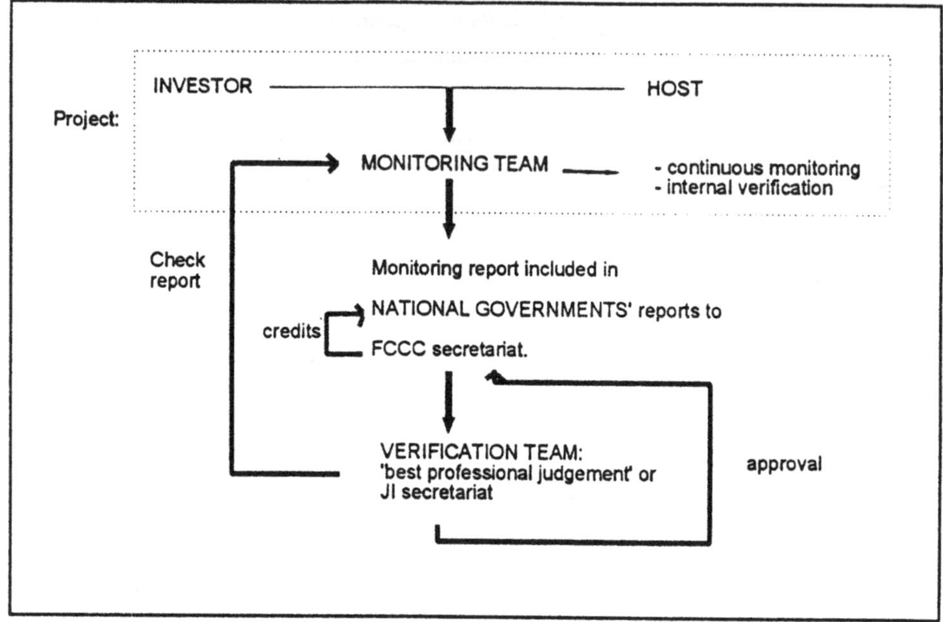

FIGURE 5. A potential framework for monitoring, verifying and reporting on AIJ/JI

5. Joint Implementation versus trading

The discussion about trading of the difference between national GHG emissions and the national quantified emission limitation and reduction obligations (QUELRO's) on the one hand versus JI on the other hand is one of the key points in today's debate. Two important aspects in this context are: the creation and trading opportunity of 'hot air' and the possibility of individual countries to participate in the Protocol system.

5.1 CAPS AND THE CREATION OF 'HOT AIR'

As mentioned above the acceptability of national GHG emissions caps is to some extent related to the introduction into the Protocol of such instruments as JI and trading. In this context there is a discussion now about the position of the Central and especially Eastern European countries. Since they are part of the Annex I Parties, they are also expected to accept some kind of QUELRO, just like the individual OECD countries.

But would Russia accept an emission reduction commitment for the period around 2010, which would really mean that it is forced to take measures, in other words would it accept a cap which is lower than its projected emissions for 2010? Compared to China, India and even to South Korea and Malaysia, countries that all have very high rates of economic growth, but do not seem to accept caps, Russia could argue that its economic performance is much worse and therefore that it would not accept any hard caps. In fact, it may be both in the interest of Russia and some Western countries if Russia's cap would actually be set above its projected emission level. If for example the projected emission level for Russia is 10, and its cap would be at a level of 12, the country cannot only just do whatever it wants, but also they now have emission rights of 2 left over, what I would call 'hot air', which it can sell, if trading is allowed.

So, if the caps for some Eastern European countries are set higher than their actual projected emissions, the difference can probably be sold to the US or Europe. Moreover, this would represent only a transfer of money going from the OECD to Russia without any real reduction of carbon emissions. That is the big risk of a trading regime. Even worse, the country buying the 'hot air' would use this to fulfil part of its commitments, without any real abatement effect!

5.2 PARTICIPATION OF INDIVIDUAL COUNTRIES

Another point about trading is that it is only feasible between countries that have accepted a cap. Hence there can, for instance, be trading between the Netherlands and Germany, or between the United States and Russia, but there cannot be trading between Annex I countries and Non-Annex countries. In opposition JI has the big advantage that it is feasible between Annex and Non-Annex countries, because it is a voluntary deal with any country worldwide based on cost-differences. It seems likely that these differences are generally greater between Annex I countries and Non-Annex countries than between Annex I countries. So, as a consequence, if one would concentrate on within Annex trading alone, not only would the main emitters of the future, the LDCs, be left out of the game, but also would an important part of the potential cost savings not be reaped.

6. Joint Implementation in the UNFCCC process

FIGURE 5 summarises the decisions on JI at the Conference in Berlin.

Because of the discussions mentioned in the earlier paragraphs the Parties decided to install a pilot phase until the year 2000 without crediting. During that time the instrument is not called JI, but Activities Implemented Jointly (AIJ). Issues to be resolved in the pilot phase are summarised in FIGURE 6..

DECISION ON ACTIVITIES IMPLEMENTED JOINTLY

FIRST CONFERENCE OF THE PARTIES, BERLIN MARCH - APRIL 1995

- START AIJ PILOT PHASE AMONG ANNEX I AND, ON A VOLUNTARY BASIS, WITH NON-ANNEX I PARTIES
- ACTIVITIES SHOULD BE COMPATIBLE WITH AND SUPPORTIVE OF NATIONAL AND DEVELOPMENT PRIORITIES
- AIJ ACTIVITIES SHOULD BE APPROVED OR ENDORSED BY THE PARTIES' GOVERNMENTS BEFOREHAND
- ENVIRONMENTAL BENEFITS AND FINANCING OF ACTIVITIES SHOULD BE ADDITIONAL
- NO CREDITS SHALL ACCRUE TO ANY PARTY FROM AIJ ACTIVITIES DURING THE PILOT PHASE

FIGURE 5. Decisions on Joint Implementation in the UNFCCC process

ISSUES TO BE RESOLVED DURING THE PILOT PHASE

- BASELINE DETERMINATION & Additionality
- TRANSACTION COSTS
- MONITORING
- VERIFICATION
- CREDIT SHARING

FIGURE 6. Issues to be resolved during the pilot phase

FIGURE 7 and TABLE 5 give an overview over the ongoing and planned AIJ/JI projects.

According to FIGURE 7 there is a regional bias of host countries, because most projects are located in Central and Middle America as well as in Central and Eastern Europe.

The list in TABLE 5 is not complete, because there are 41 projects in the pilot phase now. Similar to the regional bias of host countries there is a regional bias at the side of the investors: the majority of the projects are financed by US or the Northern European countries as well as the Netherlands and most of in any case the European projects are still subsidised by the governments.

FIGURE 7. Planned and ongoing AIJ/JI pilot projects

TABLE 5. Planned and ongoing JI/AIJ pilot projects

	Project	Host country	Investing country
1	Dec:n	the Czech Republic	USA
2	Rio Bravo conservation and forest management	Belize	USA
3	CARFIX	Costa Rica	USA
4	Plantas Eolicas S.A. Wind Facility	Costa Rica	USA
5	ECOLAND	Costa Rica	USA
6	Rural Solar Electrification project	Honduras	USA
7	RUSAFOR-SAP	the Russian Federation	USA
8	Krkonose	the Czech Republic	The Netherlands
9	Profafor	Ecuador	USA
10	Energy saving	Hungary	The Netherlands
11	Compressed Natural Gas Fuel Engine	Hungary	The Netherlands
12	Landfill project	the Russian Federation	The Netherlands
13	Horticulture Tyumen	the Russian Federation	The Netherlands
14	ILUMEX	Mexico	Norway
15	Coal-to-Gas Conversion	Poland	Norway
16	Reforestation Vologda	the Russian Federation	USA
17	Klinky forestry	Costa Rica	USA
18	El Hoyo-Monte Galan Geothermal	Nicaragua	USA
19	Bio-Gen Biomass Power Generation Phase I	Honduras	USA
20	Dona Julia Hydroelectric	Costa Rica	USA
21	Tierras Morenas windfarm	Costa Rica	USA
22	Aeroenergia	Costa Rica	USA
23	Biodiversifix forest restoration	Costa Rica	USA
24	RUSAGAS-Fugitive Gas Capture	the Russian Federation	USA
25	Micro Hydroelectricity	Bhutan	The Netherlands
26	Uganda National Park FACE project	Uganda	The Netherlands
27	Renewable Energy Systems	Indonesia	Germany, Japan
28	Ainazi windpower project	Latvia	Germany
29	Virilla river basin	Costa Rica	Norway
30	Skoda-Mlada Boleslav	the Czech Republic	Germany
31	Bel/Maya Biomass Power Generation	Belize	USA
32	Bio-Gen Biomass Power Generation Phase II	Honduras	USA
33	District heating improvement Zelenograd	the Russian Federation	USA
34	Halophyte Cultivation Sonora	Mexico	USA
35	Noel Kempff M. Climate Action Project	Bolivia	USA
36	Reforestation Chiriqui Province	Panama	USA
37	Sustainable Energy Management	Burkina Faso	World Bank
38	Renel-SEP, energy efficiency	Romania	The Netherlands
39	Bilsa-Reserve forest conservation	Ecuador	USA
40	Scolel TJ, sustainable land management	Mexico	USA
41	Reduced Impact Logging	Indonesia	USA
42	Consolidation of biological reserves	Costa Rica	USA
43	Ushgorod Corridor Wolgotransgas	the Russian Federation	Germany
44-59	16 boiler conservation/district heating projects in Estonia in the framework of the Swedish NUTEK programme		
60-75	16 boiler conservation/district heating projects in Latvia in the framework of the Swedish NUTEK		
76	Dry coke quenching	China	Japan
77	15 fossil fuel fired Power Units	Ukraine	The Netherlands
78	Glassworks factory	Ukraine	The Netherlands
79	Beet sugar plant	Ukraine	The Netherlands
80	Byczyna district heating	Poland	The Netherlands
81	Bacstej dairy factory	Hungary	The Netherlands

Source: JI Quarterly, 1998, Volume 4, p.14

Since the First Conference of the Parties in Berlin in 1995 some progress has been made with the AIJ pilot phase. This is summarised in FIGURE 8.

REVIEW OF PROGRESS WITH AIJ AT THE SECOND CONFERENCE OF THE PARTIES: GENEVA: JULY 1996

- INFORMATION RECEIVED ABOUT 32 PILOT PROJECT (PROPOSALS) -- DOCUMENT: FCCC/CP/1996/14
- 5 COUNTRY REPORTS (NUMBER OF PROJECTS IN BRACKETS): AUSTRALIA (2), GERMANY (7), THE NETHERLANDS (6), NORWAY (2), USA (15)
- NUMBER OF AIJ PROJECTS THAT HAVE RECEIVED HOST AND INVESTING COUNTRY APPROVAL: 28

FIGURE 8. Progress with AIJ until 1996

7. Conclusion

JI/AIJ is a project oriented instrument that reduces the costs of global warming mitigation by exploiting interregional differences in the marginal abatement costs.

The main difficulties of JI/AIJ are of technical nature and can be solved by a national and international system of stepwise monitoring, verification and reporting backed up with best professional judgement systems of independent experts.

JI/AIJ can be carried out in combination with trading of the differences between national QUELRO's and actual emissions, although the latter system may be inflated insofar as 'hot air' is created in defining the QUELRO's for some economically vulnerable countries in Central Europe.

A major advantage of JI is that it opens the opportunity to include countries that not yet have a national cap on their emissions, and thereby to take the developing countries on board in the Protocol.

PRACTICAL EXAMPLES OF ACTIVITIES IMPLEMENTED JOINTLY (AIJ) IN COSTA RICA

Dr. FRANZ TATTENBACH
Oficina Costarricense de Implementación Conjunta
P.O.Box 7170-1000, San José, Costa Rica

This is a transcript of the tape recorded during the lecture of Mr. Tattenbach at the Advanced Study Course. After moderate revision by the editors to make contents and format compatible with this volume and to include the relevant parts of the discussion it has been authorised by Mr. Tattenbach.

1. Introduction

The purpose of this paper is to present the perspective of a developing country on Joint Implementation (JI) and Activities Implemented Jointly (AIJ). Therefore a brief introduction to the situation of Costa Rica is given, before describing the different generations of AIJ projects in Costa Rica and their benefits.

2. Situation of Costa Rica

Costa Rica is located in Central America between the Caribbean Sea and the Pacific Ocean. It has an area of 51,000 km^2 and a population of 3.5 million. Costa Rica has a relatively high GDP per capita of US $ 2,610. It distinguishes itself by a very high human development index. As a Non-Annex country Costa Rica has no obligations to reduce its emissions, but the government created an economic condition that favours investments in projects that reduce GHG emissions.

FIGURE 1 gives an overview over the relation between GDP per capita and CO_2 emissions per capita for different countries. Note that the lines are only indicating a range of possible relations. The figure highlights different possibilities of how economic development brings about higher emissions per capita. Nonetheless there seem to be differences in the per capita emissions even for countries with similar per capita GDP. These could be explained by consumption habits, production patterns and resource endowment. The relatively low emissions of Costa Rica are achieved

J. Hacker and A. Pelchen (eds.), Goals and Economic Instruments for the Achievement of Global Warming Mitigation in Europe, 273–281.

by a high share of hydro power in power generation. The great differences in the current level of per capita emissions between developed and developing countries are the foundation of the common by differentiated obligations embodied in the United Nations Framework Convention on Climate Change (UNFCCC).

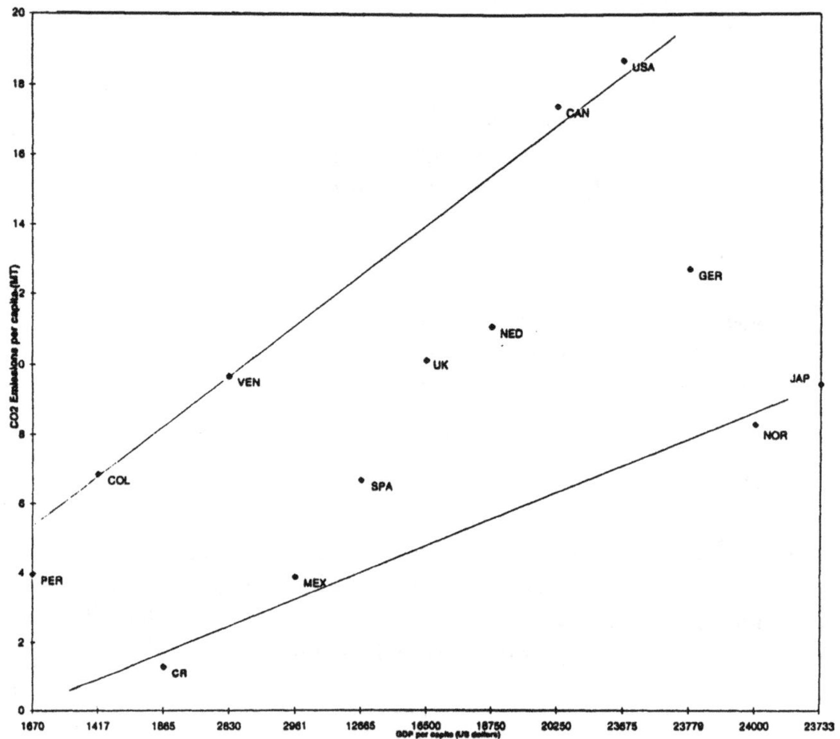

FIGURE 1. CO_2 emissions per capita versus GDP per capita
Source: Internet Sites of World Bank and IPCC

3. Three generations of JI projects in Costa Rica

Costa Rica went through three generations of AIJ projects (FIGURE 2).

COSTA RICA: AIJ Generations

- First Generation of Projects, 1994
  ☞ Project by project
- Second Generation of Projects, 1996
  ☞ Portfolio of projects ➔ CTOs
- Third Generation of Projects, 1997
  ☞ Portfolio of projects ➔ CTOs
  ➔ International Commodity/Stock Exchange Market

FIGURE 2. AIJ project generations in Costa Rica

The first generation was a project by project approach in which it was hard to distinguish direct private investment from the project's AIJ additional investment. Nonetheless Costa Rica dealt with that by selecting sectors (renewable energy and forestry) where there was no doubt that is was justifiable from a "need-to-increase-profitability" and "strategic-for-Costa Rica's sustainable-development-agenda" point of view to give 100% of the credits to the project equity owners. If the project equity owners were half Costa Rican and half Americans, it was up to them to decide how they will going to split the offsets. If the American equity owners were the end users of the offsets, then it was among equity owners that the two bases and economic values had to be negotiated. On the other hand, if the American equity owner were not the end user of the offsets, then it was treated as a commodity belonging to the project and an increment of the final value project. At that stage we had several projects with high transaction cost and some more difficulties, but we had success in this phase. The projects are listed in TABLE 1 and 2.

TABLE 1. Costa Rican AIJ projects: Energy sector

Project Name	Type of Project	Install Capacity (MW)	Annual Production (GWh/yr)	%Total Production	Total Cost (US$ millions)	Avoided Emissions (mt C)
Plantas Eólicas	Wind	20.0	98.0	2.10	30.40	17,000
Tierras Morenas	Wind	20.0	90.0	1.90	27.00	85,000
Aeroenergía	Wind	6.4	30.0	0.63	8.85	10,048
Doña Julia	Hydroelectric	16.0	85.0	1.80	27.00	85,000
CNFL	Hydroelectric	22.4	110.6	2.37	41.50	486,982
TOTAL		84.8	413.6	8.80	134.75	684,030

TABLE 2. Costa Rican AIJ projects: Land-use sector

Project Name	Type of Project	Area (ha)	Total Cost (US$ million)	Project Life (years)	Emissions Avoided/Sequestered (mtC)	(mtCO2)
ECOLAND	Conservation	2,340	1.0	15	345,548	1,267,124
CARFIX	Conservation Regeneration Reforestation	110,000	21.4	25	5,939,000	21,778,313
BIODIVERSIFIX	Regeneration	58,500	64.7	50	5,040,000	18,481,680
KLINKI	Reforestation	6,000	3.8	40	1,968,000	7,216,656
CNFL	Conservation Regeneration Reforestation	4,000	3.3	25	313,646	1,150,139
TOTAL		180,840	94.2	155	13,606,194	49,893,912

Joint Implementation has helped Costa Rica to balance the problem brought by the increasing globalization of the economies that led to an increase of the private sector investment decisions on the technology that is chosen in power generation. It is no secret that there is a tendency towards fossil fuel at the expense of hydro electric and wind power as it has much shorter pay off times and less risk. Here Joint Implementation seems to create an impetus for investment in hydroelectric and wind power. Costa Rica has a share of around 85% hydroelectric power today and about 9% of the installed capacity is due to Joint Implementation projects. Likewise, in the land use projects there are considerable amounts of hectares, as a percentage of the country's total, that are put into the different types of sustainable forestry and conservation projects.

The second generation was called portfolio project phase. This phase allowed a lot more project implementators in Costa Rica to participate in an ambitious nation-wide project which is called private forestry project. It was first launched with Norway and allowed Costa Rica to combine many small projects into a portfolio and to treat this portfolio as a single project for the investor. The basic principle is depicted in FIGURE 3. The main difference to the first generation is that the investor, who still agrees in advance to finance a certain portfolio, does not directly deal with the farmer who maintains the forest or plants the trees. Instead he deals with the Costa Rican Greenhouse Gas Fund that sells Certified Tradeable Offsets (CTOs) to the investor and pays the farmers via an independent organisation, that organises the day to day supervision of the projects.

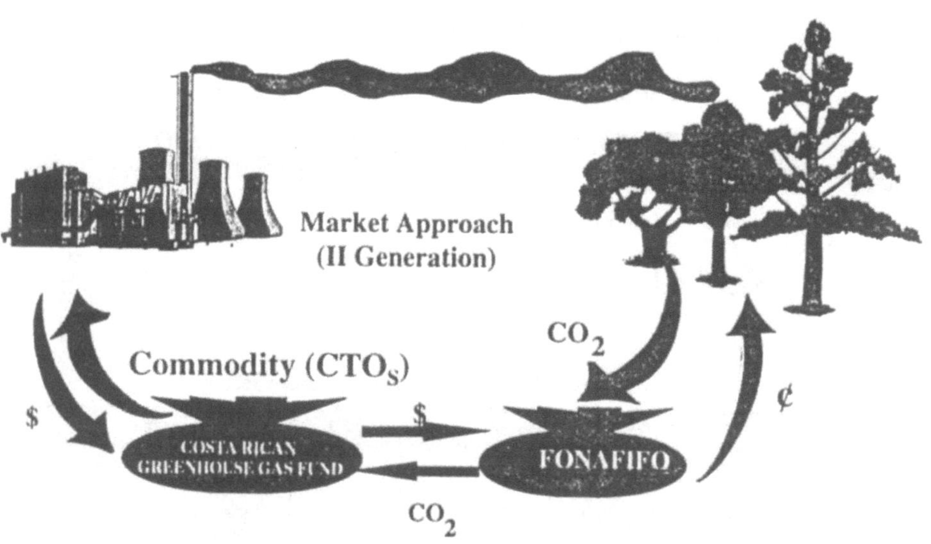

FIGURE 3. Basic mechanism of second generation AIJ portfolio project

The third generation is now being launched and has just been accepted by the United States Initiative on Joint Implementation. These portfolio projects not only pool projects in the country, but they will also be totally integrated in the market of tradeable offsets. The CTOs of this project will be handled by a world recognized broker, namely Centre Financial Products from Chicago. Additionally these offsets will be certified by a credible third party, namely by SGS, the Societe Generalle de Survillance, which is a recognized European Company that certifies different kinds of commodities as an auditor of the process. Hopefully this will increase the validity of the CTOs and also the trust of the investors in this kind of projects. The main difference to the second generation is the fact, that the CTOs are offered at the international markets without the need for a predetermined investor. FIGURE 4 summarises this approach.

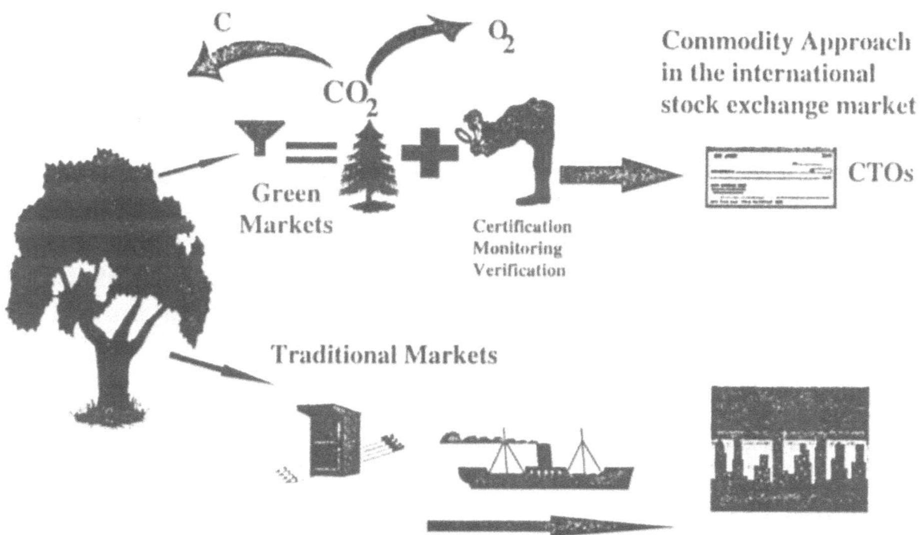

FIGURE 4. Basic mechanism of second generation AIJ portfolio project

This new approach is linked to a range of benefits both for the owner, the country and the world. A matrix showing the distribution of the benefits is presented in TABLE 3.

TABLE 3. Benefits from forests in Costa Rica

Type of environmental service	Beneficiary		
	Owner	Country	World
Sustainable Wood	✔		
Water for different uses		✔	
Scenic beauty		✔	
Biodiversity			✔
Carbon sequestration	✔	✔	✔

The benefit of some of the forests' services like the value of the wood is accrued by the owner. Because of externalities water and scenic beauty mainly benefit the country as a whole. On the other hand biodiversity in the sense of preserving genetic banks for the humanity and greenhouse gas sequestration was up to now more of a global benefit than of a local benefit. Now the third generation JI projects allow us to bring these global benefits to the country in case of national parks and to the forest owners in case of private sector forestry.

How does this work in practice? The payment of the individual farmers is carried out by the National Forestry Financing Fund. This was established a long time ago for the distribution of subsidies to forestry. It has therefore the advantage of knowing how to deal with large scale financing. Right now this programme is transformed into a programme that is based on the environmental services of different forestry activities like forest conservation, forest management and reforestation. In either case the farmer gets a yearly payment for environmental services that his forest provides and for his collaboration to start, keep or preserve his plantation. The Fund itself will then deal the carbon offsets to the Ministry of Natural Resources and Energy, Greenhouse Gas Fund, which in turn will deal directly with the AIJ investor, whose money will be passed to the National Forestry Financing Fund for redistribution. The seed capital for the programme has been levied by a 5% tax on fossil fuel that was created in the Forestry Law. Additional funds come from AIJ investors and voluntary payments by private hydroelectric plants. This is a nice example for collaboration and the role the government can play in internalising different externalities. The programme can involve more than 20.000 farmers in Costa Rica and makes the AIJ mechanism very transparent to AIJ investors. We have designed a close loop from receiving the money from private sector investments via producing offsets to returning CTOs to the investors. This mechanism is visualised in FIGURE 5.

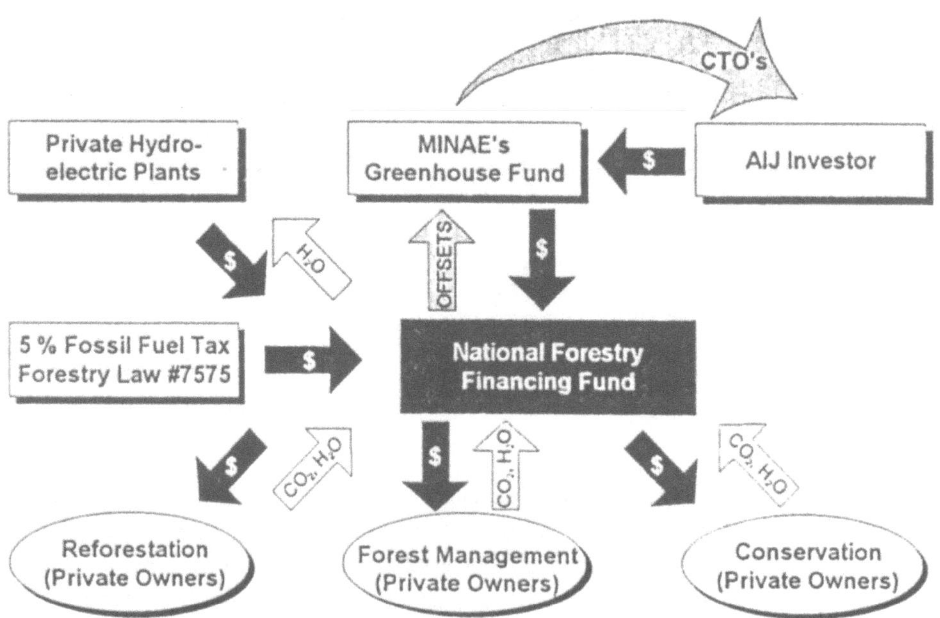

FIGURE 5. Forestry environmental services payment in Costa Rica

With this mechanism two Forestry Umbrella Projects have been set up, which include 0.55 million ha of national parks and 1.66 million ha of private forests. The details are outlined in FIGURE 6.

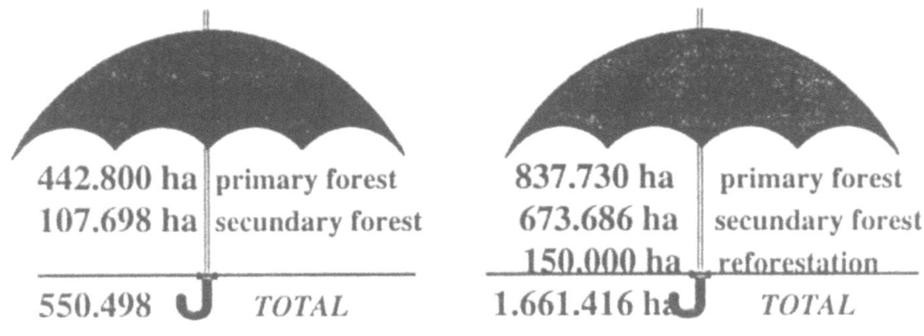

FIGURE 6. AIJ - Forestry Projects Umbrella

In the context of these projects the question of additionality arises: If a country like Costa Rica invests some of its own money to start the AIJ process via the National Forestry Finance Fund it is creating an economic way of thinking of reducing emissions and this is the only way to keep the process host country driven. But on the other hand a project started in this programme is different from a project started completely independent of the host country and all its elements are carried out by an international NGO. Strictly speaking this project started without a predetermined AIJ investor. Therefore Costa Rica is a little bit nervous about all the discussion on strict financial additionality. A "commodity approach" to AIJ, using CTOs, simplifies the issues and guarantees financial additionality.

Another important question is how to guarantee the carbon sequestration of the forest projects and to insure them against forest fires etc. Costa Rica has not yet approached insurance companies, but institutions like the World Bank to see, if they are interested in the concept of insuring Costa Rican offsets. The problem is that Costa Rica would not like to pay insurance fees, but rather keep some extra offsets to back up the CTOs that are sold. In this sense Costa Rica is going to insure itself by only selling a certain amount of the CTOs that could be claimed from a certain project. The amount that has to kept is determined in an independent risk assessment. Therefore in the moment Costa Rica will be selling only 80% of the total offsets that can be claimed.

To give an impression FIGURE 7 shows a Costa Rican Certified Tradeable Offset (CTO).

REPUBLIC OF COSTA RICA
MINISTRY OF ENVIRONMENT AND ENERGY
OFFICE FOR JOINT IMPLEMENTATION

SAN JOSÉ, JANUARY, 1997

GREENHOUSE GAS EMISSIONS MITIGATION CERTIFICATE

The Costa Rican Office for Joint Implementation of the Ministry of Environment and and Energy of the Republic of Costa Rica certifies that the bearer of the Certificate has contributed to the improvement of the global climate by providing additional financial support to specific national sustainable development projects in Costa Rica that have mitigated a quantity of greehouse gases in carbon dioxide (CO_2) equivalent units of.

ONE METRIC TON OF CARBON

FIRST CERTIFIED AND TRANSFERABLE GHG OFFSET TITLE

COMMEMORATIVE ISSUE 1/200

Through the emission of this Certificate, the Government of the Republic of the Costa Rica commits itself to maintain the validity of the amount of greenhouse gas emissions offsets specified in this Certificate during the next 20 (twenty) years, and guarantee replacement offsets if it is demonstrated that the offsets here certified have not been produced in the amount indicated on the Certificate.

The commitments and activities written above follow from the principles and obligations established under the United Nations Framework Convention on Climate Change and its annexes I and II of May 9, 1992, ratified as the high Law of the Republic by means of Law No. 7414 of June 13, 1994 and, within the principles and obligations of the Convention on Biological Diversity and its annexes I and II of June 13, 1992, ratified as the high Law of the Republic by means of law No. 7433 of September 7, 1994 and in conformity with Executive Decrees No. 25066-MINAE: Creation of the Costa Rican Office for Joint Implementation and 25067-MINAE: Creation of a Specific, National Fund for the Conservation and Development of Sinks and Deposits of Greenhouse Gases, published in the official Costa Rican Report, La Gaceta, No.76 of April 22, 1996.

This Certificate documents an activity that is additional to the obligations of the Republic of Costa Rica under the United National Framework Convention on Climate Change.

René Castro Salazar
Minister
Ministry of the Environment and Energy

Franz Tattenbach Capra
National Coordinator, Costa Rican
Office for Joint Implementation

Made with 90% recycled paper and 10% banana paper

FIGURE 7. Greenhouse Gas Emissions Mitigation Certificate

6. Conclusions

The following four points should be kept in mind, when talking about AIJ from a host country's perspective:

- Host country approval should be maintained:
  This is important to insure, that AIJ does not lead to eco- or neo-colonialism. Costa Rica for example has decided, that there will only be AIJ projects in the forestry and renewable energy sectors, because the country would like to see more direct private investment in these sectors and their financial returm is not enough to drive these investments. We need to value their environmental benefits via CTOs to make them competitive. Other sectors have enough economic returns and are therefore not (yet) allowed.
- Commodity approach should be applied to GHG benefits:
  This approach is applied, because it is the best way to deal with the savings that are created for Annex I countries. One way to share the benefits, is to create a commodity that can be traded. It can be compared to mining a certain resource. Mining is an example where a country can decide on the policies for extracting this commodity: either a royalty or part of the profits are asked for. The commodity approach is very much the same, because emission reductions in Non-Annex countries can only happen there by definition. Hence host country governments can decide on the modalities of their 'extraction' as well.
- Governments should enhance the financial incentives for AIJ:
  Up to now the investment in AIJ has not been as much as anticipated. One of the reasons is that not enough financial incentives are given by Annex I country governments. Even the clarity that some portion of the risk will be shared by governments is lacking, if companies are investing in AIJ today. No investor can be sure that the credits will be recognised by the investor's country government. There is even the risk of AIJ not being passed in Kyoto. Investor country governments have to give some guarantee to the potential AIJ investor, that his investment is not completely lost in case of changing political conditions.
- Crediting after the pilot phase:
  In the context of incentives for potential investors it is of course vital that crediting is allowed after the pilot phase.

PROBLEMS AND LIMITATIONS OF AIJ AND THE JI POTENTIAL FROM THE PERSPECTIVE OF A GERMAN PROJECT BROKER

JÜRGEN HACKER
UMB Environmental Management Consultancy Hacker GmbH
Selchowstrasse 1, D-14199 Berlin, Germany

1. An Introductory Parable about Joint Implementation

1.1 THE PROBLEM

Somewhere there is a jungle with a broad biodiversity and with many kinds of predators. Several kinds of predators are becoming a problem for this jungle and they are killing too many other animals, so that there is a danger that the equilibrium of the jungle will be lost.

1.2 THE PROPOSAL FOR A SOLUTION

Let us erect a big hall in the middle of the jungle, put several boxes with meat in the middle of this hall and open the hall to all predators of the problematical kinds, so that they have food without the need to hunt other animals.

1.3 ESTABLISHING A PILOT PHASE

As you don't know the exact behaviour of these predators, there may be a problem of unfair use of this free food, so you decide to establish a pilot phase:

As proposed above, erect the hall and put in several boxes with the inscription "meat boxes" and open the hall to the relevant predators. But in contrast to the proposal, you put no meat in the boxes but only salad.

Then you go around in the jungle and shout to the predators:
"Please, problematical kinds of predators, come and use our cost-free meat boxes in the hall in the middle of the jungle instead of hunting other animals outside".

Now you will be very excited about the results you will get from this pilot phase for the final design of the proposed solution.

J. Hacker and A. Pelchen (eds.), Goals and Economic Instruments for the Achievement of Global Warming Mitigation in Europe, 283–291.

1.4 THE EVENTS IN THE PILOT PHASE

- Most predators won't come in the hall because they have not heard you.
- Most predators having heard your announcement won't come, because they heard too that in spite of the inscription there will be no meat in the boxes.
- Nevertheless some predators are very inquisitive and they come in the hall looking into the boxes for the following reasons:
  - some think that they can nevertheless find some meat in the boxes
  - for some "green" predators, salad may be a useful additional nutrient
  - some think that if they already learn the way to the hall and to the boxes, they will have advantages in future, when you put real meat in the boxes
  - few predators want to please you for some other reasons and e.g. to be stroked by you
  - finally some think this may be an easy way to find some weaker predators to eat

1.5 SOME PRELIMINARY EVALUATIONS OF THE PILOT PHASE

There is a lot of criticism of your general proposal because nobody can see a change in the behaviour of the predators in killing other animals during the pilot phase, not even by the predators which visited the hall. Instead, loss of the equilibrium has accelerated. Reasons discussed are:

- Maybe the results of the pilot phase would have been better if you had shouted louder to the predators.
- Maybe the number of boxes are not appropriate for the number of animals.
- You have identified the problem that you cannot precisely calculate the number of animals a predator is killing less or more after having visited the hall.

The argument that predators need meat not salad, is refused as being too friendly to predators. But at least you have made several administrative agreements e.g.:

- About the arrangements of the report sheets for the door keepers, which should register the number and type of predators entering and leaving the hall.
- That there should be used in future only boxes of the latest technology and of the most modern style!

1.6 THE DIFFERENCE TO JI?

It would be absolutely unfair to compare the majority of private enterprises and households in the industrialized countries with predators and the ones in developing countries with vegetarians. But more important than this is that the concept of the Joint Implementation of measures for the reduction of CO_2 emissions by industrialized countries together with countries in transition and developing countries is a real solution with a huge economic and ecological potential to the climate change problem, in contrast to the stupid proposal in the parable which assumed that a jungle is a zoo.

2. The Credit Problem

Obviously, the decision by the First Conference of the Parties in Berlin in 1995 that CO_2 credits are not allowed for Activities Implemented Jointly (AIJ) projects is an important defect in the design of the pilot phase for Joint Implementation *(like not putting meat into the boxes for the predators).*

This leads to the fundamental but widely held misunderstanding:

«AIJ projects cannot offer economic incentives for an investing private company!»

This sentence is not true, at least it need not be true. It is up to the governments of the Annex I countries, the industrialized countries, to establish direct economic incentives for their own private sector by exemption from national duties like

- the possibility of reducing the CO_2 tax basis by the reduced CO_2 emissions of the AIJ project,
- exemptions from local duties for expensive measures for energy conservation,
- acceptance of the reduced CO_2 emissions of AIJ projects towards the fulfilment of voluntary commitments of industries,
- treatment of the reduced CO_2 emissions of AIJ projects as equivalent to CO_2 emission permits in a national system of tradeable permits.

The only limitation from the convention is that the countries cannot use the reduced CO_2 emissions of AIJ projects as fulfilment of their country's obligations to stabilize their emissions of CO_2 by the year 2000.

The convention does not stop the industrialized countries establishing direct incentives for their private sector, but the governments of the industrialized countries, especially in the EU, have not yet done their homework. And it does not seem that they will do it during the pilot phase.

Even if, after the pilot phase, there is an acceptance of a JI phase with credit of CO_2 to parties, which is not certain, that does not automatically mean direct economic incentives for private companies if they then invest in a JI project.

Why should a private company invest in a project which only helps its government to reach its obligations? Again, governments must first establish at the national level a system to transfer the credits from the state level to the level of the private sector with a certain economic value for these transferred credits. But there is no light at the end of this tunnel.

- No CO_2 tax is in sight, especially not with a linkage to JI either in Germany or in the different tax proposals of the EU (see pages 231-237).
- We never heard of planned exemptions from local duties of measures for energy conservation.
- There is only the statement of the German Environmental Ministry of a "may-be" willingness to accept AIJ or JI projects in the fulfilment of the targets of the voluntary commitments on climate change by German industry. But German industry does not need any credits of AIJ or JI projects to fulfil their current commitments as they can reach them more or less by business-as-usual (see pages 177+180). Therefore the Ministry demands more ambitious renegotiated targets as a precondition for recognition, but this is rejected by industry.

So, you cannot see any preparation of governments in the EU worth-mentioning for establishing economic incentives for their private sector after the pilot phase. Only in the case of some non-European industrialized countries does there seem to be work under way in connection with the establishment of a tradeable permit system.

But such systems will not be installed before the year 2005, or more realistically before 2008. What will happen till then?

3. The "soft" Incentives of AIJ

Though legal crediting is not allowed in the pilot phase *(no meat in the boxes)*, there are still four so-called indirect or "soft" incentives for the private sector to finance AIJ projects *(like reasons for very inquisitive predators to visit the hall)*:

1) Image enhancement in home-country (eco-sponsoring)

2) Public attention and image enhancement in the host country

3) Improvement of starting position for joint-ventures in a host country

4) Experience and securing cost-effective projects for the time after pilot phase

But these "soft" incentives are of very limited value and justify only small financial involvements. It has to be taken into account that especially the public relations and marketing incentives decrease with the number of projects. Which journalist will report about the umpteenth AIJ project?

For instance for a country like Germany, we assume that no more than about 10 to 20 AIJ projects seem to be justified by the above-mentioned "soft" incentives. But even this number has not been reached in Germany; by mid-1997 only four projects could be reported:

TABLE 1. German AIJ Projects announced till mid-1997

	German Company	Host Country	Type of Project
1	Preussen Elektra AG	Latvia	2 wind power generation facilities
2	Bayernwerk AG/RWE AG	Czech Republic	combined heat and power plant
3	RWE AG among the E-7 Initiative	Indonesia	decentralized rural electrification
4	Ruhrgas AG	Russia	reducing methane losses in part of Russian pipeline system

These projects are not really good examples of AIJ projects. The motivation for these four big utilities are mostly, it seems, the appeal of the German Environmental Ministry to them: "Please, please do an AIJ project. We, as the German Government, also want something to report about in the relevant international climate change conferences and not only the US."

It is not necessary in this context to go into more details of these projects, except to note that:

To projects 1+2:

Both projects are ordinary project investments, which proved to be economically profitable. The German companies simply earn a lot of money with these investments. The AIJ aspects played no role in the investment decision, but can be characterized as a "windfall profit". This seems to be an important reason why the second project is not yet approved by the Czech Government and may never be.

To project 3:

The specific CO_2 reduction costs are calculated at about US $ 110 per metric tonne CO_2. To get such performance you need not go abroad. There are thousands of potential projects to reduce CO_2 emissions with lower specific costs in Germany itself. But the philosophy of JI is to find projects with lower costs abroad than at home.

To project 4:

There is still little information available about this project. But it is expected to be very cost-effective, because it seems to consist mainly of a transfer of specific software know-how, (how to run the pipeline compressors in an optimum energy-saving way) and not in costly technical upgrading.

What are the reasons for these disappointing results of the pilot phase so far in Germany?

First, at the beginning of the pilot phase, there were no attractive project proposals by host countries, who very often lacked quantitative data concerning potential CO_2 reduction and especially specific CO_2 reduction costs. Even if this information is provided by a host country, it is insufficient to attract a German investor.

Although specific CO_2 reduction costs are an interesting and important criterion, it is wrongly believed to be decisive. This will only be the case if CO_2 crediting is possible to the investor. Without this possibility the decisive motives for a commercial investor can only be the "soft" economic incentives mentioned.

To make use of these "soft" incentives, it is necessary to develop and offer AIJ projects that are geared towards the specific interest of potential investors concerning their business planning and marketing strategies. Most project ideas or proposals did not take these "soft" incentives properly into account and consequently did not attract backers to invest in them.

But for that, one cannot blame the potential host countries. How should they know who could be a potential investor from an industrialized country and especially how can they know their business planning and marketing strategies?

This is already very difficult for governments of the industrialized countries and even for us as a German consultancy and potential project broker. It has been a challenge for us as a German Environmental Management Consultancy, but we are facing the following problems:

i) There is no more public pressure on industry to support AIJ as a result of the media success of the voluntary commitments and general economic difficulties in Germany combined with increased unemployment.
ii) Due to mostly unfair criticism of green NGO's, AIJ now has a problematical image in Germany.
iii) As a result of the first experience of the pilot phase, the requirements for baseline definitions have risen enormously *(problem of how to calculate precisely the number of animals a predator is killing less after having visited the hall and the meatless boxes)*.
iv) There is a big uncertainty about future policy of the government towards JI (no preparation of a legal incentive system).

Problems ii) and iii) could be solved by us, but problem i) makes it very difficult to gain the ear of the decision-making managers of bigger companies, although in some cases we have been successful.

But then we are faced with problem iv) as a new big one. There are two alternatives:

- the optimistic alternative is that after the pilot phase there will be a JI phase with transferred crediting to the private sector. Automatically the question was raised what will then happen to AIJ projects? Will at least the CO_2 reduction of AIJ projects be accepted from that time on for crediting? If not, why should a company invest now, not getting a credit, instead of waiting another 2-3 years, investing then and getting credits?
- The pessimistic alternative is that after the pilot phase there will be no JI phase with transferred credits to the private sector. Why would a company then invest at all today?

Both alternatives lead to the board members of our clients saying:

"Let us wait for the results of Kyoto. Maybe then, we will decide about your project proposals!"

And we can hardly argue against this, or can we?

4. The Vision of "hard" Economic Incentives of JI projects Independent of Formal Governmental credits

As we are no clairvoyants, we cannot predict the outcome of the Third Conference of the Parties in Kyoto in December 1997, but let us assume the following optimistic results:

The Conference will agree upon binding time-tables for individual quantitative limitations of greenhouse gas emissions for the industrialized countries and, as a precondition to be acceptable to the USA, the Conference will also allow the concept of Joint Implementation with credits at state level in addition to other flexible economic instruments. In this case it nevertheless remains very unlikely that the Conference will already decide about detailed modalities about accepted JI project types, approval procedures, credit calculation, verification and monitoring. This will be the task for upcoming working committees, expert hearings and conferences etc. for several years. As described in section 2, there will still be the need for national governments to establish internal systems for transferring these credits from the state to the private sector. This will last some additional years.

So even with this optimistic outcome, there will be no state-related credits combined with direct economic values to private investors for the next 3 to 10 years. But at least the direction of the political development would be clarified:

- additional efforts to reduce GHG emissions in the industrialized countries would be necessary
- the instrument of Joint Implementation would be generally accepted by the international community

What alternatives for activities exist in industrial countries?

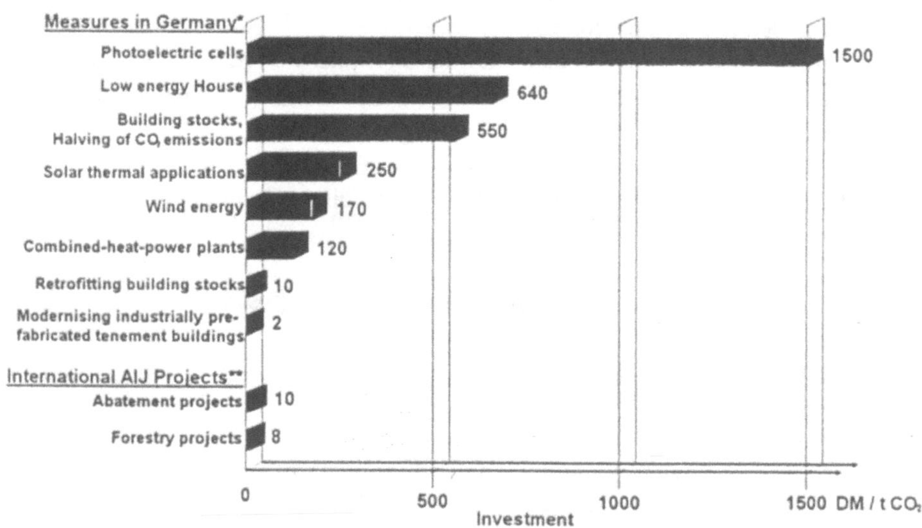

FIGURE 1. Abatement or Compensation Costs for CO_2 Emissions in Germany

* Data from the Ministry of Urban Development, Environmental Protection and Technology of Berlin, 10/1997; ** UMB Price Index 9/1997

Figure 1 shows a selection of alternatives with their specific costs for the example of Germany. It demonstrates the attractive economic potential of JI.

The data of the measures in Germany are based on calculations of realized measures in Berlin. Though it would be more correct to indicate price ranges than exact prices, the tendency of the results would be unchanged, maybe even worse for the photoelectric cells (the Berlin utility calculates more than DM 2,000/t CO_2). As you can see there are very cost-effective activities in Germany too, but if you look more closely you see that their potential is very limited. For instance, more than 60% of the east Berlin industrially pre-fabricated tenement buildings have already been modernized since the German reunification in 1990. The remaining

ones tend to be more cost intensive.
The UMB price index for AIJ projects is based on the available cost data of about 75 announced projects and about 100 additional proposals and indicate the maximum specific price for new AIJ or JI projects guaranteed by UMB for practically unlimited amounts.

The availability of cost-effective JI projects opens the possibility for companies with products or services connected with high CO_2 emissions to compensate their emissions by costs of only 2-3% of the product or service prices.

For instance, burning of one litre of petrol will produce about 3.15 kg of CO_2. The compensation by JI projects with the maximum price of the UMB price index of DM 10/t CO_2 will cost only about DM 0.03 per litre petrol, which is even less than 2% of the actual product price.

For an average car driver in the EU with a performance of about 12,600 km a year and a fuel consumption of 9.6 litre petrol per 100 km, the surcharge of DM 0.03 per litre will sum up to a total of only about DM 36 per year. We believe that nearly every German car driver would accept this surcharge.

Similar calculations exist for other climate sensitive products like electricity or air transportation services.

We strongly believe that companies offering these kinds of environmentally improved products or services could significantly increase their market share and their turnover. In some markets it even seems possible to gain acceptance of over-proportional price increases, so that JI projects may be a direct source of profits. This correlation is shown in table 2.

TABLE 2.

Company	Product/ Service	CO_2 emissions	Costs	Prices
A	a	e_a	c_a	p_a
A with JI project	a*	$e_a - e_{JI} = 0$	$c_a + c_{JI}$	$p_a + p_{JI}$
JI with DM10/t CO_2	a*	$e_a - e_{JI} = 0$	c_a + 2-3% p_a	p_a + 2-3%
	a**	$e_a - e_{JI} = 0$	c_a + 2-3% p_a	p_a + 4-5%
B	b	e_b	c_b	$p_b \sim p_a$

SECTION IV

TRADEABLE EMISSION PERMITS -

A PRACTABLE EXAMPLE: TRADEABLE EMISSION PERMITS FOR SO_2

U.S. EXPERIENCE WITH TRADEABLE SO_2 ALLOWANCES

BRIAN McLEAN
Acid Rain Division, Environmental Protection Agency
401 M Street, Washington, D.C. 20460, U.S.A.

The following pages represent the original transparencies of the lecture held during the Advanced Study Course. A full text is not available due to the time constraints of the author. Nevertheless the editors felt that most of the transparencies are self-explanatory and worth of being reproduced in this volume to give an overview over the Acid Rain Programme of the United States of America.

Further information on the programme is available in the World Wide Web under the following URL:

http://www.epa.gov/acidrain

Additionally the paper by K. A. Brockmann et al. on 'A European model for tradeable SO_2-emission permits' also contains some information of the American Acid Rain Programme (see pp. 314-316)

- Background on SO2 Control
- SO2 Allowance Program
- Lessons Learned

J. Hacker and A. Pelchen (eds.), Goals and Economic Instruments for the Achievement of Global Warming Mitigation in Europe, 295–307.

SO2 CONTROL: TRADITIONAL MEASURES

- Switch to lower sulfur fuels
- Reduce sulfur in fuel
- Disperse sources and emissions
- Employ control technology

SO2 CONTROL: TRADITIONAL GOALS

- Protect public health and welfare
- Meet ambient air quality standards
- Prevent degradation of clean areas

GOAL OF SO2 ALLOWANCE PROGRAM

To reduce SO2 by 10 million tons (8.5 million tons from power generation) as cost-effectively as possible in order to protect public health and the environment

MAJOR POWER PLANTS

HISTORIC REGULATION OF EMISSIONS

- Traditional air pollution control requirements stated as technology requirements, emissions rates, concentrations, percent removal; not as allowable emissions
- Early emissions trading programs -- bubbles, offsets, ERC's -- added flexibility to, but did not alter, existing regulatory structure

EXISTING APPROACHES TO EMISSIONS TRADING

- Conversion to a tradable commodity (emissions) required determinations of:
  - present and future utilization and emission rates
  - appropriate credit / intent
  - emissions quantification protocols
  - compliance agreements
  - assessment of air quality impacts
- Case-by-case development and approval of trades were resource-intensive and time-consuming
- High transaction costs discouraged trading

SO2 ALLOWANCE PROGRAM: A NEW APPROACH

- Not built upon existing programs
- Made allowable emissions (output) the implementation objective
- Required measurement of emissions (output)
- Capped total emissions - ensured maintenance
- Allowed flexibility to minimize cost
- Made penalties & offsets automatic
- Established default limits to assure reductions & discourage delay

SO2 ALLOWANCE PROGRAM: KEY FEATURES

- "Allowance" = authorization to emit one ton of SO2
- EPA allocated allowances to utilities based on desired performance rates and representative utilization levels
- Total allowances per year = half of 1980 SO2 emissions
- Permits are simple, flexible
- Utilities continuously monitor emissions
- At the end of each year, utilities must hold enough allowances to cover emissions
- Also, all utilities must comply with health-based state or local limits

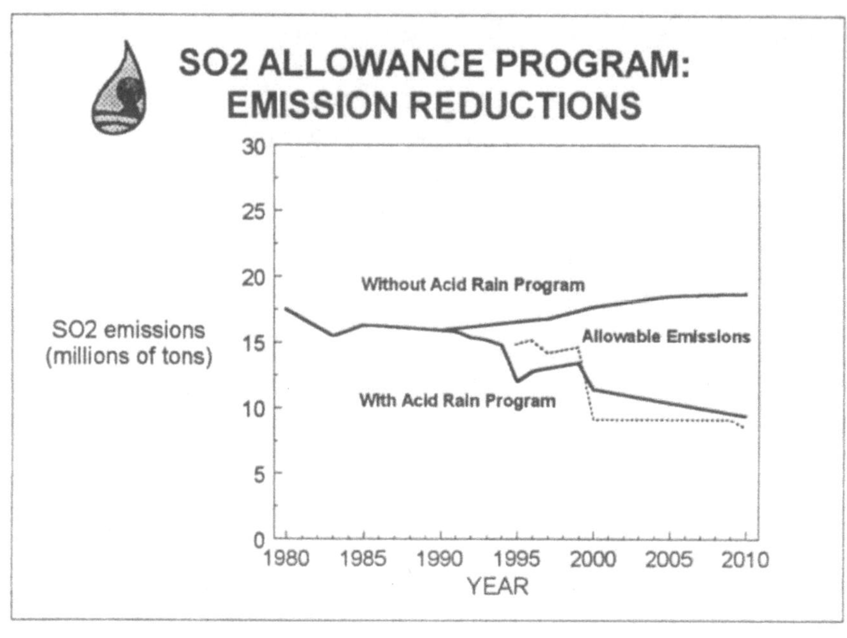
SO2 ALLOWANCE PROGRAM:
EMISSION REDUCTIONS
SO2 emissions
(millions of tons)
Without Acid Rain Program
Allowable Emissions
With Acid Rain Program
30
25
20
15
10
5
0
1980 1985 1990 1995 2000 2005 2010
YEAR

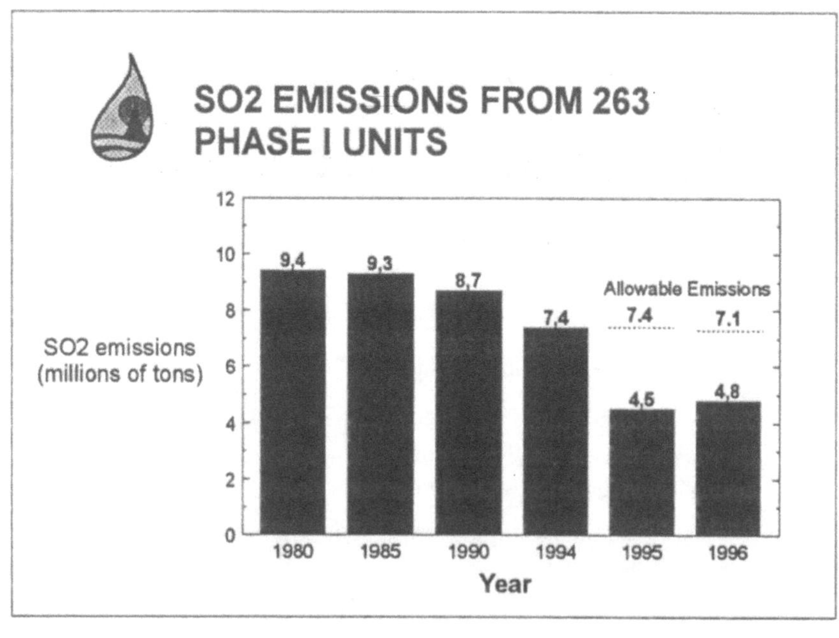
SO2 EMISSIONS FROM 263
PHASE I UNITS
SO2 emissions
(millions of tons)
12
10
8
6
4
2
0
9.4
9.3
8.7
7.4
Allowable Emissions
7.4
7.1
4.5
4.8
1980 1985 1990 1994 1995 1996
Year

This Figure is reproduced additionally as Figure I in the colour section of the book.

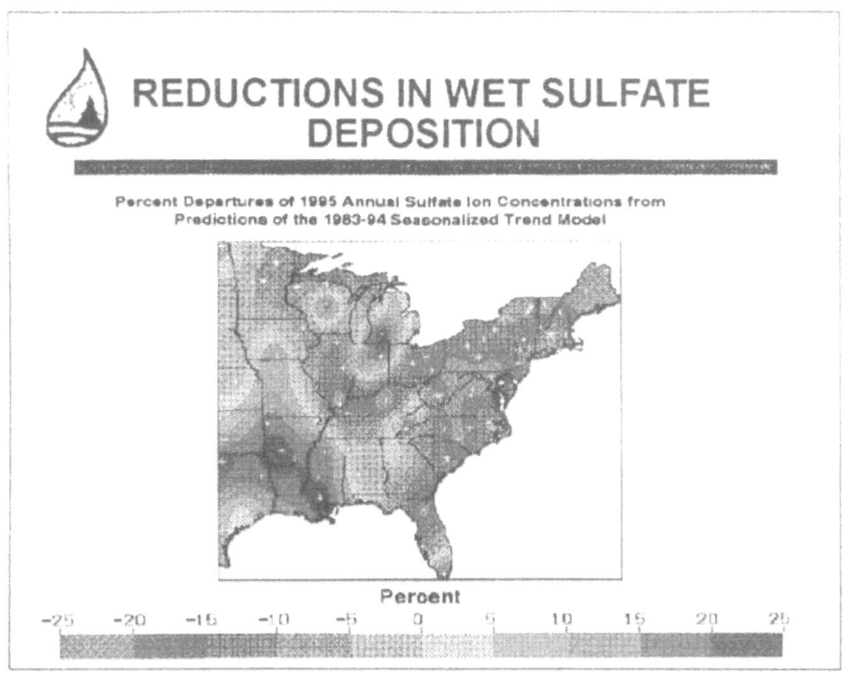

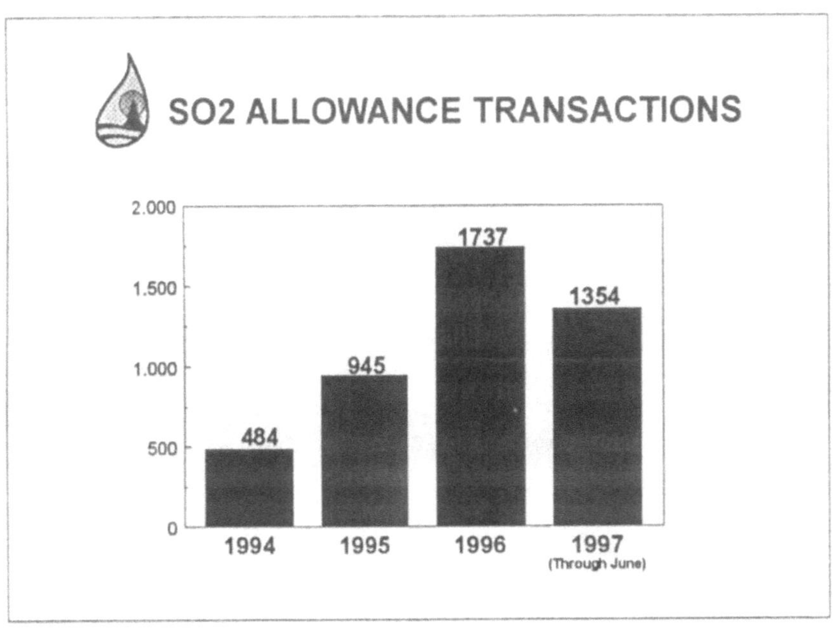

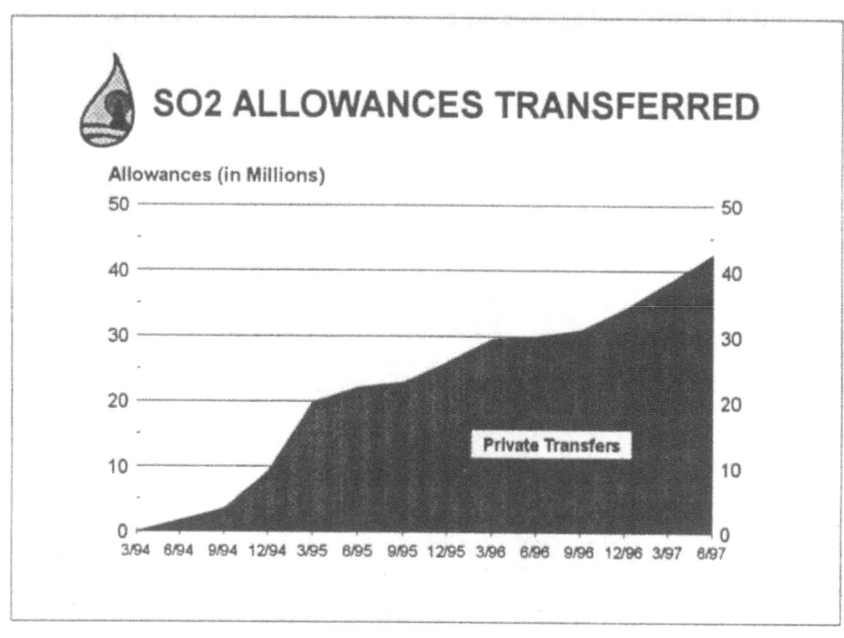
SO2 ALLOWANCES TRANSFERRED
Allowances (in Millions)
50
40
30
20
10
0
Private Transfers
3/94 6/94 9/94 12/94 3/95 6/95 9/95 12/95 3/96 6/96 9/96 12/96 3/97 6/97

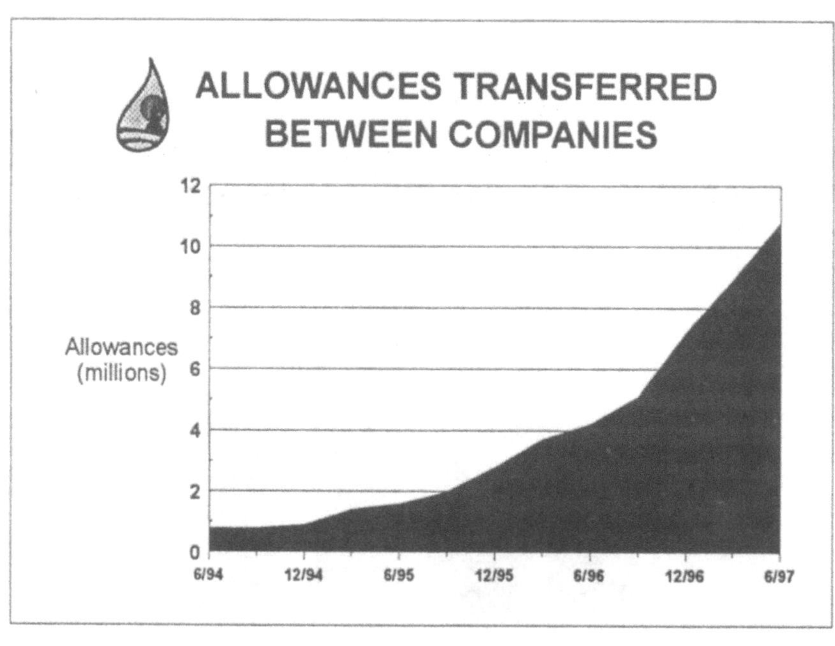
ALLOWANCES TRANSFERRED
BETWEEN COMPANIES
Allowances
(millions)
12
10
8
6
4
2
0
6/94 12/94 6/95 12/95 6/96 12/96 6/97

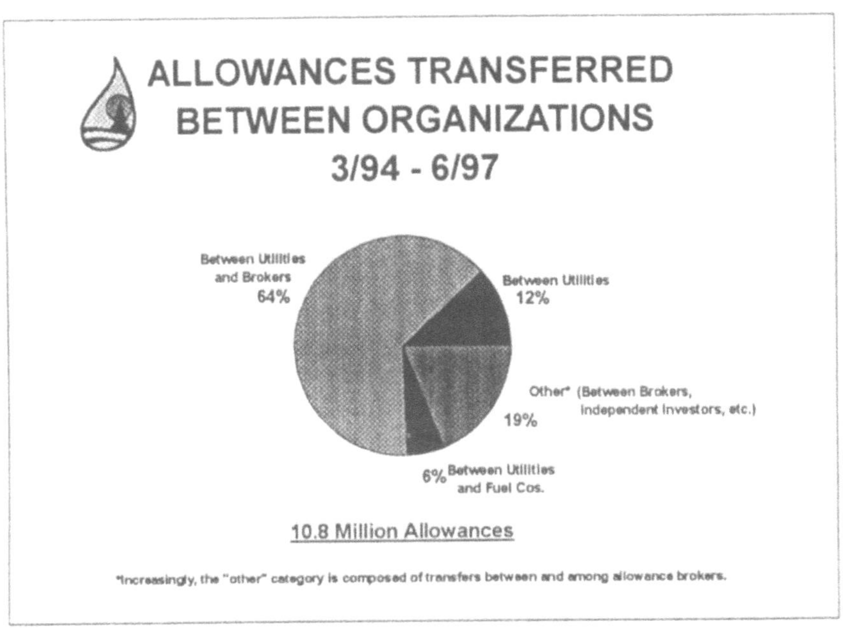
ALLOWANCES TRANSFERRED
BETWEEN ORGANIZATIONS
3/94 - 6/97
Between Utilities
and Brokers
64%
Between Utilities
12%
Other* (Between Brokers,
Independent Investors, etc.)
19%
6% Between Utilities
and Fuel Cos.
10.8 Million Allowances
*Increasingly, the "other" category is composed of transfers between and among allowance brokers.

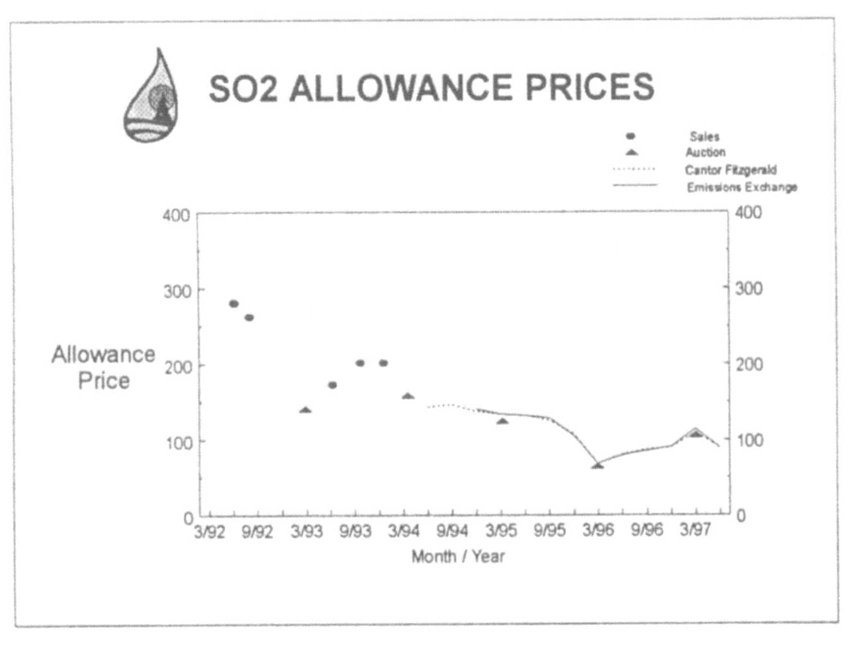
SO2 ALLOWANCE PRICES
Sales
Auction
Cantor Fitzgerald
Emissions Exchange
Allowance
Price
400
300
200
100
0
3/92 9/92 3/93 9/93 3/94 9/94 3/95 9/95 3/96 9/96 3/97
Month / Year

SO2 TRADING PROGRAM: EXPECTED BENEFITS

- Health: $12 - 40 billion per year by 2010
- Visibility: $3.5 billion per year by 2010
- Fewer acidic lakes & streams
- Reduced damage to buildings & monuments

SO2 TRADING PROGRAM: EXPECTED COSTS

- Estimated annual cost using "policies and measures": $5 billion per year
- In 1990, projected annual cost by 2010 of $4 billion
- By 1994, projected annual cost by 2010 of $2 billion
- Government costs are small

WHY HAS THE COST OF COMPLIANCE BEEN SO LOW?

- Competition across all emission reduction options
- Markets provide continuous incentives for innovation
- Banking provides timing flexibility for emission reductions
- Markets reveal true costs

ADVANTAGES OF ALLOWANCE TRADING

- Provides an asset rather than a restriction
- Allows flexibility to achieve lowest cost emission reductions
- Facilitates trading through low transaction costs
- Provides accountability

LESSONS LEARNED

- Keep the system as simple as possible
- Avoid partial participation by a source category within the same geographic area
- Match flexibility with accountability

DIFFERENCES BETWEEN SO2 and GHG TRADING

- Geographical impacts
- Gases, sources, and sinks
- Institutions

KEY FEATURES FOR A GHG PROGRAM

- Legally binding target, allocated to Parties
- Accurate emissions measurement & consistent and timely reporting
- Mechanism for recording and verifying trades
- Public availability of all emissions & trading data
- Consequences for noncompliance

HIGHEST PRIORITY FIRST STEPS

- Highest priority is to set standards for measurement/estimation methods
- Credible national measurement systems with consistent & public data reports are critical to fostering an effective market

A EUROPEAN MODEL FOR TRADEABLE SO_2-EMISSION PERMITS*

KARL LUDWIG BROCKMANN
HENRIKE KOSCHEL
TOBIAS F.N. SCHMIDT
Zentrum für Europäische Wirtschaftsforschung
Kaiserring 14 - 16, D-68161 Mannheim, Germany

* The paper summarizes the results of a study commissioned by the German Federal Ministry of Economics. It will be published at the end of 1997: K.L. Brockmann, H. Koschel, T.F.N. Schmidt, M. Stronzik, H. Bergmann, Handelbare SO_2-Zertifikate für Europa - Konzeption und Wirkungsanalyse eines Modellvorschlags, Heidelberg: Springer/Physica.

1. Introduction

In the context of SO_2-emission reduction the United Nations Economic Commission for Europe states, that

„In order to deal with this inherent uncertainty [on abatement costs], countries could be given some flexibility in how they meet their commitments. One way is to give countries the option to exchange emission reduction commitments. [...] The possibility for exchange will contribute to cost-effectiveness in the allocation of abatement measures over time ...“.[1]

In the following, a practicable SO_2-emission permit scheme is developed for the EU-15, which could support the realization of the reduction targets set by the Protocol of Oslo in 1994. The Protocol of Oslo is related to the 1979 Convention on Long-Range Transboundary Air Pollution on Further Reduction of Sulphur Emissions. For 28 European countries as well as for Canada, the Protocol establishes sulphur emission ceilings for the year 2000 (for some countries for 2005 and 2010, too).

---

[1] UN-ECE (1991): Exploration of Economic Instruments for Implementation of Cost-Effective Reductions of SO_2 in Europe Using the Critical Loads Approach. EB.AIR / WG.5 / R.22, dated 26.6.1991, § 22.

J. Hacker and A. Pelchen (eds.), Goals and Economic Instruments for the Achievement of Global Warming Mitigation in Europe, 309–328.

2. Theory

Environmental permits, partly rooted in the property-rights based Coase-concept, constitute a solution to environmental problems which is highly conformant to the market system. The state merely establishes an ecological framework, and permits trade allows for an efficient allocation of the distributed „rights to pollute“. Thus, the costs of meeting the given target are minimized. Moreover, the price signals of the permits markets do not only assure a short-term efficient allocation of abatement measures but also enhance competition between existing abatement options and give incentices to the invention and diffusion of new abatement options („dynamic efficiency“).

Economic theory suggests a lot of tradeable permits models with different spatial dimension (see TABLE 1). But, the models with the highest appeal in terms of ecological effectivity are characterized by a lack in practicability. This is due to the fact, that, the higher the degree of spatial differentiation, the higher the ecological goal-conformity and the higher the transactions costs related to the gathering and evaluation of information (see Figure 1 for ambient permits). Furthermore, the parallelism of different permit markets - which is an inherent feature of spatially differentiated schemes - raises the risk costs of emission sources not to obtain all necessary permits, or only at prices that are difficult to forecast. This again raises the opportunity costs of relying on permits and thus the attractiveness of technical solutions (e.g. the installation of a flue gas desulphurization process).

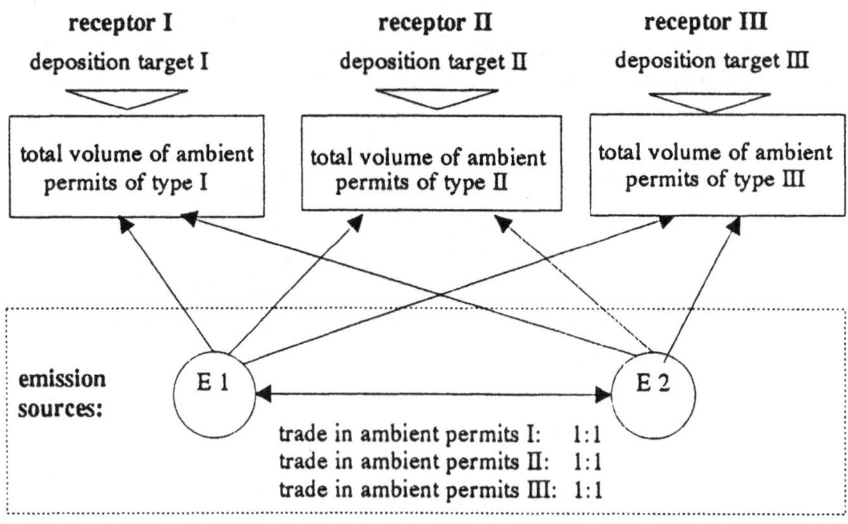

FIGURE 1. Ambient Permits (AP)

TABLE 1. Evaluation of Emission Permit Schemes

permit scheme / criteria	AP (ambient permits)	UDP (undiffe-rentiated discharge permits)	ADP (ambient differentiated permits)	EDP (emission discharge permits), without interzonal trade	Trade with fixed exchange rates between different zones	LDP (local discharge permits)
Number of independant permit markets	one market for each immisson target	one market for each emission target	one market for each immission target	one market for each zone and immission target (thin market problem!)		one market for each zone and immission target (thin market problem!)
Information requirements on behalf of the emission sources	own emissions and diffusion matrix	own emissions	own emissions	own emissions		own emissions
Information requirements for the central authority	—	cost functions and ex post market allocation	diffusion-relationship between market zones and receptor areas	difusion-relationship between market zones and (Rezeptoren)	diffusion-relationship between market zones and receptor areas, and cost functions	relation of marginal abatement costs while operating at minimal cost, in order to fix cost optimal exchange rate between countries
Ecological effectivity	assured	only assured if total emissions are reduced by a large margin	relatively well assured (problem: establishing zones simplifies the diffusion relations)	not assured as inter-zonal diffusion relations are not taken into consideration	not assured	not assured
Cost efficiency	assured	low because the overall reduction is high	relatively high, but lower than with an ADP	relatively high, but lower than with an ADP	efficieny gains possible versus cost minimum plus current reduction plan	not assured

Transaction costs	very high	very low	high	low	very low	very low
Practical experiences	—	Acid-Rain- and RECLAIM-Program, Basler emissions trading	--	—	—	—
Practicability for SO_2	not practicable: relations of diffusions and thus transactions are too complex	potentially practicable if used in combination with measures of ecological fine tuning	potentially practicable, yet institutions for lowering transaction costs are necessary	potentially practicable, yet the problem of thin markets arises	possibly practicable if modified adequately	less practicable than UDP

The models with the highest appeal in terms of practicability are the model of Klaassen as well as a system of locally undifferentiated discharge permits (UDP).

Klaassen et al.[2] suggest a system of tradeable permits with fixed exchange rates between the countries involved. This could be considered as a simple and practicable solution. In comparison to a situation without any permit trade, cost savings are obtained. On the other hand, this system is linked to ecological deficits. As the problem of information connected to the fixing and the regular revision of the fixed exchange rates cannot be solved completely, it would be necessary in practice to combine such a system with measures preventing a violation of the deposition targets. In that case specific advantages in comparison with a UDP-system appear to be minor.

In the European context, a UDP-system establishes one market of discharge permits and allows all emission sources to trade permits 1:1 (see FIGURE 2). Different from a CO_2-emission permits scheme, location issues cannot be ignored in the context of SO_2. Theoretically, an unrestricted permit trade could result in deleterious emission concentrations. Thus, a UDP-system has to be combined with measures of ecological fine tuning („hot spot measures“).

---

2 G.A.J. Klaassen, F.R. Førsund, M. Amann (1994): Emission Trading in Europe with an Exchange Rate. In: Environmental and Resource Economics, Vol. 4, No. 4, S.305-330.

In summary, the theoretical considerations lead to the conclusion that a good compromise between the existing economical and ecological demands would be a system of undifferentiated discharge permits with additional measures to prevent hot spots. The UDP-scheme allows a maximum market size and offers therefore a maximum degree of functioning of the market as well as stable price signals. In addition to this, it is administratively simply to implement and the addressees of the scheme can easily grasp its mechanisms and its economic implications.

Even though the ecological effectiveness of a pure UDP-scheme might not work effectively for all receptors, the cost efficiency of the tradeable permit system eases the carrying through of more ambitious emission reduction goals in the future and would therefore increase the dynamic efficiency of the entire set of environmental policy measures. An ecologically accurate but not functioning and therefore inefficient tradeable permit system would be more likely to run contrary to the long-term aim of a sustainable management of emissions in the sense of the critical loads concept pursued by the Protocol of Oslo.

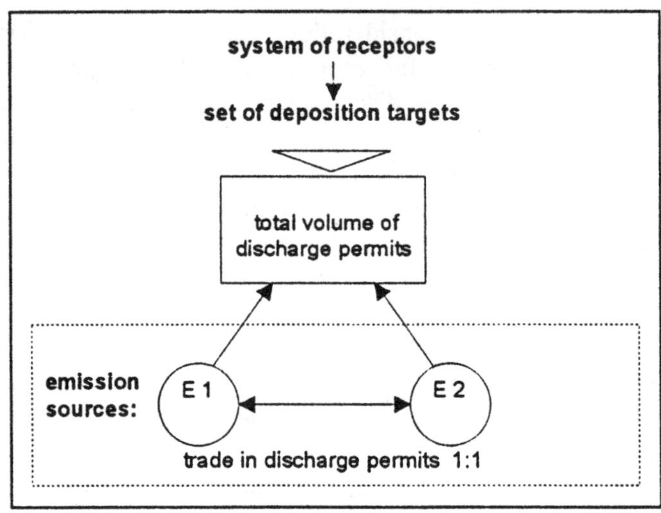

FIGURE 2. Undifferentiated Discharge Permits (UDP)

3. Lessons Learned from the U.S.

First experiences with tradeable permits in pollution control were already gained in the U.S.A. The Regional Clean Air Incentives Market (RECLAIM) was implemented in 1994 in the region of Los Angeles, the Acid Rain Program (ARP) was implemented in 1995 on a national scale. Both programs can be classified as UDP-systems.

So far, in both programs neither indications for negative ecological effects nor for market power in the permit market could be observed. The volume of trade and the allowance prices stay below prior estimates in both, ARP as well as in RECLAIM. Definite conclusions regarding the reasons cannot be made so far because of the early program phases. Nevertheless some preliminary thesis can already be put foreward.

Following a period of relatively low prices in early 1996, meanwhile the allowance prices have risen, as can be seen from FIGURE 3. In order to explain the all in all low permit prices (see FIGURE 4), one can firstly refer to the generous placing of excess allowances (under the ARP) as well as to a relatively generous initial allocation of permits to the companies concerned, which went beyond the real need (under the RECLAIM). Secondly, it seems that during the discussions about the design of both programs the abatement costs were overestimated by far. One reason can be the fact that under the predominant politics of command and control, the companies had an incentive to overestimate the abatement costs. Thirdly, the relatively low prices can be

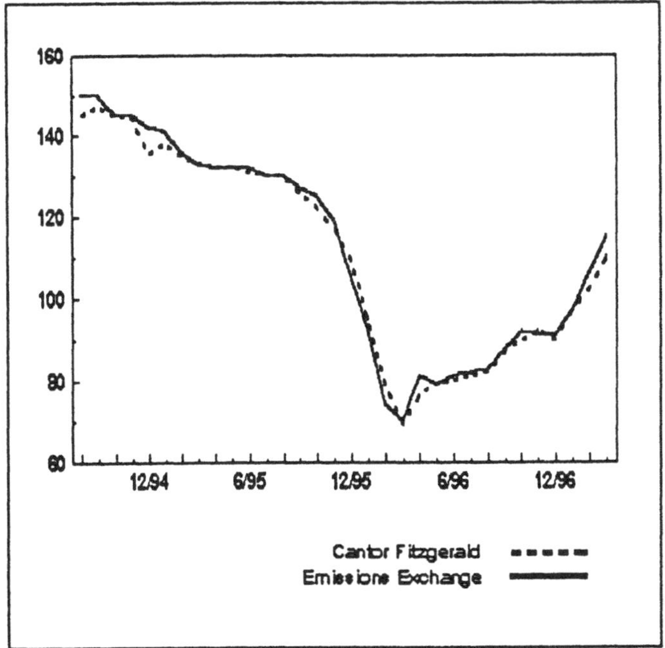

FIGURE 3. Monthly Average SO_2-Allowance Price under the ARP [US$/t SO_2]]
Source: U.S. Environmental Protection Agency: Acid Rain Division's Home Page, http://www.epa.gov/acidrain/ardhome.html.

attributed to the dynamic efficiency of a permit market, i.e. the potential of innovation and the resulting long-term cost reductions of a permit scheme were underestimated.

Despite lower permit prices as expected, the volume of trade in both programs lies below the initial estimates, too. Again, the reasons can be seen in the dynamic efficiency of a tradeable permit system. To a higher degree than within the traditional command and control politics, the technical emission control options start competing with each other if economic instruments are being used. This competition was supported by deregulation in the gas industry and in the transport sector. While under RECLAIM there has not been any indication for this phenomenon yet, the indications at the ARP are evident.

Seen altogether, especially the ARP can be valued as being a successful, practicable implementation of the idea of emission permits for the pollutant SO_2. The volume of trade is sufficiently high to generate a steady and stable price signal. The relatively low permit price is not so much a sign for a non-functioning of the market as a hint to the dynamic efficiency of the permit market. Thus, the theoretical results are being confirmed by these experiences.

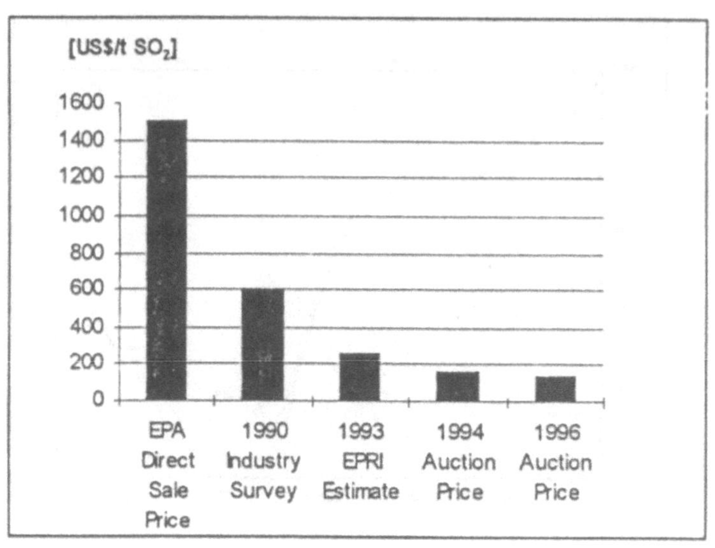

FIGURE 4. Estimated and realized SO_2-allowance prices under the ARP
Source: U.S. Environmental Protection Agency: Acid Rain Division's Home Page, http://www.epa.gov/acidrain/ardhome.html.

4. The Technical and Economic Setting in Europe

Technical setting: In EU-15, big thermal power plants in Germany, Belgium, Austria and Finland are already fitted with flue gas desulphurisation processes to a high degree. In Greece, Portugal, France, Italy, Sweden, Ireland and the Netherlands more than 50% of the installed capacity are based on low sulfur fuels. A relatively low coverage of plants with sulfur control technologies can be recognized for Greece, France, Portugal and Denmark (between 40% and 50% of the net capacity), as well as for the United Kingdom and Spain (between 80% and 90%). All in all, there is a high potential for SO_2 emissions reductions in EU-15 electricity production. A big share of existing plants (representing 31% of total net capacity) has no control technology. Another 28% of the capacity are presently fired with gas or other low sulfur fuels and could be combined with additional control technologies.

An analysis of the *economic setting* reveals, that an SO_2-emissions trading scheme for public utilities would be characterized by a sufficient number of agents; this works against possible oligopolistic structures. A grandfathering procedure would result in a distribution of emission permits where 18 public utilities cover approx. 78% und 39 public utilities approx. 92% of all permits. Nonetheless, the number of agents seems small enough to keep transactions costs of measuring emissions and of supervising the agents low. These conditions are at least as good as the conditions in Phase I of ARP. A higher number of big thermal power plants (EU-15: 671; original ARP Phase I units: 263) with altogether roughly the same volume of SO_2-emissions (and thus permits) would be included into the scheme (EU-15 1990 SO_2-emissions from power production: 8.6 million tons; ARP Phase I units 1990 SO_2-emissions: 8.7 million tons). Only with full information about the site-specific marginal abatement costs, the question of the potential trading volume under the proposed scheme can be answered precisely. The general equilibrium modelling calculations made in the study for public utilities in EU-11 indicate that a share of 31% of all permits may be traded. For ARP, as well, a volume of 30% had been estimated, while actually only approx. 17% to 24% can be realised. Nevertheless, as can be seen from FIGURE 3, this volume is enough to guarantee a steady development of prices and thus the functioning of the permit market.

5. A European SO_2-Emissions Trading Scheme

Based on theory and the lessons learned from the U.S.A. a concrete SO_2-emissions trading scheme for public utilities is elaborated. Public utilities were chosen as addressees of the proposed scheme in order to achieve a certain degree of simplicity in the early phase of the program. They cause approx. 50% of all anthropogeneous SO_2-emissions in the EU-15 region. Other sources are intended to join the scheme in a later phase.

There are several **principles** underlying the scheme's design:

- The mode of the initial allocation should allow for early price signals, in order to make full use of the instrument's potential to discover static and dynamic efficiency gains.
  Consequently, and considering the vested interests of existing emissions sources a long-term transition from a mode of grandfathering to the mode of a free auction was chosen for the initial allocation („limited grandfathering").
- A functioning permit market needs a sufficient number of agents and a sufficient volume of trade.
  This principle led to the selection of a system of undifferentiated emission permits (UDP) which provides the possiblility of banking.
- All mechanisms should be simple and reasonable in order to improve the confidence in the effectiveness of the system.
  Therefore, it is necessary that the emission reduction path (i.e. the number

of present and future permits) as well as all rules governing the permits trade are fixed definitely and in advance, in order to guarantee long-term planning security for all agents involved.

- The maximum capture of the cost savings potentials involves ecological risks. The limitation of the risks of hot spots has to be assured by a catalogue of optional measures with different degress of intensity of intervention.
  Size and point in time of the intervention have to be chosen in a manner which restricts the degrees of freedom of as few economic agents as possible and only as far as it is necessary for ecological reasons.

During the scheme's conception the Acid Rain Program functioned as a model. The major differences lie in the long-term transition in the initial allocation of emission permits from grandfathering to an auction and in a broader set of options to avoid hot spots. The details of the proposed scheme are sketched in the following (for a summary see the appendix).

As the model shall be obligatory for all EU-15 countries, it is necessary to set up a single *European certificate agency*. For the certificate agency a structure should be chosen, which is divided into an administration unit and an emission council. The emission council would be authorised to initiate measures which keep the permit system functioning, for example measures to counteract possible hot spots, or measures for the fine tuning of the market. The administration - which may be a private company - will manage the „daily business" of the permit trade, of booking, etc..

As *criterion for the initial allocation* of the permits the product of fuel consumption and an emission factor typical for this fuel will be selected. An assignment according to the actual emissions is rejected, because „pioneers" of emissions control would be penalized through a low assignment. The emission factor will have to be *inter*nationally differentiated. This is necessary for two reasons. Firstly, the national emission caps of the Protocol of Oslo differ - this is a major difference to the ARP where the allocation of permits does not have to reflect different obligations of the States to reduce SO_2-emissions. Secondly, each country is going to select a different contribution of its public utilities to the overall national reduction target - again this is a major difference to the ARP as no EU country can be forced to assign specific reduction requirements to its power producing sector.

*Intra*nationally a single emission factor for each fuel should be applied to all national public utilities. A differentiation into different fuels is necessary, as, for example, gas has a significantly lower sulphur content than coal. If such a differentiation did not take place and an average value for all fuels was chosen, the allocation of permits would be unrealistically low for example for coal-fired plants and unrealistically high for gas-fired plants. A differentiation in terms of more or less sulphur containing coal, oil or gas, following the example of the ARP, is denied. Regarding desulphurised fuels, again the „pioneers“ would be penalized. A differentiation according to the natural sulphur content is rejected because it would imply a lower potential volume of trade: The initial allocation would more and more approach the actual need, and for any plant the average demand respectively supply of permits would decrease.

Legal construction: The permits assign the right to discharge one ton of SO_2 in a stated year; banking is allowed but emission debts will not be granted. The permits are freely tradeable, the owner can keep them, sell them, lease them, or may give them away. Does one of the utilities shut down a plant, it definitely keeps the permits assigned to him for the years coming. A withdrawal of the permits if a plant is closed down would give an incentive to delay an investment into modern and ecologically friendly technologies. In contrast, the continuing validity of the permits for future years creates an incentive to replace old plants earlier than planned.

In order to hand over the control of emissions to the permit scheme, all *emission standards* for existing plants or units have to be frozen on a status quo basis. In particular, all dynamic standards have to be either anullated or substituted by explicit standards. New more strict standards, including those of annex V of the Protocol of Oslo, should not be introduced as they constrict the scope of the permit scheme. At least should the standards of Annex V be postponed from July 1st, 2004 to the starting date of Phase II of the proposed scheme.

In consideration of the traditional patterns of using the environment, and in order to avoid abrupt changes, the *initial allocation procedure* of grandfathering is chosen. This principle is going to be softly broken at two points. On the one hand a limited grandfathering is preferred, on the other hand in the first years a small amount of approx. 3% of all permits is distributed by the centrale certificate agency through auctions and fixed price sales and is not being booked on the accounts of the public utilities. The initial allocation of permits to existing sources covers only the expected remaining regular lifetime of the respective plant type. To reach a higher acceptance, a generously defined „regular technical lifetime“ of a plant is to be considered. It seemes adequate to take a value of 35 to 40 years. The reasons which support this procedure of a *limited grandfathering* trace back on the one hand to distributional aspects and aspects of competition, and on the other hand to aspects of the functioning of the permit market. Firstly, the transfer of capital to the existing sources, which is connected with the allocation of tradeable permits, should be minimised, because otherwise a distributionally not justifiable preferential treatment of existing sources would take place. Secondly, in order to promote competition in the energy market, it is useful to provide for a long-term rise of the share of permits which can be auctioned; this makes it easier for new sources to purchase permits. Thirdly, a long-term growing amount of auctionable permits improves the price signals which come from each auction.

It is important for the achievement of the ecological targets that the permits, which rest at the agency because of the limited grandfathering, will not stay there, but will distributed to new and old sources by spot- and advance-Auctions. Only in this way the emission reduction path announced at the beginning of the program will be met. In a system of limited grandfathering, a rising share of permits stays with the certificate agency the longer the program runs, becauser more and more old plants overrise the regular technical lifetime. As just mentioned, these permits will be recycled to the agents via auctions in order to maintain the announced emission reduction path. At these auctions growing revenues will be achieved during the years. These can be used for the coverage of administration costs of the certificate agency and for adjustment programs for extremely affected plants or regions, as well as for the promotion of renewable energies or as a reserve pool for the fine tuning of the permit market.

The auctions should be carried out by an established *stock exchange*, which has experiences with futures markets, in the order of the certificate agency. Because of the European context one of the more important European stock markets should be selected as the main stock exchange. Other parallel stock markets should not be allowed in order to maximize the market volume, liquidity and efficiency of the main market place. One may think of the establishment of an automatic computer supported stock exchange, because of the more likely small frequency of trade in permits in comparison to other futures markets.

In the U.S., *bonus permits* were used as an instrument to avoid too much friction. It is possible that such a need for political cushioning also comes up in the heterogeneous economic landscape of the EU-15 countries. But, other than in the U.S. this need should not be managed via special and additional assignments of permits, but via other instruments, as, for example, financial transfers. A too generous allocation of permits bares risks concerning the functioning of the permit market.

The binding prior announcement of the emission reducion path and the distributed permits are major parts of a permit policy which is orientated towards planning security and therefore towards functioning and efficiency of the permit market. For the same reason, an ongoing *rough tuning* of the volume of circulating permits is not planned; it would impair the planning security of the public utilities considerably. But, it could well be that, e.g. due to progress in natural sciences, the emission caps decided in Oslo would mean a fall below or an exceeding of the emission caps now considered necessary. Such a unique rough tuning is acceptable, but one should bear in mind and take care for the vested interests of the public utilities. For example it could be decided that rough tuning can only be implemented as part of new negotiations of the Protocol of Olso or via a decision of the Council of Ministers.

A permit system can only correspond to the receptor-orientated approach of the Protocol of Oslo if *precautions against hot spots* are taken. This need can be found in Art. 2 (1) of the Protocol, which admittedly does not set binding measures to avoid hot spots, but asks for long-term keeping of the critcal loads. One can imagine different measures in this context, which to a certain extent add up to the permit system without destroying its basic flexiblity and still looking useful to avoid hot spots:

1. Promotion and extension of the estimative capability of existing monitoring systems about the effects of acid rain in critical regions.
2. Documentation of all permit transactions and emissions of „critical sources“ (power plants whose emissions may contribute to a hot spot problem).
3. Obligation to register for all bilateral permit trades.

This set of informational instruments establishes an early warning system for hot spots and prepares adequate countermeasures. If it indicates the coming up of hot spots, out of the following spectrum of instruments the one should be selected with the lowest intensity of intervention and which least interfers in the mechanisms of the permit market. The instruments are command and control measures, but, different to former politics, they will be implemented on a case by case basis only for single „critical sources“ and only as long as the hot spot problem goes on.

4. Integration of an assessment of risk concerning the effects on imissions into the approval procedure for new sources in regions whose emissions cause a major problem.
5. Permit trade restrictions for „critical sources“.

Measure 4 is more of a preventive nature, whereas the 5th Measure is relatively put into action. But the precondition is, that these limitations, which mainly concern the spot market and the spot-auction, are accompanied by preparatory informational measures concerning all market transactions taking place earlier (bilateral and organized futures markets and advance-auctions). Consequently, this measure has to be linked to the 2nd Measure.

In case the former shown instruments are not sufficient to avoid the appearance of hot spots, in the end one has to fall back on a short-term and restrictive command and control instrument:

6. Restrictions in the validity of permits held by „critical sources".

A bubble within this group should be allowed. But ultimately, the affected „critical sources" will have to implement technical solutions to restrict emissions. Within the traditional command and control approach this option is compulsory for all sources and from the beginning. The proposed permit scheme just takes it as a „lender of last resort" and thus offers a considerably higher degree of flexibility.

6. Macro-Economic and Sectoral Impacts of an EU-Wide Tradeable Permit System for SO_2

To estimate the overall and sectoral effects of a European SO_2-emission trading policy two *scenarios* were simulated using a computable general equilibrium model for eleven EU-member states:[3]

- In scenario 1 the emission reduction targets, which were derived from the Protocol of Oslo for the national public utilities, are put into action for the electricity sector on a national level using national tradeable permit systems (non-coordinated policy).
- In scenario 2 the emission reduction targets, which are laid down in the Protocol of Oslo, are aggregated for the eleven countries covered in the model. This single reduction target is achieved by imposing a single, EU-wide permit market. In this type of politics the opportunities of emission reduction will be exhausted in the different countries according to their cost-effectiveness (coordinated policy).

The *environmental and economic effects* of the non-coordinated and the coordinated policies are minor. The variation in the rates of all aggregated macro variables lies in both scenarios in general (with the exception of Spain)

---

[3] The GEM-E3 model was developed on behalf of the European Commission, DG XII, by P. Capros, T. Georgakopoulos (NTU-Athens), S. Proost, D. Van Regemorter (CES/KU-Leuven), K. Conrad, T.F.N. Schmidt (Universität Mannheim, ZEW-Mannheim), N. Ladoux, M. Veille (CEA/GEMME-Toulouse) and P. McGregor (University of Strathclyde-Glasgow). Further references and informations are given in K. Conrad, T.F.N. Schmidt (1997): National Economic Impacts of an EU Environmental Policy - An Applied General Equilibrium Analysis, ZEW-Discussion Paper No. 95-22.

within a range of ± 0,5%. Due to transboundary air pollution, the development of national depositions of sulphur is not proportional to the national abatement efforts in both scenarios. In an EU-wide scheme of permit trade Germany, Denmark, United Kingdom and the Netherlands observe higher depositions than in the non-coordinated scenario 1, whereas depositions are lower in Spain, Portugal and Greece. The scenarios do not include any hot spot measures as contained in the proposed SO_2-permit scheme. Thus, when implementing the scheme, possible exceedances of targeted deposition levels will be counteracted by the set of hot spot measures described in the previous section.

Both scenarios differ in their effect on *relative prices*. Scenario 2 leads to a single permit price of 1,419 ECU/ton of SO_2 in all EU-11 countries. Plants that already undertook large emission reduction efforts before 1990 prefer to buy permits than to invest in further abatement measures. Hence, there is a net-demand for permits in Germany, Denmark, France, Italy, the Netherlands, and the U.K.. In the non-coordinated case of scenario 1 permit prices vary between 250 ECU/ton of SO_2 in Spain and 3,195 ECU/ton of SO_2 in Germany. This outcome reflects the differences in marginal reduction costs and reduction targets of countries. A model run taking into account actual abatement efforts after 1990 may change the results. In particular German power plants located in the New Bundesländer, which carried out large investments, can expect to become net sellers of permits and may thus enjoy a partial financial compensation.

Both scenarios imply a loss in *GDP*. At the EU level this loss is smaller under the co-ordinated regime of scenario 2 (-0,08%) than under the regime of scenario 1 (-0,12%). Under the coordinated policy the distribution of total economic burden is altered in disadvantage of those countries that sell permits, i.e. those countries that take the higher reduction efforts in comparison to scenario 1 are worse off. This is due to the fact that the latter observe higher prices (in particular the price for electricity) when a coordinated approach is taken. The negative effect of this rise in national prices is not fully compensated by the (net) sale of permits to other countries as the decision on buying permits or abating more is taken by the electricity companies which are interested in minimizing their individual costs. The economy-wide spill-over effects of higher electricity prices on the rest of the economy are not incorporated in their calculation and are therefore not reflected in the permit price.

According to *sectoral effects* (on an aggregate European level), with the exception of the three sectors oil, gas and energy intensive industry, all sectors have to accept a reduction of their gross production under a non-coordinated as well as under a coordinated permit policy. The following statements match scenario 1 as well as scenario 2: 1) The winners of a policy of SO_2-permits are (i.a.) the sectors oil and gas; they show absolute growth in production, resulting in a growing share of the European gross production. 2) The sectors agriculture, consumer goods industries, transport, services and non market services show absolute decreases of production in comparison to the reference scenario, but nevertheless as well a growth of their shares in the whole of the European gross production. 3) The loosers of a SO_2-permit policy are the sectors coal, electricity as well as equipment goods industries. Their gross production falls more than the average aggregated about all sectors, which means that their competitive position gets worse relative to the other sectors.

To *conclude the findings* of the simulation: The economic costs of a system of tradeable permits in the electricity sector are within acceptable limits. A coordinated EU-wide market could lower the total costs, but the differences to the non-coordinated policy are again rather small.

7. Summary and Conclusions

A main finding of the study is, that, considering the setting in the EU-15 countries, the proposed scheme of undifferentiated emission permits seems to be economically advantageous and legally feasible. But, there is an inherent conflict between ecological effectivity and economic efficiency in a UDP-scheme. Due to the missing differentiation regarding the receptors compared to other emissions trading schemes, the risk of hot spots increases in the short run. On the other hand, this risk is lower in the long run, as the cost efficiency of this type of scheme facilitates the carrying through of new and ambitious emission reduction plans in the future.

The ecological risks of the scheme seem to be justifiable and should be accepted, in order to make full use of the gains in static and dynamic efficiency. Nevertheless, a differentiated bundle of measures counteracting possible hot spots should take a central role within such a scheme. The use of the measures should be graded according to their intensity of interference, in order to keep the degrees of freedom for all agents in the market as high as possible.

The cost saving potentials - a lesson learned from the American experiences - of an EU-wide permit scheme may increase the acceptability on the side of the public utilities and may thus contribute to an acceleration of the decision making process. The simplicity and the high practicabilitiy of the proposed scheme will support this process, too.

The mode of primary allocation of permits should allow for early price signals, in order to make full use of the instrument's potential to discover static and dynamic efficiency gains. Under additional consideration of the vested interests of the existing emissions sources, a long-term transition from a mode of grandfathering to the mode of a free auction was chosen. This „limited grandfathering“ is implemented by a primary allocation of permits to existing sources which only covers the expected remaining regular lifetime of the respective plant type.

The computed variation in the rates of all aggregated macro variables caused by an emission permit scheme in EU-11 lies in both scenarios in general within a range of $\pm$ 0,5%. In particular, the potential seller countries of permits bare the risk of (small) losses, in terms of GDP, as a sectoral permit system only allows for the minimization of the sectoral costs (of public utilities). The revenues from selling permits to other countries do not necessarily cover all national costs arising from the higher degree of emission reduction in an EU-wide scheme. The purchasing countries shift a part of the emission reduction costs on these. Therefore, a financial compensation mechanism compensating the divergent national costs of the scheme should be taken into consideration. This scheme would also allow for an inter-country compensation of the deviations of the actual national emissions (realized under the scheme) from the levels fixed in the Protocol of Oslo.

Appendix. A European SO_2-Emissions Trading Scheme for Public Utilities

addressees	- large combustion plants in the EU-15 countries - opt-in for (i.a.) industrial power plants and production processes with combustion - extension to Middle-East- and East-European countries in the long-run
permit design	- locally undifferentiated discharge permits, dated for a specified year - unrestricted banking for all years and phases - no emission debts
assessment basis	- SO_2 emssions to be monitored by continuous emission monitoring systems
criterion for the initial allocation	- product of fuel consumption and an emission factor typical for the fuel used - internationally differentiated emission factors - intranationally a single emission factor for each type of fuel
legal construction	- right to discharge one ton of SO_2 in the stated year - permits are not discounted - free permit trade (buy, lease, rent, or give away) - if a plant is shut down before the end of its regular technical lifetime the owner keeps the rest of the permits assigned,(even those dated for future years)
procedure over time	- preparatory phase before start of programme, incl. advance auctions and bilateral trade - constant yearly permit allocation for all years of each phase - dynamics by establishing two phases with different overall emission levels
emission standards for existing plants or units	- all emission standards are set on a status quo basis - all dynamic standards are either anullated or substituted by explicit standards - new standards, including those of annex V of the Protocol of Oslo, should not be introduced. At least should the standards of Annex V be postponed from July 1st, 2004 to the starting date of Phase II
emission standards for new plants	- are to be omitted or to be formulated explicitly and moderately
initial allocation mechanism	*Limited Grandfathering* - grandfathering for existing plants before start of programme for the defined regular technical lifetime; the liftetime is defined generously - 3% of the initial allocation are kept as reserve for auctions and fixed price sales as long as the agency itself does not dispose over a sufficiently large number of permits

Appendix (cont'd). A European SO_2-Emissions Trading Scheme for Public Utilities

<table>
<tr><td rowspan="3">initial allocation mechanism (cont'd)</td><td colspan="2">Stock-Exchange Auctions
- in the long-run the share of permits initially allocated by auction rises, as for late years the certificate agency disposes over a larger number of permits remaining at the agency due to the system of limited grandfathering
- zero-revenue auction if permits have been cut off from the sources' initial allocation</td></tr>
<tr><td>Spot-Auction
- permits dated for the present year
- once or twice a year
- approx. 0,6% of a year´s permits</td><td>Advance-Auction
- permits dated for future years
- 4, 5, and 6 years in advance
- at least approx. 0,6% of each year´s permits
- permits resting with the agency due to limited grandfathering are to be added</td></tr>
<tr><td>Direct Sales from Fixed Price Sales Reserve
- approx. 0,5% of each year´s permits
- a fair part of the revenue from those permits that were cut off from existing sources is redistributed to them</td><td>- no fixed price sales of permits dated for future years</td></tr>
<tr><td rowspan="2">secondary allocation</td><td colspan="2">Bilateral Trades
- informally by brokers or via stock-exchange</td></tr>
<tr><td>Spot
- presently valid permits</td><td>Futures
- permits valid in future years
- futures and options</td></tr>
<tr><td>choice of stock market</td><td colspan="2">- existing stock exchange experienced with futures
- large European stock exchange as central stock exchange; no parallel stock markets
- possibly automatic computer supported stock exchange</td></tr>
<tr><td>monitoring</td><td colspan="2">- continous emission monitoring systems
- spot checks</td></tr>
<tr><td>enforcement</td><td colspan="2">- automatic enforcement
- between 500 and 1.500 ECU per ton SO_2 not covered by permits
- corresponding number of permits is subtracted from the next year´s allocation</td></tr>
</table>

Appendix (cont'd). A European SO_2-Emissions Trading Scheme for Public Utilities

balance period	- one year - those months with extreme peaks or fluctuations in electricity production are not to be set at the end of the period
bonus permits	- none - other instruments to cushion too much friction, such as financial transfers
rough tuning	- no ongoing rough tuning by the central agency - an additional reduction of allocated SO_2 permits only when taking into account the plants' vested interests and only as part of new negotiations of the Protocol of Oslo or via decision of the Council of Ministers
fine tuning	- done by the certificate agency - aiming at stable prices and a stable market volume when supply bottlenecks, speculative tendencies and other misdevelopments in the permit market occur - open market operations, no discount rate policy
avoiding hot spots	- various instruments with different intensity of intervention 1. promotion and extension of the estimative capability of existing monitoring systems about the effects of acid rain in critical regions 2. documentation of all permit transactions and emissions of „critical sources" 3. obligation to register for all bilateral permit trades 4. integration of an assessment of risk concerning the effects on imissions into the approval procedure for new sources in regions whose emissions cause a problem 5. permit trade restrictions for „critical sources" 6. restrictions in the validity of permits hold by „critical sources". A bubble within this group should be allowed - in order to reach an identified ecological target the instrument having the lowest intensity of intervention and interfering least with the functioning of the permit market is chosen

SECTION IV

TRADEABLE EMISSION PERMITS - DESIGN OF A CO_2 EMISSION SYSTEM

EFFICIENCY, EQUITY AND OPTIMAL CARBON ABATEMENT

JOAQUIM OLIVEIRA MARTINS [1]
Organisation for Economic Co-operation and Development
Economics Department
2, Rue André Pascal, F-75775 Paris Cedex 16, France

Abstract. The separability between efficiency and equity has been an underlying assumption in most of the assessments of the costs to reduce carbon emissions. Chichilnisky and Heal (1994) suggested that this property does not hold anymore if carbon abatement is viewed as a public good produced in a decentralised way by private consumption activities. In that context, the optimal carbon tax, the optimal abatement level and equity considerations are deeply interrelated. This paper shows: i) in what sense equity and efficiency are not separable; and, ii) while the equalisation of marginal abatement utilities is not necessary for Pareto efficiency, the principle of an uniform carbon tax or a system of tradeable permits remains valid. Some policy conclusions are drawn from these results.

1. Introduction

Given the uncertainties concerning the net benefits from stabilising the global climate, finding an optimal level of global carbon abatement has been viewed as a exceedingly ambitious objective. Without a valuation for the damage of climate changes in the utility function of policy-makers, the problem is reduced to defining a given level of emission reduction and seeking the least costly way to achieve this constraint. This is the road that has been followed by the majority of analyses and assessments of the global warming problem that have been published so far.

In such a framework, economic efficiency is typically separable from the issue of equity. Put differently, the basic economic efficiency principles can be applied independently from the distribution of income among economic agents.

---

[1] The author would like to thank G. Chichilnisky, G. Heal, J.-C. Hourcade, R. Stavins, P. Sturm and H. Tulkens for helpful discussions. The opinions expressed in this paper are those of the author and cannot be held to represent the views of the OECD or its Member countries.

J. Hacker and A. Pelchen (eds.), Goals and Economic Instruments for the Achievement of Global Warming Mitigation in Europe, 331–337.

In the case of carbon emission reductions, the principle of equalisation of marginal abatement costs through a uniform carbon tax or a tradeable permit scheme minimises global costs, even if this overall efficiency principle may entail an extremely uneven sharing of the burden across countries (OECD, 1995). To secure an international agreement where the emission abatement effort is displaced from high to low abatement cost countries; it may be necessary to make transfers that compensate the latter for their incremental costs. Nonetheless, provided that the emission constraint is applied at the global level and the abatement price is uniform across countries, it can be shown that abatement efficiency ensures an optimum outcome (in the Pareto sense) and reciprocally. This point is particularly important for the design of a system of tradeable permits. Abstracting from uncertainty or transaction cost considerations, it implies that *any* initial distribution of allocation of permits will achieve Pareto efficiency. The considerations about income distribution can be viewed as a separate problem that can be solved, say, through a negotiation process.

The recent progress in the science of climate change is changing progressively this picture. With a better assessment of the risks and damages of climate change, the problem will probably need to be reformulated in terms of the optimal provision of a public good. Indeed, climate change is a public "bad" (or carbon abatement is public good) which affects all agents irrespective of their level of consumption of carbon-intensive goods or their efforts to decrease carbon emissions[2]. This implies that the basic first-best principles for efficiency no longer apply here. This argument was put forward in the first place by Chichilnisky (1994) and Chichilnisky and Heal (1994) in the context of the global warming issue.

It is well known that the optimal provision of a public good typically would imply a non-uniform price across agents. For example, the so-called Lindhal solution for this problem would require a system of differentiated taxes set according to the preferences of each agent. Such a system is difficult to implement because preferences have to be revealed before designing the tax schedule, which obviously raises all kinds of free-ridding incentives and moral hazard issues. However, providing that information is known an optimal tax system could be constructed.

---

[2] In addition to the public good nature of global carbon abatement, there is increasing evidence that the effects of global warming are differentiated across regions. This issue would complicate further the problem, but would not invalidate the public good argument. Even if the impact of climate is different across regions and, actually some countries may benefit from it, there is no way a country can influence alone the level of concentrations of CO_2 in the atmosphere.

The problem of the optimal carbon abatement is even more complicated. Indeed, contrary to the standard case where a public good can be produced in a *centralised* way by say, a central planner, carbon emissions are a public bad that is produced in a *decentralised* way by the consumption activities of all economic agents. As shown in the next section, this feature implies that the optimal solution depends not only on the specific preferences of each agent towards the public good, but also on the *weights* that, in a world welfare function will be attributed to the preferences of each country. This means that efficiency and equity can no longer be separated. The underlying intuition is rather simple to explain. Suppose that each agent has in his or her utility function a public good, which in turn depends on the consumption activities of all other agents. Then, each agent will also have the level of consumption of all other agents in his/her utility function. Instead of maximising its own consumption under a certain set of constraints, each region will then have to care about the level consumption of all other regions. This is perhaps one of the characteristics that makes climate change such a deeply *global* issue.

This note aims at clarifying the analytical issues under what conditions the equalisation of marginal abatement costs across regions is not a necessary condition for achieving a Pareto efficient allocation of scarce world resources and in what sense equity and efficiency issues cannot be separated. The consequences of these results for policy are briefly discussed.

2. Optimal Carbon Abatement

Ideally, the level of global carbon emissions should be set at the (global) welfare optimising level. This means that each country and the world community as whole, should determine the effective damages of climate change and in this way establish a balance between costs and benefits of a policy action aiming to reduce the risk of climate change. Considering the abatement externality directly in the utility function means that carbon emissions can be viewed as a public "bad" that is produced in a decentralised way by private consumption activities.

Along these lines, suppose that in each region there are two-goods: a carbon-free good C_k and a carbon-based good F_k, say, fossil-fuels that generates carbon emissions E_k when it is consumed. Defining an objective function having global abatement (A) as an argument, implies that in a given country i, the utility function depends on the level of consumption of the carbon-based good in all the other countries, as follows:

$$U_i(C_i, F_i; A) \quad \text{with} \quad A = \sum_j E_j = \sum_j h_j(F_j) = E_W \tag{1}$$

where function $h(F)$ represents the emission generation function associated with fossil-fuel consumption (with $h' > 0$). A Pareto optimum can be obtained by maximising the utility of each country subject to the constraints that there will be no utility losses for any other country, and their *specific* production frontiers:

$$\begin{aligned} &\text{Max} \quad U_i\left[C_i, F_i; A(F_1, F_2, \ldots, F_n)\right] \\ &\text{s.t.} \quad U_k(C_k, F_k; A) \leq \overline{U}_k \qquad \text{for } k \neq i \\ &\qquad\;\; g_k(C_k, F_k) = 0 \qquad \text{for } k = 1, \ldots, n \end{aligned} \tag{2}$$

where $g_k(.)$ represents the country-specific production frontier. Differentiating the corresponding Lagrangian with respect to all C_i and F_i, the first-order conditions for a given country i are:

$$\mu_k \cdot \frac{\partial U_k}{\partial C_k} = \theta_k \cdot \frac{\partial g_k}{\partial C_k} = p_k^C \tag{3}$$

$$\mu_k \cdot \frac{\partial U_k}{\partial F_k} = (\theta_k \cdot \frac{\partial g_k}{\partial F_k}) - \left[\sum_j \mu_j \cdot \frac{\partial U_j}{\partial A}\right] \cdot h'_k = p_k^F + t_k^F \tag{4}$$

$$\text{for } \; k = 1, \ldots, n \quad \text{and} \quad \mu_i = 1$$

where θ_k and μ_k are respectively the Lagrange multipliers associated with the resource and the Pareto constraint, respectively. Equation (3) says that the marginal social valuation of F is equal to the competitive market price[3] (p^F) plus a term reflecting the valuation of the emission externality. The latter can be interpreted as the excise tax on fossil-fuels (t^F) needed to bring the private cost of F to its social cost. Noting that, by definition, one must have:

$$t_k^F \cdot dF_k = t_k^E \cdot dE_k \tag{5}$$

where t^E is the excise tax to be levied on carbon emissions. The carbon tax in each country will then be equal to:

3 The competitive price of each good is equal to the shadow-price of the resource constraint times the opportunity cost of production (see, Varian, 1994).

$$t_k^E = t_k^F \cdot \frac{1}{h_k'} = -\left[\sum_j \mu_j \cdot \frac{\partial U_j}{\partial A}\right] = T \tag{6}$$

This implies that tax the on carbon emissions will be the same in each country (*T*) and correspond to a weighted sum of each country's marginal abatement utility[4]. This is an important point to be noted, as some confusion has emerged in the literature on whether or not an optimal solution would require a system of differentiated taxes. However, while all countries face the same carbon tax, marginal abatement costs (or the marginal utility of abatement, i.e. $\partial U_k / \partial A_k$) are not necessarily equalised across countries. As it could be expected, this is different from the standard (first-best) case where a uniform carbon tax also equalises the marginal abatement costs across countries.

A second important implication of the optimality conditions (4), is that Pareto efficiency also requires:

$$\frac{\mu_k \cdot \frac{\partial U_k}{\partial F_k} - (\theta_k \cdot \frac{\partial g_k}{\partial F_k})}{h_k'} = T \qquad \forall k \tag{7}$$

Given that countries may differ by their marginal preferences towards the carbon-based good (*F*) and the characteristics of production conditions, the above relation implies that not all combinations of utility weights will lead to Pareto efficiency. Actually, only one combination of welfare weights will be compatible with the requirement that relation (7) must be verified for all countries[5].

Indeed, it is straightforward to see that equations (6)-(7) are a set of $(n+1)$ linear relations enabling to determine *jointly* the optimal carbon tax T and n multipliers (μ_k, $k=1,\dots,n$). In the case that the set of welfare weights could be related to the initial allocations of permits in a system of tradeable permits, this would have a quite strong implication. It would imply that only *one* initial allocation of permits could be Pareto efficient. In this sense, equity and efficiency are not longer separable.

---

4 This correspond to the usual solution of the optimal tax with externalities (see Baumol and Oates, 1988).

5 Only in the case of a global production frontier, the optimal conditions will not depend from these welfare weights (see Oliveira Martins and Sturm, 1998). The latter is also equivalent to the possibility of unlimited lump-sum transfers among countries (see Chichilnisky and Heal, 1994).

3. Concluding Remarks

From the above, two main options can be envisaged. A *first* road, is to view the problem of finding an optimal level of global carbon abatement as an exceedingly ambitious objective, given the uncertainties concerning the net benefits from stabilising the global climate and income valuations across countries. In that case, the problem is to define a given level of emission reduction and seek the less costly way to achieve this constraint. Uniform carbon taxes or a system of tradeable permits under any feasible initial allocation of permits are efficient ways to achieve the objective. However, if the climate change damages are not really taken into account, the optimal carbon tax could be virtually zero as well as the level of global emission reductions. As an aside, this is probably the *ex-post* rationalisation for the present outcome of the international negotiations to reduce the risks of climate change.

A *second* road is to accept the public good nature of the global warming problem. In that case, the international negotiations should tackle jointly the equity and efficiency issues. Once the solution for the equity problem and the optimal carbon tax are found, the optimal level of global carbon emission reductions can also be found. In practical terms, this would enable to design a system of tradeable permits involving, at least, the major carbon emitters.

The present state of the negotiations can be interpreted as the search of a compromise between these two roads. Clearly, the first road is being rejected by developing countries on the grounds that it not satisfies some basic equity considerations. Developing countries do not seem keen to accept a deal where they are just compensated for the emission reductions done in place of the industrialised countries because it is cheaper to do it in that way. The implications of the second road are that the historical responsibility for past emissions being mainly attributable to industrialised countries, the latter would have to accept either a level of costs or financial transfers to developing countries that could be seen as properly unacceptable by governments and the public opinion. No easy negotiation path can be expected. The damages related to climate change could also become so forceful that there is no other alternative but a global co-operative and co-ordinated solution.

References

Baumol, W. J. and W. E. Oates (1988) *The Theory of Environmental Policy*, Cambridge University Press.

Chichilnisky, G. (1994)"*Commentary on* Implementing a Global Abatement Policy: The Role of Transfers", presented at *OECD/IEA Conference on the Economics of Climate Change*, Paris, OECD.

Chichilnisky, G. and G. Heal (1994) "Who Should Abate Carbon Emissions ? An international view point", *Economic Letters* , (Spring).

OECD (1995) *Global warming: Economic dimensions and Policy responses*, Paris.

Oliveira Martins, J. and P. Sturm (1998) "Efficiency and distribution in computable models of carbon abatement", *OECD Economic Department Working Papers*, no. 19x, (forthcoming).

Varian, H. R. (1984) *Microeconomic Analysis*, W. W. Norton & Company.

POSSIBILITIES OF AN EU-SYSTEM OF CO_2 EMISSION PERMITS

Dr. PAUL KOUTSTAAL
Nederlands Energy Research Foundation
ECN Policy Studies
P.O. Box 1, NL-1755 ZG Petten, The Netherlands

1. Introduction

It is now widely recognised that the increase in greenhouse gases in the atmosphere will have consequences for the worldwide climate system, which might lead to higher temperatures and an increasing number of extreme events. In the Framework Convention on Climate Change (FCCC) (UN, 1992), signed at the United Nations Conference on the Environment and Development in Rio de Janeiro in 1992, it was recognised that steps must be taken to curb emissions of the main greenhouse gases. Stabilization of emissions at the 1990 level by the year 2000 is mentioned as a first target of a global climate policy. A number of countries have consequently formulated emission reduction targets and implemented measures to attain these targets, but so far no effective policies have been introduced. One of the instruments which have been introduced in the last years in some countries, notably the Scandinavian countries, is a tax on CO_2 emissions. However, the levels of the introduced taxes are low and in many cases energy-intensive industries are exempted. This illustrates once again that a large and probably decisive obstacle to the introduction of such an economic instrument is the political resistance that any proposal to impose a new tax on a specific interest group calls forth. This has ,for example, been shown by the activities of the Union of Industrial and Employers' Confederation of Europe, a European business lobby based in Brussels, which has been a strong opponent of the carbon/energy tax (SkjÆrseth 1994, p. 28). For that reason it makes sense to look for instruments that do not have the disadvantage of raising the tax burden or changing the existing tax structure but yet leave emitters more flexibility than the direct regulation of emission or carbon use would do. In this paper we shall discuss such an instrument: tradeable carbon permits. Section 2 provides a short survey of the most important properties of tradeable emission permits (TDPs). Section 3 then sketches the outlines of a system of tradeable carbon permits (TCPs). Special attention is given to the design and feasibility of a system of TCPs for the European Union.

J. Hacker and A. Pelchen (eds.), Goals and Economic Instruments for the Achievement of Global Warming Mitigation in Europe, 339–355.

2. Tradeable Permits

Tradeable permits in environmental policy are a relatively new instrument, in theory as well as in actual policy. The idea was developed by J.H. Dales in 1968. The basic concept – the rationing of production and the handing out of coupons to consumers, who are allowed to trade coupons among themselves – has a much longer history. But although the concept is fairly new, the instrument has by now been applied on a number of occasions (see Klaassen 1996 for an overview). A full-blown system of TDPs consists of the following elements:

- on national, or if necessary regional, level the acceptable total release of a pollutant is determined and expressed in a homogeneous unit of measurement, for example tons of carbon dioxide;
- permits that entitle their owner to release pollutants are issued either free or in exchange for payments. The total of pollution quota distributed in this way equals the pollution ceiling mentioned in the first point;
- the pollution permits can be traded.

It should be noted that permission to pollute without payment is usually a part of existing environmental policies in developed countries. The innovative element is the possibility to transfer the entitlement to pollute.[1] In principle the tradeable pollution permit is an attractive instrument: it is effective, efficient and stimulates the development of cleaner technologies. Furthermore, by giving out permits free to pollution sources the excess burden that is typical of (pollution) charges can be avoided.

Tradeable permits are effective because the number of units of released pollutants they represent is limited and is determined by environmental policy targets. Consequently, the total amount of pollutants emitted cannot increase. Individual sources may increase their emissions and new sources may be established, but this has to be compensated by reductions in released pollutants elsewhere. The total level of emissions permitted can be reduced in the course of time as environmental necessity dictates.

The efficiency of tradeable permits arises from the ability to trade permits. Those who can reduce emissions at low cost will do so and sell their permits to emitters who could reduce emissions only at very high costs. Consequently, the opportunity to trade permits opens up the possibility to reallocate emissions and emission reduction in such a way that the total costs of emission reduction are

[1] A second important difference is that no additional permits are made available for new sources, as is the practice under direct regulation.

minimized. In a perfect market, trade would take place and reallocate pollution abatement in such a way that all sources reduced their emissions at equal marginal costs; total costs would be at a minimum and the reduction of emissions would be allocated efficiently (see Tietenberg 1988, ch. 14).

Emitters are obliged to obtain permits for every ton of a regulated pollutant that they emit. Since the permits have a price (even if they are handed out free, they have an opportunity cost), there is an incentive to search opportunities to reduce emissions and to invest in the research and development of new, cleaner technologies. In other words, tradeable permits are a dynamically efficient instrument (Downing and White 1986; Nentjes and Wiersma 1988).

The last, and certainly not the least, attractive feature of tradeable permits (from the viewpoint of existing emitters) is the possibility of distributing permits free to the emitters. This form of distribution is known as grandfathering. As a basis the environmental authority can take the 'historical rights' of established polluters: existing sources receive a number of permits relating to a given percentage of their emissions in a reference year. Compared with a system of pollution charges, polluters can make considerable savings, since the individual source need pay only for any additional permits required, whereas under the charge system a price must be paid for every unit of pollution released. Taken as a group, permit-holders who receive permits free will only have to bear the abatement costs. Compare this with the cost impact of a charge. If within a given period the emissions of CO_2 are to be reduced by 10 or 20 per cent only, the expenditure on charges for the residual emissions of CO_2 would be a multiple of the abatement costs. Even if the increase in tax revenue for the government were returned to taxpayers in the form of a lower rate for other taxes, it would not be possible to perform such an operation neutrally from a distributional point of view (see, for example, Pearson 1992). Those who would benefit from tax reductions would not fully coincide with those who bear the charge. Consequently, the resistance of industry, especially of the pollution-intensive sectors, can be overcome more easily with a system of tradeable permits with grandfathering than with a charge.

Tradeable pollution permits are a suitable instrument for reducing several forms of pollution as long as the market for pollution permits works well. The conditions are the usual ones for developed markets, such as sufficiently large numbers of buyers and sellers in order to induce 'workable competition' (Hahn 1984), certainty of entitlements, and frequent transactions (which implies reasonably low transaction costs). Competition (anti-trust) policy would apply to permit markets just as to other markets.

Another important question with regard to the permit is whether grandfathering can create barriers to entry for new firms. This might be an important issue, especially in a system which covers the whole economy, as is the case with the tradeable carbon permit scheme discussed below.

A last point concerns the grandfathering of permits. Usually, the number of permits an emitter receives is based on emissions in a reference year. Therefore, the greater the emissions in the past, the more permits are received. This favours emitters who have done the least to diminish their pollution. One way to overcome this injustice is to limit the total number of permits an emitter can receive by choosing as a point of reference, not the actual emissions in a reference year, but the emissions that would have resulted if the firm had complied with a given (minimal) emission standard.

3. Outline and feasibility of a system of tradeable carbon permits

In this section, a system of tradeable carbon permits is described. In particular, the question is whether and how such a system would work in the context of international common markets like the European Community. Attention is given to (a) the definition of the permits; (b) the issue of permits over time; (c) the initial distribution of permits; (d) the permit market; (e) compliance with the system; and (f) the EU dimension of TDPs.

3.1. DEFINITION OF THE PERMITS

For the time being fuel saving and fuel substitution are the major and almost the only economically feasible options for reducing emissions of CO_2. For that reason and also for reasons of administrative efficiency and enforcement it makes sense to implement a policy of restricting CO_2 emissions by the use of tradeable carbon permits. The use of carbon contained in fuel that is allowed in total can be calculated from the CO_2 emission targets of the government. On this base a limited number of tradeable carbon permits is issued. A carbon permit is equivalent to 1 ton of carbon: one carbon permit allows the use of a quantity of fossil fuels which contains 1 ton of carbon. The permits are not limited in any way as regards the period or the place where they can be used. This property arises from the fact that the greenhouse effect is a consequence of the accumulation of gases like CO_2 in the atmosphere and is independent of the place where CO_2 is emitted. Since the carbon permit can be used at any unspecified date, it will retain its validity until the moment it is 'used up'; that is to say until the time the carbon is released to the atmosphere. Consequently, permits are a homogeneous good that can be easily traded, at low transaction cost, among a nationwide or even larger public of potential carbon users. The importance of these properties is illustrated by the experiences with the EPA emission trading programme: trading was restricted to the geographical area in which the permits originated and every single deal had to be approved by the authorities. Transactions costs were high and the future value of permits was uncertain. These limitations have seriously restricted the number of trades and, by the same token, also the efficiency gains.

When permits are grandfathered, firms receive a number of permits each year free (a number which will decrease when the overall emission limit is reduced). The right to receive these gratis permits during an indefinite number of future years may be termed a quota. In addition to trading permits, firms can also trade quotas. For example, a firm which stops producing can sell its right to receive a number of free permits to another firm, which consequently is assured of a supply of free permits each year.

3.2. ISSUE OF THE PERMITS

Fossil fuels are an essential resource to keep the economy going. Therefore a steady supply at a reasonably stable or steadily changing price is a necessary condition for economic stability. A system of tradeable carbon permits comes very close to a system of fuel rationing. Such a system must have enough flexibility to allow the economy to adjust smoothly to changing circumstances. A system which rigorously limits the number of permits available in each single year can lead to large price variations, with negative consequences for the economy.

Therefore, special care should be exercised to avoid unnecessary bottlenecks caused by a temporary lack of permits. One of the ways to increase flexibility in the supply is to maintain a permanent stock of permits which can be drawn upon, for example, in an extremely harsh winter which drives up fuel consumption. Such a permanent stock can be created when the system is launched. Instead of permits being issued for only one year, they could cover expected use for four or five years. During the first years this stock of permits would be adequate to meet exceptional demand variations. The permits intended to cover the next period could be issued in advance in order to maintain a reserve stock of permits. For example, permits for the second five-year period could be issued at the start of the last year of the first five-year period. Such mechanisms would ensure that there are always permits available to meet changes in demand due to exceptional circumstances. It should be noted that such a system does not mean that the number of available permits exceeds the emission limit. The reserve is created exclusively through the timepath used for issuing the permits.

In addition, it is important that the permit system allows the authorities some flexibility in setting its future emission targets, because the problem of global warming is beset with uncertainties. Care must therefore be taken in issuing permits to avoid committing policy to a specific emission limit for a long period. However, this might conflict with the requirement for the supply of permits to be known in advance for a sufficient number of years. Given a known supply of permits, economic subjects can anticipate future demand and therefore form expectations about the development of the permit price. This is important not only for the development of a well-functioning permit market, but also for firms which have to make long-term investment decisions in which the permit price is a factor. For example, investments in the electricity-generating sector will be influenced by the current and future permit price. As these investments are made for periods of up to 20 years, it is important that there is some idea about the future price of permits. In order to reduce uncertainty for fuel users and at the same time to allow the government some flexibility with regard to future emission targets, a scheme can be used in which the government's emission targets are set for a certain period, during which they should not be changed. The emission limit for subsequent years need not be precisely specified. Instead, the government could announce an upper and lower limit for its planned distribution of permits, with a gradually increasing gap between the two for the more distant future. Within these margins, the authorities have room to set the exact number of permits made available, taking into account new insights into the enhanced greenhouse effect. The exact number of permits to be issued must be announced sufficiently in advance of the year in which they are distributed to assure a well-functioning market.

In addition to these schemes, the development of a forward market in tradeable carbon permits will add further opportunities for risk-averse fuel users to shift uncertainties to those who are willing to bear them.

3.3. DISTRIBUTION OF THE PERMITS

In a system of tradeable carbon permits, both grandfathering and auctioning can be used side by side, according to political expediency. Since fuel-intensive industries in particular would have to incur large expenditures if they had to obtain carbon permits in auctions it can be politically expedient to hand out permits free to firms in this category. A practical dividing line for the Netherlands would be between, on the one hand, industry, horticulture and possibly freight transport as sectors which are fuel-intensive and for that reason benefit from grandfathering and, on the other hand, consumer households, services and (personal) transport as sectors which fall under the auction regime.

The objective of grandfathering permits to energy-intensive industries is to exempt them from the additional financial expenditure of buying permits for their full fuel use. For example, reducing CO_2 emissions in the Netherlands by 10 per cent from the 2015 level would require a charge which would raise 33 billion Dutch guilders (about 16 billion ECUs, 1 ECU = 2.09 Dutch guilders). The abatement costs are only 2.3 billion Dutch guilders, or 7 per cent of the revenue of the charge (Koutstaal 1992). According to calculations of the Dutch Central Planning Bureau (CPB 1992), reducing emissions in 2015 by about 10 per cent by means of a unilateral tax on fossil fuels would result in energy-intensive industry being wiped out almost completely in the Netherlands. The proposed system of grandfathering tradeable permits would cost industry only a fraction compared to a charge. Firms would still bear opportunity costs for the permits grandfathered but their total expenditure would be far less.

The carbon permits needed for the emissions of CO_2 by the less energy-intensive sectors and consumer households are auctioned by the government. It would not be efficient for consumer households and small enterprises in the service sector to have to buy the permits at the auction themselves because transaction costs would be huge. The alternative is for distributors of fossil fuels, such as gas distribution and oil companies, to buy permits at auction. Subsequently, they can sell fossil fuels to customers, putting a mark-up on the fuel price which is equal to the price of the permits. This will motivate small fuel users to reduce their consumption (or to switch from fuels with a high carbon content, like coal, to fuels with a low carbon content, like natural gas).

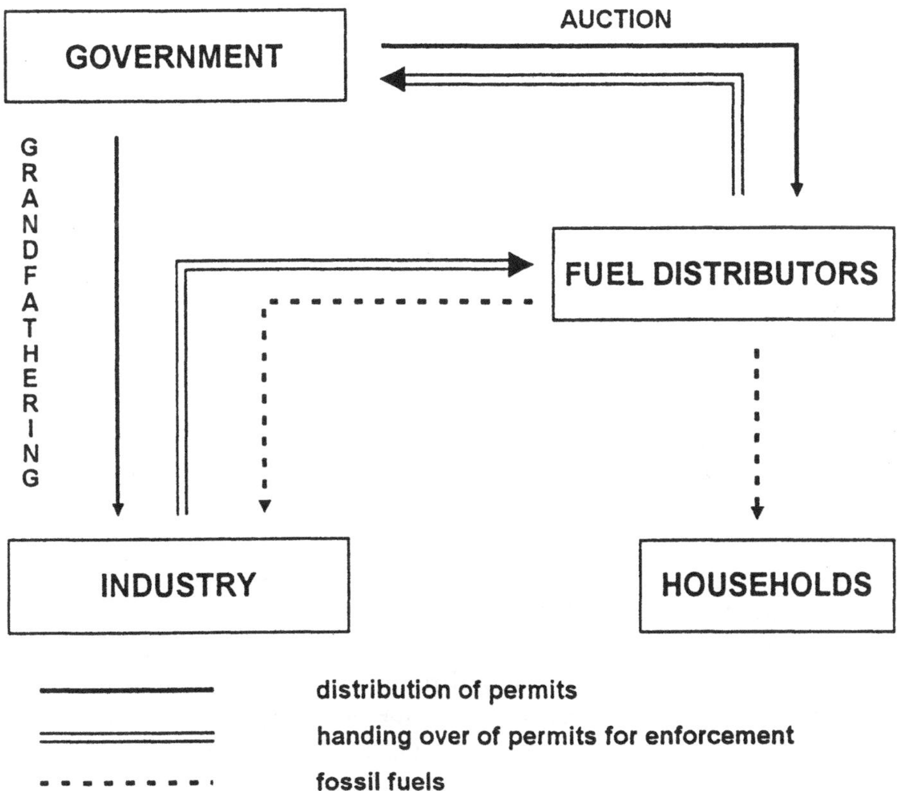

FIGURE 1. Operation of the system of tradeable permits.

It should be noted that in such a system it would not be consistent to hand out permits free to the distributors because they would be able to collect the scarcity rent without incurring any costs for reducing carbon use, because only their customers can reduce CO_2 emissions.

FIGURE 1 shows schematically the way in which the system functions. Permits are distributed by the government, both through grandfathering (to industry, horticulture and transport) and by auction (to distributors of fossil fuels). Distributors deliver fossil fuels without a mark-up on the fuel price to those who have already acquired permits themselves through grandfathering (for example, industry); the buyer pays the price and transfers permits equal to the carbon content of the fuel to the distributor. The other trade channel for distributors is to buy permits at auction or from other firms. Those who have not acquired permits in one way or another (for example, consumer households) buy fossil fuels from distributors with a mark-up on the price.

3.4. PERMIT MARKET

In the system outlined above two markets can be discerned. First, there is the auction of some of the permits. This can be called the primary market. In addition, firms can trade permits between themselves on the secondary market. In theory, the permit price will be the same on both markets; otherwise arbitrage would occur which would equalize prices.

An important condition for a system of TDPs is the development of a well-functioning (secondary) market. A well-functioning market implies that permits are traded in sufficiently large numbers to facilitate stable price-making. Whether this will be the case in the system of TCPs will depend on the number of potential actors in the market and on the supply of and demand for permits. TABLE 1 provides data for the Dutch economy on emissions per sector, the number of firms per sector and average emissions per firm. Grandfathering permits to industry, refineries, transport and horticulture implies that about 50 per cent of the permits are grandfathered. The other 50 per cent would be sold at the auction, guaranteeing a large supply on the primary market.

The number of potential actors on the secondary market is large. In industry alone there are more than 45,000 sources and in addition energy suppliers such as gas distribution companies and power generators can also be expected to trade.[2] However, presumably not all industrial firms will trade actively on the market. Transaction costs might be relatively large for smaller firms and for those whose energy costs are only a small fraction of production costs. Instead, these firms can

---

[2] In a European-wide system of tradeable carbon permits the number of (potential) actors on the market would be much larger.

arrange with their suppliers of fossil fuels to supply them with the fuels they need and to acquire the permits which these firms might need in addition to those they received through grandfathering (and presumably pass on the price). The suppliers might also get a mandate to sell permits which firms do not need. In essence, suppliers would take on the role of a broker, bringing together supply of and demand for permits.

TABLE 1. CO_2-emissions, Dutch economy, 1989.

sector	CO_2 emissions [mln. ton]	number of sources	average emissions [1000 ton/source]
Food and beverages	3.9	7258	5
Textiles	0.3	1515	2
Paper	1.5	362	41
Fertilizer	6.6	12	5500
Other chemicals	3.8	838	44
Building materials	2.2	684	32
Base metal	8.8	119	740
Other metal	1.8	14563	1
Other industries	0.5	21358	< 1
Refineries	12.1	7	17285
Power companies	36.8	88	4182
Transport	26.4		
Households	19.3		
Horticulture	6.4		
Others	11.3		
Total	141.7		

Source: Koutstaal 1992, p.74.

The potential for trade depends not only on the number of actors but also on the supply of and demand for permits and relates to differences in abatement costs. The larger the differences, the larger is the potential for trade because the gain from trade increases when cost differences increase. Studies indicate that the costs of reducing CO_2 emissions differ considerably between various sectors (see Velthuijsen 1995; Blok *et al.* 1990). Consequently, there are sufficient incentives for trade.

A last point concerning the permit market is the form which the secondary market will take. Ideally, permits are traded on an exchange similar to stock markets or exchanges for raw materials or agriculture products. However, this will only be possible if the volume of trade is large enough. Given the potential for trade and the number of actors, such an exchange might develop in time in the TCP system. Until then, trade can take place through brokers, a role which might be filled by suppliers of fossil fuels. In the CAAA sulphur-trading scheme, brokers facilitate trade on the secondary market (Klaassen and Nentjes 1995). In the EU tradeable milk quota programme, a number of firms of solicitors which were already active in agriculture have specialized in milk quota brokerage (personal interview, Scottish Agricultural College dairy test farm 1993).

3.5. MONITORING, ENFORCEMENT AND ADMINISTRATIVE COSTS

As has been shown, distributors acquire permits for the fossil fuels they sell to consumer households and other small CO_2 emitters. The permits industry receives through grandfathering can be handed over to the distributors in exchange for the carbon contained in the fossil fuels they buy from the distributors (see FIGURE 1). In this way, all permits will end up in the hands of the distributors, who to a large extent are the same as the producers and importers of fossil fuels. Those distributors who do not produce or import fossil fuels themselves can in turn hand the permits over to their suppliers. In the end, all permits will land up with the producers and importers. This property of the system can be used to set up an efficient system of monitoring and enforcement. Producers and importers of fuel are placed under the obligation once a year to turn over to the environmental authorities carbon permits for the carbon contained in the fossil fuels they have sold on the market (see FIGURE 1). They have either received the permits from their clients or bought them at auction.

The advantage of supervising compliance with the tradeable permit system in this way is that it fits in with existing institutions for levying excise duties on fossil fuels, which exist in most western countries. In the Netherlands, traders and suppliers of mineral oils must have a licence. They are obliged to report each month how much they have supplied to the market and turn over the excise tax to the authorities. This administrative system of self-reporting is supplemented by occasional physical checks. The system operates satisfactorily (Dutch Parliament 1991, nr. 21368, p. 21). Instead of handing over the excise, suppliers and producers of fossil fuels hand over permits, as described above. In addition, suppliers of other fossil fuels, such as natural gas and coal, should be brought into the system.

In addition to the administrative monitoring system, other sources of information could be used for double checking. Victor (1991) mentions four other sources of information:

- *direct monitoring of emissions* This might be an option for some large stationary sources. However, it is not practical for smaller and mobile sources.
- *data from other reports by either the source itself or from third parties* For example, data reported to the fiscal authorities from buyers of fossil fuels.
- *data from environmental annual reports* Reports of this kind, which are sometimes verified by accountants, are published by an increasing number of companies.
- *data generated by modelling exercises* Starting from known input data, estimates can be made of emissions. These data are probably not very accurate, but they might serve to detect large-scale frauds.

One way of direct monitoring is by the use of tamper-proof metering devices fixed to machinery and pipelines (Cnossen and Vollebergh 1992). Using the excise system will guarantee a high level of compliance and will make a system of TCPs just as feasible as a carbon charge as far as compliance is concerned. Only the limited number of firms producing or importing fossil fuels has to be checked. In the Netherlands, for example, there are between 40 and 50 such firms.

Not only must there be an effective system of monitoring, enforcement must also be assured. For effective compliance two elements are of importance. First, sanctions should be such that the expected costs of fraud exceed the costs of sticking to the rules.[3] Second, the environment should not suffer from fraud. Consequently, in addition to a fine, firms which have failed to supply permits for the fossil fuels they have used should be forced to acquire permits to cover these emissions (this is the approach taken in the tradeable sulphur allowance scheme; see Kete 1992).

The administrative costs of the system of TCPs will consist of several elements. The two main cost factors are the costs of monitoring and enforcement and the costs of registration of the ownership of permits. Furthermore, data will have to be collected for determining the quota which firms receive when permits are grandfathered. This will also entail costs which, however, only have to be incurred once. Last, a yearly auction will have to be set up. The first cost factor, monitoring and enforcement, will be comparable with those of implementing a charge on carbon in fossil fuels. They will not differ very much from the costs of the current system for levying excise duties on fossil fuels. In Allers (1994, pp. 71–75) the administrative costs of existing environmental taxes in the Netherlands are given. The environmental taxes which are levied under the WABM law *(Wet algemene bepalingen milieuhygiëne)* consists mainly of a tax on fossil fuels. In 1990, the administrative

---

[3] The expected costs of fraud equal the chance that fraud is discovered times the level of the sanctions.

costs ran to 40.5 million guilders, about 20 million ECU. Compliance costs of this environmental tax are difficult to ascertain due to the fact that only data on compliance costs for all environmental taxes in the Netherlands are available (Allers 1994, p. 179).

Registration of the ownership and trade in permits is characteristic of TDPs; no comparable arrangement is necessary when a charge is levied or standards are applied. The most efficient way to register ownership is probably to use a giro system comparable to those used to register ownership of and trade in shares and certificates traded on stock exchanges. In the Netherlands, NECIGEF (the Dutch central institution for trade in shares by giro) registers the ownership of shares. Yearly operating costs of the institute were 9.5 million guilders in 1990 (4.6 million ECU) (NECIGEF 1990, p. 29). The daily average number of trades registered in that year was 4,532. It is not to be expected that trade in CO_2 permits will exceed this number, therefore the costs of this institution serve as an example of the costs of registering the carbon permits.

A rough estimate of the yearly operating costs of the system of TCPs described here is about 50 to 70 million guilders if implemented in the Netherlands: around 29 million ECU. In addition costs must be incurred to grandfather permits (which, however, will only happen at the start of the system) and an auction must be set up.

In the next sub-section, compliance at the European level in a EU-wide system of TCPs will be addressed.

3.6. EU DIMENSION OF TRADEABLE CARBON PERMITS

An important question is whether a system of TCPs would have to be confined by national boundaries. The Member States (MS) of the EU have delegated part of their power to the European Union. Consequently, the freedom for independent policy is restricted by primary legislation (the Treaty of the EU) and secondary legislation (e.g. directives and regulations).[4] A national system of tradeable carbon permits restricts the use of carbon contained in fossil fuels. Consequently, TCPs imply a restriction on the quantity of fuel that can be produced at home and imported from abroad. It might therefore be seen (by the European Commission) as a restriction of the free movement of goods between MS and consequently as a violation of article 30 EG. Art. 30 EG states that trade in goods between MS may not be restricted. In a system of TCPs the import of fossil fuels is restricted by the total amount of carbon permits available and therefore it might be contrary to the article.

However, there are exemptions on article 30 EG (Pisuisse and Teubner 1994,

[4] For an overview of European Law, see Pisuisse and Teubner 1994 and Teubner 1993. The discussion of TCPs in European law is based on personal communications with A.M.M. Teubner (1995).

p. 97). Restrictions of imports are allowed under article 36 EG and under the 'rule of reason'. The rule of reason applies if the following conditions are met:

1. There must be no Community measure regarding the policy area in which the MS wants to implement the policy measure which could restrict free movement of goods.
2. The proposed policy measure must apply to domestic and imported products alike.
3. The proposed policy measure must justify certain interests which are accepted by the Court of Justice. Examples of these interests are environmental protection and consumer protection.
4. The policy measure must be reasonable and is subject to the proportionality test. This means that trade between MS must not be restricted any more than is necessary to achieve the policy purpose. Moreover, there must not be another policy measure which would be as effective (and efficient) but which would not restrict trade to the same extent.

Concerning the first condition, there is currently no substantial greenhouse policy in the European Union. There has been considerable discussion in the last five years on the use of a EU tax to limit carbon dioxide emissions. However, on 15 December 1994, the Council agreed that no Community tax measures will be taken (Europe Documents no. 1918, 6 January 1995). Instead, it was agreed that those MS which wish to do so may introduce a CO_2/energy tax themselves. Such unilateral carbon taxes should follow a Community framework which would be developed in further discussions by ECOFIN (the Council of Finance Ministers). They have asked the Commission to develop a proposal for such a framework, which will probably consist of minimum levels of excise duties on a number of fuels. It is not clear by what time the framework will be developed and implemented. However, it will probably leave room for MS to use other additional policy measures to limit CO_2 emissions like TCPs. In the mean time, it seems plausible to assume that as long as there is no Community policy nor a proposal for a regulation, MS can take their own policy measures.

The tradeable permit scheme described above does not discriminate between fossil fuels sold by national firms and firms from another MS. For all fossil fuels brought on to the market, regardless of their country of origin, carbon permits have to be acquired. The second condition is therefore fulfilled.

Environmental protection has been recognized as an interest which justifies a restriction of free movement of goods, therefore a system of TCPs meets the third condition.

A system of tradeable emission permits is both an efficient and effective instrument to control CO_2 emissions. Other instrument types are either not as effective (for example, a carbon tax) or not as efficient (for example, emission standards). The last condition appears to be met as well.

Another question regarding the possibility of implementing a system of TCPs in one country in the EU is the issue of enforcement. As long as differences in excise taxes are allowed, the administrative system which is used to enforce the system remains in place.

Enforcement might also become more difficult when national frontiers within the EU are abolished. A start has been made in the Schengen agreement, which abolishes border controls between seven countries of the EU. Without such border controls, fraud is more difficult to discover. End-users could import fossil fuels directly from suppliers in other MS without procuring carbon permits and without running the risk of discovery at border controls. However, this problem should not be overstated. Large-scale evasion of the obligation to acquire permits will remain difficult because of the properties of fossil fuels: movement of fossil fuels is by bulk transport which, moreover, needs large installations for unloading. The obligation to acquire permits will be hard to evade for large quantities of fossil fuels because it will be difficult to escape detection.

It can be concluded that it is probably possible to implement the national system of TCPs sketched in this chapter in one MS of the EU. A system of TCPs might conflict with free movement in goods within the EU but it will probably fall under the rule of reason which allows exemptions to article 30 EG. One possible problem is that there will be no room for other instruments once it is decided that MS can introduce a carbon/energy tax unilaterally. However, it is not yet clear if and when such a decision will be taken and how it will be formulated. Enforcement might become more difficult when border controls are abolished. However, fraud on a large scale will still be difficult.

Although TCPs might be realized within one country, it would be preferable to introduce a system across the whole of the EU; it would be more effective to limit emissions in all MS of the EU instead of only one. Moreover, the problem of firms in MS which have implemented TCPs being at a disadvantage *vis-à-vis* firms from MS without CO_2 reduction policies is avoided.

As a first step in the introduction of the system the Council of the EU will have to decide on: a time path for total carbon use within the EU; the sectors that are selected for grandfathering; and the basis on which permits are grandfathered. Another question to be decided at EU level is whether and how the available permits should be distributed among the different MS. Permits can be grandfathered and sold at EU level or they can be allotted to the MS which in turn distribute them. With regard to this last option, it should be realised that MS would not be allowed to use their permits to support specific sectors or firms by grandfathering permits in excess of those allowed by the rules for grandfathering. Such behaviour would be contrary to articles 92 – 94 EG. These articles prohibit governments from supporting sectors if this would reduce competition and obstruct trade between MS. Furthermore, a member state would not be allowed to sell the permits exclusively to firms registered within its own borders, thereby favouring its own industry, as this would be discrimination. Hence, the only rationale for allotting quotas to MS would be as a method of distributing the revenue generated by the auction of (part of) the permits among MS. The other option is to decide on an allocation rule for the revenue and to auction the permits centrally.

When the system is introduced, a distinction must be made between the activities which should be undertaken at the central EU level and those which can be delegated to the MS. For the execution of the various tasks, a Brussels bureau should be set up (or alternatively the European Environmental Agency in Copenhagen could perform the task) as well as a network of national bureaux. The tasks of the national bureaux include:

- registration of the ownership of the permits;
- grandfathering of permits to designated sectors;
- monitoring and enforcement.

One of the tasks of the national bureaux is to set up and operate the above described giro system for registration of the permits. Furthermore, they should implement the rules on the grandfathering of permits to industry. For this purpose, the national bureaux must draw up a list of all firms eligible for grandfathering and issue to them each year their allotted quota.

An important task of the bureaux is to enforce the tradeable carbon scheme. They should collect the permits handed in by importers and producers of fossil fuels registered in their MS. In addition, the national bureaux should make periodical inspections to check whether firms accurately report the amount of fossil fuels they have brought on to the market. The task of the Brussels bureau would be threefold:

- to supervise (the performance of) the national bureaux;
- to act as a clearing house for transactions between permit owners registered at the various national bureaux;
- to evaluate the programme.

The Brussels bureau should supervise the national bureaux on a number of points, the most important of which is enforcement. The Brussels bureau should check with great care whether all the national bureaux enforce the carbon permit scheme equally accurately and collect all the permits due. If some MS do not enforce the system an EU-wide system of tradeable carbon permits would suffer from lacunae which would detract from its effectiveness (and efficiency). This would also be true of any other instrument (charges or regulation) that had to be applied under such awkward conditions, but there is one difference: under TCPs, firms which operate in a MS without adequate enforcement can emit carbon dioxide without handing over permits. Consequently, they can sell their permits to firms in other MS and, as a result, pollution would increase in these other MS. When taxes or regulation are used, firms which defraud cannot sell permits to sources in other countries. With these instruments pollution only increases above the allowed level in the country in which enforcement is inadequate. In a European system of TCPs insufficient monitoring and enforcement in one or more MS will lead to higher overall pollution levels compared with instruments like charges or regulation.

As has been described above, MS are not allowed to favour specific firms or sectors by allocating them more permits than is allowed under the grandfather rules. The implementation of these rules should therefore be monitored at the European level. This task could be delegated to Directorate General IV of the European Commission, which deals with competition. Another field for supervision is competition between firms: firms would not be allowed to use carbon permits to limit competition. This task can also be undertaken by DG IV.

The evaluation of the programme can be made in the form of an annual report. Subjects to be dealt with are, among others: the number of permits issued and permits used; the volume of trade; and the occurrence of fraud.

4. CONCLUSIONS

Tradeable permits are an instrument of pollution control that can be used to tackle a large class of pollution problems. It is particularly suitable if the problem is created by emissions from a large number of sources and the pollutant is spread more or less evenly throughout the environment. This is the case with emissions of CO_2; the greenhouse effect occurs worldwide and the pollution sources range from large stationary sources like power plants to small mobile ones like cars. With such a large number of different sources, it will be impossible to reduce emissions in a cost-effective way by means of direct regulation. The instrument of tradeable permits has the important advantage that emission reduction will be cost-effective.

TDPs have other attractive features. The instrument is effective in the sense that emission targets are realized: the total amount of polluting emissions is limited by the number of permits issued. With regulation and taxes, the level of emissions can increase, although it has not been planned: for example, as a consequence of economic growth or sectoral shifts. Furthermore, permits can be grandfathered to polluters. This will considerably reduce their costs compared with the auction of permits or with emission charges; they will only have to bear the abatement cost. Especially in the energy-intensive sectors of industry, expenditure on permits would be several times larger than abatement costs. Therefore, tradeable carbon permits would be politically more acceptable than a tax.

In this paper we have studied in detail how a system of tradeable carbon permits should be designed. Elements on which the analysis has focused are: the definition and the issue of permits; the distribution of permits; the permit market; monitoring and enforcement; and the specific characteristics of an EU-wide system of TCPs. As regards distribution of the permits, it seems most practical to grandfather permits to industrial sources and to sell the permits which cover the other emissions. As it is not practical to compel consumers to buy (and trade in) permits themselves, a government agency can sell these permits to their suppliers of fossil fuels, who can subsequently mark up their prices by the price of the permits.

Compliance in a national system of TCPs does not pose greater problems than compliance with a carbon tax. Under both instruments producers and importers of fossil fuels can be obliged to hand over either the tax payments or the permits for the carbon contained in the fossil fuels which they bring on to the market. The existing mechanism for levying excise duties on fossil fuels can be used to levy the carbon tax or carbon permits. In addition, other information sources can be used to supplement monitoring, such as periodical checks and data on other taxes. The administrative costs do not have to be excessive as long as a giro system is used for the registration of trades and the ownership of permits.

So far there is no concerted EU policy on carbon dioxide reduction. There has been much discussion about introducing a European carbon/energy tax but in the end it was decided to leave it to individual countries whether to introduce a tax or not. In the absence of a European policy, it seems possible for one MS to introduce a national system of TCPs. Although a national system might be seen as a restriction on the free movement of goods within the EU, it will probably fall under the 'rule of reason' which allows exemptions to article 30 EG which deals with the free movement of goods. The disappearance of border controls within the EU might make enforcement more difficult, but this does not seem to be a large problem, due to the bulk character of fossil fuels.

Although a system of TCPs can probably be introduced in one MS, it would be preferable to implement it at the European level. The instrument would be much more effective and there would be no consequences for the competitiveness of firms in different MS. Within a European system of TCPs all MS would have to use the same grandfathering rules. They would not be allowed to use the permits to support specific sectors or firms. Special care must be given to enforcement within an EU-wide system. TCPs would not be effective (or efficient) if firms could evade the system in one or more MS. Although a system of TCPs might be just as sensitive to fraud as other instruments such as standards and taxes, the consequences for the overall pollution level are greater under TCPs because firms which can evade the obligation to hand over permits can sell them to firms in other countries. As a result, pollution would increase not only in the MS where enforcement is not adequate but also in the other MS.

LAUNCHING A PLURILATERAL GREENHOUSE GAS EMISSIONS TRADING SYSTEM - THE UNCTAD-EARTH COUNCIL INITIATIVE*

FRANK T. JOSHUA
Greenhouse Gas Emissions Trading
Division on Globalization & Development Strategies
United Nations Conference on Trade and Development (UNCTAD)
Palais des Nations, CH-1211 Geneva 10, Switzerland

* This contribution is an updated version of the presentation to the Advanced Study Course, taking into account the development of the second half of 1997 especially the results of the Climate Change Conference in Kyoto. This contribution is dated 14th January 1998.

1. Aims and Objectives of the Policy Forum

In June 1997, UNCTAD and the Earth Council, supported by Centre Financial Products Limited, established the Greenhouse Gas Emissions Trading Policy Forum. The aim of the Policy Forum is to provide timely support to interested governments, corporations, and non-governmental organizations in their efforts to design and implement a plurilateral international greenhouse gas emissions trading system, in accordance with the Kyoto Protocol to the United Nations Framework Convention on Climate Change. The goal of the Forum is to launch a plurilateral market for trading in greenhouse gas emission allowances and reduction credits by the year 2000, thus contributing to the early and effective implementation of the Kyoto Protocol.

The Policy Forum is dedicated to facilitating a dialogue among a core group of government policy makers, corporate executives, and leaders of non-governmental organizations for the purpose of identifying feasible steps to implement the trading market. This includes (a) assisting the Parties to the Kyoto Protocol in their efforts to establish a comprehensive regulatory framework for emissions trading (including with respect to defining the tradeable commodity, accounting, monitoring, certification, reporting, non-compliance, and enforcement); and (b) assisting national authorities and market makers in their efforts to develop efficient trading rules, trading instruments and supporting institutions.

This initiative is an outgrowth of several years of research by UNCTAD into the

J. Hacker and A. Pelchen (eds.), Goals and Economic Instruments for the Achievement of Global Warming Mitigation in Europe, 357–366.

feasibility of a global greenhouse gas emissions trading market, and the Earth Council's Global Environmental Trading System initiative, launched in 1995. Furthermore, the success of the United States sulfur dioxide allowance trading programme in dramatically reducing SO_2 emissions well ahead of schedule, and at significantly lower cost than had been predicted, provides a proven model that emissions trading can bring early and quantifiable environmental, economic and social benefits.

2. Main Elements of The Work Plan

In line with the above, the Policy Forum's Work Plan focuses on matters which are essential to the design and early implementation of an initial-phase international plurilateral greenhouse gas emissions trading system, and which require close consultation and coordination among participating countries, corporate entities, and other institutions. UNCTAD and the Earth Council, as cosponsors of the Policy Forum, will ensure regular coordination with the Climate Change Secretariat, and other international and corporate trading initiatives including, among others, the work of the OECD and IEA, the World Bank, the International Panel on Climate Change (IPCC), United Nations Environment Programme, United Nations Development Programme, the Global Environment Facility, the North American Commission for Environmental Cooperation, the Center for Clean Air Policy, World Resources Institute, Environmental Defense Fund, British Petroleum, and Resources for the Future.

3. The Policy Framework Working Group

3.1. IMMEDIATE PRIORITIES:

3.1.1. Review and analysis of the Kyoto Protocol and related COP-3 decisions
To ensure a common starting point for all participants, and a policy framework for the work of the Forum, an analysis of the results of Kyoto will be commissioned. The review and analysis will outline the key implications of the Kyoto decisions as related to the international emissions trading project, as well as pull together an overview of what the Subsidiary Bodies to the UN FCCC and other international organizations are being asked to do in relation to emissions trading. The analysis will be kept up to date as negotiations continue and as the development of the market proceeds.

3.1.2. International Participants' Agreement

The Policy Framework Working Group will develop a draft international agreement that outlines the rules to govern the plurilateral trading programme. The agreement would be an instrument of international law established in accordance with the UN Framework Convention on Climate Change. Depending on the overall design, the agreement could be open for signature by both governments and corporate entities. The agreement would include, *inter alia,* the articles required to establish the trading system, technical guidelines on domestic allowance allocation, trading, monitoring, certification, reporting, enforcement, dispute settlement, amending procedures, and rules for admissions of new members. Details of elements to be addressed in the International Participants' Agreement include:

Technical guidelines on domestic allowance allocation practices. How emissions allowances are allocated to emission sources participating in a cap and trade programme is likely be an important issue not only because of the potential environmental, economic and social consequences at the domestic level, but also because of the potential that resource allocation mechanisms have for competitive manouvering at the international level. The Working Group will prepare a set of technical guidelines on the allocation of allowances to domestic emission sources in order to provide a framework within which all countries participating in the plurilateral trading system can operate. The guidelines would assist governments in designing acceptable allowance allocation methodologies and at the same time help to ensure a level playing field at the international level.

Guidelines on trading system expansion (opt-in provisions). A major purpose of establishing a plurilateral trading system is to demonstrate its efficiency in controlling greenhouse gas emissions and thus provide an incentive for other countries, including non-Annex I countries, to participate. To ensure the orderly and timely expansion of the trading system to other participants, the Working Group will develop guidelines for opting in to the market. Every new member would presumably need to meet the requirements of the International Participants Agreement. Included in the guidelines, among others, could be commitments to set, monitor, and enforce emissions limitation caps, to respect free holding and trade in allowances and credits, to implement appropriate domestic legislation for the distribution of allowances to entities within the country, to coordinate trading through the appropriate international institutions and to be subject to the dispute resolution procedures envisaged as part of the trading system.

Technical guidelines on emissions and sinks monitoring. The value of the commodities being traded in the system will be tied to the rigour with which emissions and sinks are monitored. These monitoring functions will likely be carried out domestically in accordance with agreed guidelines. The Working Group will develop technical guidelines that may include requirements for record keeping and reporting by sources and, if included in the Kyoto Protocol, sequestration projects. Also included could be guidelines on data collection, emissions estimation and reporting, taking into account the work of relevant organizations such as the IPCC and the guidance provided by the SBSTA of the UNFCCC.

Technical guidelines for certification of commodities. Pre and post certification systems are essential for the credibility of the inter-national emissions trading market. There can be no approximate standards for certification. Auditing procedures need to be simple and transparent to facilitate the comprehensive application of common practices. The guidelines can be used by both public and private sector entities charged with certification of the commodity.

Technical guidelines on transfers and accounting. To ensure an open and accountable trading system, the Working Group will develop guidelines for Members to transfer allowances among themselves in accordance with agreed accounting principles.

Technical guidelines on reporting. Taking account of already-approved guidelines for Communications by Annex I Parties to the UN Framework Convention on Climate Change, the Working Group will identify any additional reporting requirements that member countries to the plurilateral trading system will need to include in their communications to the international community as well as the reporting requirements they will need as part of their domestic tracking of trades and transactions.

Guidelines on non-compliance and enforcement. The Working Group will develop guidelines to assist governments in dealing with domestic cases of non-compliance with agreed emissions limits. Certain minimum requirements could be articulated for domestic implementation and enforcement by the domestic authority with regard to emissions sources within its territory. For cases in which an emitter exceeds it allowed limit or in cases where the trading rules are violated, a variety of enforcement options exist ranging from allowing the market place to devalue the commodity through to imposition of sanctions such as suspension of trading and monetary fines. Such decisions would be reached in accordance with the decision making procedures associated with the International Participants' Agreement. The guidelines will, however, be helpful in setting certain thresholds for non-compliance before sanctions would be imposed.

3.1.3. *Survey of best practices for domestic legislation*
To assist participating member states in their domestic work related to international emissions trading, the Working Group will commission a survey of best practices for domestic legislation. The survey will address, at a minimum, certification, monitoring, data-gathering and reporting Protocols, public access to information, and rules for transferring and recording ownership rights in allowances or credits. Also covered could be the legal and administrative measures for ensuring compliance by sources with national emissions targets or budgets, in accordance with the international participants' agreement.

3.2. OTHER PRIORITIES

3.2.1. Allowances and credits tracking system
The International Participants' Agreement will outline the need for tracking of trades of allowances and credits throughout the life of the project. It is anticipated that this will be the responsibility of each domestic government. The Working Group will develop guidelines for the establishment of the appropriate system along with software packages designed to facilitate the tracking of both allowance and credit trades. This work will be conducted in close cooperation with the work of the Market Design and Operations Working Group (see below).

3.2.2. Public data systems
To be effective as a learning experience, and to ensure public accountability for the trading system, the Working Group will develop and/or recommend for participating governments the types of data systems that will provide wide access to the public about the trades being conducted within the plurilateral system.

3.2.3. International greenhouse gas development fund
The potential volume of new and additional financial resources which might be associated with the development of an international greenhouse gas emissions trading system appears to be enormous. Various possibilities exist in the design of the market mechanisms to mobilize private financial resources as a new source of development funds for investment in emissions reduction opportunities and related sustainable development in developing countries (including through market auctions, interest on guarantees, financial penalties for non-compliance, voluntary contributions from the private sector, etc.), and generally assist developing countries to contribute to international efforts to deal with climate change. The Working Group will explore the possibility of encouraging and supporting the establishment of an international greenhouse gas development fund, and if necessary, take appropriate steps to ensure its timely establishment.

3.2.4. National Training Programmes
Organizing and conducting national training programmes for participating countries will constitute an essential part of the activities of the Policy Forum. The training programmes will aim to assist both policy makers and market makers to put in place an efficient institutional and regulatory framework and adequate market mechanisms for international emissions trading. The Policy Forum will be responsible for all external aspects of the training programmes, including the preparation of the training modules and a comprehensive training manual, and the provision of trainers and experts. Host countries will be required to provide local support to the training programme including such support services as conference facilities, logistical support for local participants and visiting experts, and the provision of technical equipment where necessary. Priority will be given to the needs of countries with economies in transition and interested developing countries.

3.2.5. Quarterly Newsletter and Publications
As part of the Forum's commitment to an open, transparent, and educative process, quarterly newsletters, proceedings of the Policy Forum and the results of working groups and task forces will be published and disseminated widely.

4. Working group on market design and operations

4.1. IMMEDIATE PRIORITIES

4.1.1. Definition of the Tradeable Commodity
n accurate understanding of the properties of the commodity is the key to the creation of a new tradeable asset. Bearing in mind the guidance from the UN FCCC and the ongoing work of the Convention Secretariat and the work of the Convention's subsidiary bodies and the IPCC, the working group will provide a detailed technical definition of the tradeable allowance (including relevant gases, intergas exchange rates, etc.); emissions credits; and joint implementation credits.

4.1.2. Trading rules - transfers and registration
In addition to and in conformity with any relevant provisions of the Kyoto Protocol, the Market Design & Operations Working Group will prepare model rules and regulations for traders with respect to the transfer and acquisition of allowances and credits, recordation procedures, transferring titles, delivery, performance warranties, financial guarantees, transfer agents, and dispute settlement. The rules should be broad enough to apply to all trades and traders, including institutional trading houses (brokers, hedgers, speculators) and individual traders.

4.1.3. Model trading contracts

The Working Group will prepare model contracts for trades in CO_2 emission equivalents, in line with the decisions of the Parties on this matter. Contracts should cover all relevant aspects of spot, forward and futures trading such as the unit of trading; standards; price basis; delivery; trading months; trading hours; trading limits; last day of trading; position limits; inter-gas exchange rates, etc.. Swap trading documentation should also be developed. Trade documentation should also include transfer forms, confirmation notices, and relevant standardized accounting documents.

4.1.4. Market auctions

The Working Group will prepare guidelines for auctions to be conducted by organized markets. The purpose of these auctions could be to assist the market in price discovery, encourage trading, raise revenues, and provide new access to allowances and credits by various interested parties.

4.1.4. Institutional support - clearinghouse; commodity exchanges; financial exchanges

Clearinghouse. Transactions that result in delivery and transfer of allowances and credits could be cleared through an approved international clearinghouse, selected by competitive bidding. The clearinghouse would net all transactions daily and send confirmations to all appropriate parties, country regulators, buyers, sellers, and participating exchanges. The Working Group will prepare guidelines for the selection by compe-titive bidding of an emissions market clearinghouse to clear and settle all clearing members' accounts.

Commodity exchanges. Organized exchanges can facilitate emissions trading in a variety of ways, including through lowering search costs for matching willing buyers and sellers, by reducing the administrative cost of trading, by reducing the risk of trade failure (through contract standardization and trading rules), and by disseminating transaction prices to all. The Working Group will oversee the selection of international commodity exchanges to participate in the emissions trading programme. The designated exchanges may also be required to develop and sign product offset facility agree-ments, information-sharing agreements, and other appropriate arrangements for market oversight and international regulatory cooperation.

Financial exchange services. International financial institutions provide essential services to both the commodity exchanges and the clearinghouse with respect to financial transfers and settlement of payment orders. A number of highly effective electronic payments and transaction systems exist (including The Society for Worldwide Interbank Financial Telecommunications (SWIFT), GLOBEX, and the

Fedwire System). The Working Group will oversee the selection of appropriate international financial exchange services to complement other institutional mechanisms of the emissions trading system.

4.2. OTHER PRIORITIES

4.2.1. Allowances and credits tracking system

The Working Group will oversee the design and development of an operational Allowance Tracking System in support of recordation of international transfers and acquisitions of allowances and credits. The Tracking System would be one of the central mechanisms available to both the private sector and governments to ensure proper keeping of accounts, provide a record of all transfers, the documentary basis for tracking trades, such as allowance transfer forms for authorized account representatives, both transferors and transferees. The tracking system would explore all possibilities for computerization and automation. The Working Group will advise participants on the choice of appropriate institutional arrangements for the operation of the tracking system.

4.2.2. Voluntary trading groups

Voluntary trading groups will assist the smooth functioning of organized markets, in particular through self-regulation, credit support, and membership in the clearinghouse and on the commodity exchanges. The Working Group will encourage the formation of voluntary corporate trading groups, as well as other trading groups at regional and municipal levels.

4.2.3. Legal, tax and accounting practices

Confidence in the credibility of the international emissions market requires clear legal underpinnings. Participants in the trading system will need to agree a clear legal definition of the property rights of the allowance/credits holder, the transferability of allowances and credits, the transferability of title, and the tax treatment status of allowances and credits trading. Internationally comparable practices would facilitate cross-border transactions. The Working Group will prepare model guidelines on these issues.

4.2.4. Transactions information system

Various electronic possibilities exist for market institutions to provide essential transactions information (prices, volumes, bids, offers, buyers, sellers, etc.) to the public, including real-time information via the Internet. The Working Group will encourage relevant information providers to develop real-time electronic transactions information systems in support of the international greenhouse gas emissions market.

4.2.5. *International Emissions Trading Association*
A not-for-profit international association (grouping a wide cross-section of interests in emissions trading including industry groups, investors, financiers, traders, NGOs, etc.), could play an important role in assisting the long term development of an international emissions market by providing independent analysis, discussion and oversight of the trading system and its institutions, by developing and operating market support services such as accounting and tracking systems, public data systems, transactions information systems, etc., and generally by promoting the interests of its members. The Working Group will encourage and support the establishment of an international emissions trading association, which in turn will facilitate the creation of similar associations at the national level.

5. Task Forces

5.1. TASK FORCES (2-3 PERSONS EACH) OF THE POLICY FRAMEWORK WORKING GROUP

I. *Country Participation Task Force*
Main Tasks:
Draft international participants' agreement
Draft model domestic legislation
Draft technical guidelines on domestic allowance allocation practices

II. *Common Standards Task Force*
Main Tasks:
Draft protocols on monitoring, certification, accounting, reporting, and enforcement

5.2. TASK FORCES (2-3 PERSONS) OF THE MARKET DESIGN AND OPERATIONS WORKING GROUP

I. *Market Services & Practices Task Force*
Main Tasks:
Draft definition of the tradeable commodity
Draft trading rules - transfers and registration
Draft trading contracts
Draft protocols/guidelines on market auctions, clearinghouse services, commodity exchanges, financial services, transactions data, and other services)

II. *Market Support Task Force*
Main Tasks:
Draft guidelines on legal, tax and accounting practices
Draft protocols on compliance guarantees
Draft guidelines on dispute resolution
Draft protocols on voluntary trading groups; international emissions trading association

6. The Steering Committee

The Steering Committee will oversee the implementation of the Work Plan, including the activities of the Policy Forum, Working Groups and Task Forces. It will:

- Facilitate communications among Working Groups and the Policy Forum
- Facilitate communication among the sponsors of the Policy Forum and other relevant institutions.

7. Market Development Timetable

1998 Design and implementation of the policy and regulatory frameworks
Design and implementation of the market mechanisms
Training, institution and capacity-building

1999 Design and implementation of the policy and regulatory frameworks
Design and implementation of the market mechanisms
Training, institution and capacity-building

2000 Implementation of policy and regulatory instruments
Implementation of market instruments
Training, institution and capacity-building
Market preparation and market launch.

TRADEABLE EMISSIONS PERMITS AND THE WTO

SCOTT VAUGHAN*
World Trade Organization
Rue de Lausanne 154
CH-1211 Geneva, Switzerland

* Views expressed in this paper do not necessarily represent the position of the WTO Secretariat or its members.

This presentation[1] is divided into two sections:

1. A brief overview of the World Trade Organization (WTO);
2. An overview of some issues before the WTO Committee on Trade and Environment, as well as other WTO areas, of potential relevance to an international tradeable emission scheme.

Before beginning, let me state that the WTO Secretariat does not have a "perspective" or opinion on emerging tradeable emission schemes for greenhouse gas emissions. The WTO has addressed extensively various aspects of environmental policy, including market or price-based measures such as taxes, charges, eco-labelling and certification schemes, and while past discussions may be instructive in identifying possible links between the WTO and an international tradeable emissions schemes, this discussion remains speculative.

1. The World Trade Organization

As the successor to the General Agreement on Tariffs and Trade (GATT) the WTO was created in 1994 upon the completion of the Uruguay Round negotiations. The WTO agreement constitutes the legal and institutional foundation of the multilateral trading system, providing contractual obligations determining how governments design and implement domestic legislation and standards which affect trade. There are

[1] A list of abbreviations used is added at the end of this presentation.

J. Hacker and A. Pelchen (eds.), Goals and Economic Instruments for the Achievement of Global Warming Mitigation in Europe, 367–376.

currently 131 Members countries to the WTO, and 28 countries negotiating accessions, including China and Russia.

The WTO differs in several ways from the GATT. The GATT successfully addressed a relatively narrow set of trade obstacles, primarily tariffs applied to the import of merchandise goods. Over successive trade rounds, tariffs rates have gradually been reduced, whereby the post Uruguay Round average tariff rate is 3.8 percent. However, while tariffs progressively decreased, new forms of grey-area protectionism emerged. Accordingly, the WTO embraces disciplines which reach significantly beyond tariff reduction. The WTO administers and implements several multilateral agreements including the Agreement on Technical Barriers to Trade (TBT) and the Agreement on Subsidies and Countervailing Measures (SCM). The WTO also covers trade in services. This is a new and highly dynamic area, as evidenced by agreements signed this year covering telecommunications, as well as current negotiations covering trade in financial services. In addition to goods and services, what is new about the WTO is its coverage of trade in "ideas" or what is known as trade-related intellectual property rights.

It is important to emphasize that the WTO is changing and expanding as markets themselves change through technology, communications and knowledge-based comparative advantage and other factors driving economic globalisation in the objective of borderless international trade. In this the WTO weaves the old with the new: the old in that core GATT principles remain the foundation of the multilateral trading system. These include non-discrimination, most favoured nation (MFN), national treatment, and the elimination of quantitative restrictions - non-discrimination and trade liberalization remain the pillars of the rule-based multilateral trading system, and should be considered closely in the design of an international trading system regardless of whether it would or would not fall under WTO disciplines. The WTO does not set standards. However, it does encourage in several agreements the adoption of relevant international standards, while allowing scope for individual countries to adopt national standards.

Another new element of the WTO is its revised dispute settlement procedures: the WTO differs from other international legal agreements because its dispute settlement procedure is binding upon Members. Panel findings must be adopted, through either the change in measure that led to the complaint, or payment of compensation for trade damages incurred through the distorting measure. This improved and binding legal dispute process is new, and governments - in particular developing countries - are using the levers of a rule-based trading system to offset unfair economic policies. Since its creation, there have been over 100 requests for consultation under the WTO dispute settlement procedures.

As noted, there are strong ties between the GATT and WTO. One is the fundamental recognition of the vital importance of clear, transparent and predictable rules as a means to promote a stable trading environment. This is an important general observation of significance in the creation of an international trading system. Another is the provision of preferential trading benefits for developing countries, to ensure that they become full partners in international trade. The role of developing countries in a trading system will obviously be crucial.

2. The WTO and the Environment

The WTO has been working on trade-environment since its inception, with the objective of ensuring that environment and trade policies are "mutually supportive". Various agreements contain environmental provisions, including the TBT, SPS, Subsidies, Services and Investment agreements. The principal focus of the WTO's is the Committee on Trade and Environment, hereinafter referred to as "the Committee". In issuing its report to the first WTO ministerial meeting in Singapore in December 1996, the Committee met with criticism from several sides - from environmental groups arguing that the WTO did not integrate environmental concerns into trade rules, including GATT Article XX on General Exceptions, and that accordingly trade rules threaten the effective implementation of environmental policies. And criticism from developing and smaller market economies which cautioned that linking trade with environment is akin to green conditionality, in which market access commitments will be eroded because the protectionist abuse associated with some environmental standards. As neither side appears happy, the Committee's work will continue.

Of the Committee's agenda, six areas may be of relevance in discussing the design of a tradeable emissions system: (i) price or market based instruments; (ii) environmental certification systems; (iii) Multilateral Environmental Agreements; (iv) subsidies; (v) services, investment and competition policy; and (vi) trade-related intellectual property rights. Before turning to these questions, let me reiterate that it remains unclear if WTO provisions would at all cover an emissions trading scheme, in part because no interpretation exists on whether a legal definition of emissions trading would be interpreted as either trade in a good, or trade in a service. A paper by UNCTAD (Richard Stewart et al. *Legal Issues Presented by a Pilot International Greenhouse Gas Trading System,* 1996) notes that such a scheme may not be covered by the WTO: trade in capital is an obvious example of activities falling outside of WTO disciplines.

There are a number of considerations that ought to be raised here at the outset. The first is that the climate change negotiations demonstrate a blurring of lines between economic and environmental policy. The divisions between the European Union and the United States at the June Denver Summit of the Eight on climate underlined the economic stakes involved. Notice has already been given that, regardless of the outcome in Kyoto, governments will initiate a range of measures to tackle greenhouse gas emissions. A carbon tradeable scheme will not likely be of foremost importance to the WTO. Instead, other measures, in particular the possible increased use of carbon and energy taxes and ensuing questions related to border tax adjustment rules, or the use of minimum energy efficiency standards and accompanying labelling or certification schemes may raise important issues in the WTO.

Turning in general to the tradeable emissions issue, two questions can be asked. First, is such a system covered under the WTO? As noted, a definitive answer is unknown. The second question is, regardless of the question of coverage of such a scheme per se, the economic and competitiveness effects of the system are likely to be significant. There is no doubt that these effects will have as yet undetermined WTO consequences, given their projected impacts on terms of trade, competitiveness, the potential for price distortions, etc.

With that cautionary note in mind, past discussions in the Committee, while not addressing explicitly tradeable emissions, may nevertheless be instructive.

2.1 MARKET-BASED INSTRUMENTS

As a point of departure, there should be an assumption of support from or compatibility between a tradeable emissions systems and the WTO. The Committee has for some time discussed the relationship between various price-based or market creating instruments for environmental purposes and the WTO, and there is a general recognition that such measures can be more cost effective and efficient, and less trade distorting, than command-and-control regulations.

2.2 CERTIFICATION AND LABELLING SYSTEMS

The Committee has discussed under agenda item 3(a) various types of environmental labelling and certification schemes, with analysis for the most part focusing on labelling of traded products. Recently, discussion has also looked at environmental certification systems. The Committee has addressed trade-related aspects of certification systems based on different types of criteria, different thresholds, or different methods of verification. Reference to mutual recognition and equivalency may be instructive in this regard. In addition to criteria itself, elsewhere in the WTO an agreement on mutual recognition of professional services signed in 1996 (for financial auditors) may also be of relevance in designing how to recognize different certification systems. Finally, several governments has underlined the importance of transparency and notification as means to inform trading partners of proposed changes in standards, regulations, labelling or certification criteria. Notification is required usually prior to any changes to allow comments from trading partners to be taken into account.

2.3 MULTILATERAL ENVIRONMENTAL AGREEMENTS (MEA)

An important emphasis of the Committee's work has been the relationship between the approximately 185 MEAs in existence and the WTO. As a point of departure, the WTO reaffirmed its support for cooperative, multilateral solutions to shared environmental problems. Governments have focused on first defining what constitutes a multilateral approach. Suggestions include whether an MEA has been negotiated under the auspices of the UN; whether decisions are based on scientific evidence; whether a balance is struck between developing and developed countries; and whether the MEA is transparent and inclusive in its governance and decision-making structure. Clearly, in all these considerations the Climate Change Convention and whatever action is adopted under the Convention - including the possible creation of an international tradeable emissions scheme - more than meets the general WTO criteria of an MEA.

The reason the Committee has focused on MEAs is because of the possible legal incompatibility between roughly 10 percent of all MEAs which contain trade measures - usually in the form of a trade restriction - to help achieve their environmental objectives. Discussion has concentrated on CITES, Montreal, Basel and the Convention on Biodiversity thus far, and two major groupings of trade measures have emerged: (i) trade measures taken between parties or signatories to an MEA; and (ii) measures taken between parties and non-parties.

The first category includes import and export permits, other licensing requirements, prior informed consent procedures, quotas or other measures. These are applied to products clearly specified in the MEA. For example, different categories of hazardous wastes under the Basel Convention, or 37,000 species listed as endangered in CITES. Although trade restrictions applied in an MEA between parties might in theory be WTO inconsistent, in practice the WTO is of the view that problems should not arise because governments have explicitly agreed to such restrictions, and if problems do arise then a dispute should be addressed within the MEA.

The second category involves discriminatory trade measures applied against non-signatories to an MEA. For example, this is in the Montreal Protocol, whereby parties are forbidden from exporting or importing ozone-depleting substances, as well as substitute products or technologies, to non-parties to the Protocol. In addition, the Protocol includes the option, although never invoked, to ban trade in any product produced with ozone-depleting substances. The negotiating history of the Protocol suggests that discriminatory trade measures were used as a means to coerce or force governments to become parties. They were used together with the additional financing, technical cooperation and other positive measures as a kind of "carrot and stick" regulatory approach to compliance.

Although there is clear recognition of the effectiveness of the Montreal Protocol in the Committee, the use of discriminatory trade measures against non-parties has raised serious concerns, and it is improbable that the WTO would welcome their replication in the Climate Convention or a tradeable emissions scheme for obvious reasons. Such measures would appear to violate GATT/WTO principles of MFN, national treatment and non-discrimination. A theoretical problem arises whereby if a country remains a non-party to an MEA or its protocol's for any number of reasons - disagreement over scientific evidence, different risk management or risk acceptance thresholds, higher marginal adjustment costs, etc. - and as a consequence faces discriminatory trade measures, then the question might well be asked as to what might be the economic or commercial consequences of such discriminatory measures? In the Montreal Protocol, this question is moot given the wide scope of participation. However, if discriminatory trade measures were used in climate, the economic implications could be severe. One recourse to economic discrimination could be for the non-participating country subject to the discriminatory measures to bring a legal dispute under the WTO against either the MEA or parties to the MEA. Although this has never happened in the past, it clearly would create considerable instability in both the trade and environment regimes, so much so that recently several international business groups have called for the WTO to undertake a formal legal amendment to accommodate some forms of trade measures between parties to an MEA, as well as to exclude accommodation of any discriminatory measures against a non-party.

Questions which might arise in light of the Committee's discussion centre around the treatment of countries that refuses to participate in a tradeable emission system. I think the obvious question of non-participation itself in relation to non-discrimination is moot, since a tradeable emissions scheme would on assumption be open to any interested party. However, the question does arise as to whether any non-neutral policy, such as sanctions, could be considered as means of forcing or encouraging a non-participant to join the system. In that case, would the non-participating country face less favourable commercial or other treatment than participating countries? Could discriminatory treatment negatively influence the commercial opportunities of a non-participant that is a WTO Member? How will an emissions trading system deal with a country which refuses to sell surplus quotas for whatever reason? Will non-parties face either indirect economic discrimination, or direct discriminatory trade measures? Might options include some form of identifying participants through a labelling or certification scheme, which could have the effect of identifying by omission non-participation countries? In addition, given the open nature of the system, could an investor from a non-participating country nonetheless purchase permits in member country? Such questions need to be considered in looking at how to design a tradeable emissions scheme.

2.4 SUBSIDIES

The issue of subsidies in relation to a tradeable emission scheme is complex. The Agreement on Subsidies and Countervailing Measures (SCM) contains disciplines identifying actionable subsidies which affect exports. An additional category of non-actionable subsidies is included related to the environment, in which 20 percent of capital costs for environmental retrofitting of older plants may be exempt from WTO subsidies disciplines (SCM, Article 8.2(C)). This might be of relevance to more general climate change policies given the recent emphasis on retrofitting existing plants with energy-efficient technologies. Subsidy disciplines may also be of indirect relevance in a tradeable emissions scheme during the permit allocation period, that is after a global or other quantitative cap has been established, and before trading actually takes place within the market. The allocative process ostensibly represents the creation and distribution of private property rights over emissions, and would lie outside of WTO provisions. However, given the economic implications of such a system, the distribution of quotas might be designed in such a manner as to advantage a particular sector or industry, and have a similar price distorting effect as a subsidy. Past GATT and WTO panels have found that violation of national treatment provisions include measures affecting the "competitive opportunities" of enterprises. The distribution of quotas may be designed in such a way so as to grant sectoral preferences, or to raise questions about non-violation related to the nullification and impairment of WTO rights and obligations.

2.5 SERVICES, INVESTMENT AND COMPETITION POLICY

Little discussion has thus far taken place in the Committee regarding services and the environment. It remains an open question as to whether an international tradeable emission scheme would constitute a service and therefore to some extent fall under the WTO General Agreement on Trade in Services (GATS). The best generalists description of a service is one run by the *Economist* which said a service is something you can buy or sell, but you can't drop it on your foot. The GATS Agreement, not surprisingly, has a more complicated definition. The GATS does not cover trade in services per se, but rather measures affecting services crossing borders, sold by a commercial presence, etc. It is unclear if some aspects or all of an emissions trading scheme would be defined as a service. The activity represents trade in quota entitlements, which to some extent has precedent in the Multi Fibre Agreement, as well as the provision of a brokerage or other fee-based service. The Annex on Financial Services contained in the GATS defines financial services (5) as:

"(x) Trading for on account or for account of customers, whether on an exchange, in an over-the-counter market or otherwise, the following:

(A) money market instruments (including cheques, bills, certificates of deposits);

(B) foreign exchange;

(C) derivative products including, but not limited to, futures and options;

(D) exchange rate and interest rate instruments, including products such as swaps, forward rate agreements;

(E) transferable securities;

(F) other negotiable instruments and financial assets, including bullion."

The relationship between the wide scope of financial services instruments, including trade in securities (e.g. underwriting) and a tradeable emissions scheme is unclear. The scope may fall only in ensuring that financial service institutions like banks or investment houses are subject to national treatment provisions, and are thereby entitled to provide financial services within another country's borders.

This question becomes more complex because of the relationship between a permit trader and the government authority initially involved in the permit allocation. The GATS exempts all services "except [those] applied in the exercise of government authority". This is an important area that will become more clear once a system begins operation.

Another allocation or permit distribution issue concerns the role of monopolies or exclusive service suppliers. There has been consideration of ways to ensure that allowances or savings can be widely held, that any person or enterprise can engage in trading, and that other steps to address abuse of market power will be included, whereby disciplines coupled with the diversity of interested players in carbon trading, cartelization would be unlikely. Nevertheless, when a monopoly supplier of a service is in place, there is a legal obligation under the GATS that most-favoured nation provisions be followed, and that the monopoly supplier shall "not abuse its monopoly position". Monopolistic abuse is open to different interpretations based on individual panel proceedings[2], but if a permit allocation process is operated by either the government or a monopoly, then this general provision against monopolistic abuse may be of relevance.

Let me flag briefly two additional areas - investment and competition policy - as being potentially instructive. Again let me underline that the WTO has not drawn any links between the climate talks and these new trade policy areas. Discussions in the OECD draft Multilateral Agreement on Investment would, however, suggest that in the transfer of permits from a public authority to private entity, no discrimination can take place between domestic and foreign firms operating within a country's

---

[2] An additional point of reference is the role of state trading enterprises in either the allocation, or internal distribution of permits, or a possible role in the purchasing of reserve emission permits in order to suppress permit cartelization. GATT Article XVII requires that state trading enterprises - that is any enterprise that formally or in effect grants exclusive or special privileges including marketing boards - upholds the principle of non-discrimination.

borders. This assumes not only the right of a foreign entity to buy or sell permits, but it also assumes that any policy measure related to the marketing of such permits could not be designed to advantage a domestic entity. Given likely shifts in foreign direct investments associated with an initial quota regime or subsequent international transfers, the implications of non-discrimination applied to investment is likely to be very complex.

The issue of cartelization has already been raised. In the context of a tradeable emissions scheme, even though most proposed plans call for an open auction of permits, economic experience would suggest that domestic enterprises - through lobbying, regulatory capture or other means - can exert market distorting pressures to privilege some sectors or enterprises. Energy-intensive sectors which are publicly owned - such as some utilities - or private energy-intensive sectors which often are oligopolistic in nature - such as steel manufacturing - could conceivably benefit from various forms of uncompetitive behaviour. For example, could permit allocations further entrench pre-allocation imperfect market competition? Put another way, could permit allocation or patterns of trading reinforce oligopolies, and therefore afford protection to inefficient or sunset industries? Could pre-existing imperfect competition lead to either price collusion during permit purchasing, or to hoarding of permits because of an abuse of a dominant or monopoly market position? Can a dominant market position suppress the role of others in the market? Can predatory pricing be used, as in capital markets, to distort tradeable emission schemes? To these questions, consideration of national treatment and non-discrimination in permit bidding and allocation will be vital, and lessons from competition and anti-trust policy may be instructive. The question has already been raised about the possible free allocation of some permits under a grandfathering system, and this could create distortions in competition policy.

2.6 TRADE-RELATED INTELLECTUAL PROPERTY RIGHTS (TRIPs)

A final area of the Committee s discussions that is unlikely to have direct implications for the design of a tradeable emission scheme, but which may have considerable importance for a climate regime in general is the role of the Agreement on Trade-Related Aspects of Intellectual Property Rights. In the last three weeks, several governments, including the US, Norway, Japan and Korea, have underlined the importance of the development of energy-efficient process technologies as well as the transfer of such technologies to developing countries. An important ingredient of technology transfer relates to patents and intellectual property rights (IPR): some developing countries have argued that a more flexible patenting and IPR system is needed to allow deve-loping countries to acquire on an affordable basis environmentally-sound technologies. Little data exists at present on the extent to which larger scale energy technologies, for example those used in oil-fired utilities, are covered by long-term patents. However, within the context of a tradeable emission scheme itself, one could foresee a situation in which energy-intensive sectors, in order to meet quota allocations, will have an incentive to maintain existing patent schemes for their duration to protect R&D for highly energy-efficient technologies, while developing countries may press for greater IPR and patent flexibility in order to acquire such technologies on preferable commercial terms. In the Committee, India has proposed that patents and IPRs associated with the implementation of an MEA should be more flexible than other IRP Schemes, and that a formal amendment to the TRIPs Agreement is needed in this regard.

3. Final Remarks

Let me close by reiterating that these are issues to be flagged. The WTO is changing and evolving, just as international environmental policy is changing. The climate agenda, more than anything else, drives home the blurring of delineations between economic and environmental policies. Having said that, many of the questions and issues raised here will likely not be raised formally in the WTO, or in the context of the design of a tradeable emission scheme. However, it is important to be mindful of the importance which the WTO continues to assume in international commercial relations, and of the economic importance of the climate change discussions. Ensuring the two regimes work together will become an increasingly important issue.

Abbreviations

CITES	Convention on International Trade in Endangered Species of Flora and Fauna
GATS	General Agreement on Trade in Services
GATT	General Agreement on Tariffs and Trade
IPR	Intellectual Property Rights
MEA	Multilateral Environmental Agreements
MFN	most favoured nation
OECD	Organisation for Economic Co-operation and Development
R&D	Research and Development
SCM	Subsidies and Countervailing Measures
SPS	Agreement on the Application of Sanitary and Phytosanitary Measures
TBT	Technical Barriers to Trade
TRIPs	Trade-Related Intellectual Property Rights
UNCTAD	United Nations Conference on Trade and Development
WTO	World Trade Organization

SECTION IV

TRADEABLE EMISSION PERMITS -

PERSPECTIVE OF A CO_2 EMISSION SYSTEM

STATUS OF DISCUSSION AND NEGOTIATION FOR A SYSTEM OF TRADEABLE CO_2 EMISSION PERMITS WITHIN THE UNITED NATIONS FRAMEWORK CONVENTION ON CLIMATE CHANGE

JOANNA DEPLEDGE*
Climate Change Secretariat (UNFCCC)
Martin-Luther-King-Strasse 8, D-53175 Bonn, Germany

* The views expressed are of a personal nature and do not necessarily represent those of the UNFCCC secreatriat.

• Since this paper was presented in July 1997, the Kyoto Protocol to the United Nations Framework Convention on Climate Change was adopted at the third Conference of the Parties to the UNFCCC (COP 3) in Kyoto, Japan, on 10 December 1997. After lengthy negotiations, Article 17, which opens the way to emissions trading among Parties with quantified emission limitation or reduction commitments, was included in the Kyoto Protocol. The provisions of Article 17 are to be further elaborated, beginning at the fourth session of the Conference of the Parties (COP 4) in Buenos Aires in November 1998. The relevance of the paper which follows lies in the snapshot it provides of the negotiation process which led to this eventual outcome on emissions trading.

1. Introduction

• Emissions trading is an issue which has received relatively limited attention to date in the United Nations Framework Convention on Climate Change (UNFCCC) process, in contrast with the extensive work carried out on the issue in certain other intergovernmental bodies, in particular the United Nations Conference on Trade and Development (UNCTAD), the International Energy Agency (IEA) and the Organisation for Economic Co-operation and Development (OECD).

• Recently, however, a number of proposals have been tabled by Parties advocating the inclusion of provisions on emissions trading in new legal instrument to the Convention currently being negotiated under the Ad hoc Group on the Berlin Mandate (AGBM). The new instrument - which in practice will be either an amendment or protocol - will be adopted at the third Conference of the Parties (COP 3) to the Convention, which will take place in Kyoto, Japan, in December 1997. Proposals on emissions trading have provoked some discussion in the AGBM, and are included in the negotiating text for the new instrument (document FCCC/AGBM/1997/3/Add.1, dated 22 April 1997), but the debate is still in its very early stages.

J. Hacker and A. Pelchen (eds.), Goals and Economic Instruments for the Achievement of Global Warming Mitigation in Europe, 379–384.

2. Background

- The idea of establishing an emissions trading system as part of international efforts to address climate change is not new. Some delegations, the US for example, put forward such a suggestion during the negotiations on the Convention itself.

- Emissions trading does not feature in the text of the Convention. Certain Articles, however, are relevant to emissions trading and could be interpreted as providing justification for its elaboration. These include Article 3.3, which states that "Policies and measures to deal with climate change should be cost-effective so as to ensure global benefits at the lowest possible cost".

- However, the present emission reduction commitment to which Parties are subject under the Convention - to aim to return emissions of greenhouse gases to 1990 levels by 2000 - is neither specific nor legally-binding. This means that it would be difficult to elaborate a tradeable emissions system on the basis of the Convention alone.

- The issue has recently been revived for two main reasons:

  - Firstly, the launch of negotiations on possible legally-binding, specific emission reduction targets as part of the protocol or amendment to the Convention means that, if these were agreed, the basic conditions necessary for an emissions trading system to be set up would be established;

  - Secondly, the difficulties which most governments are facing in meeting their emission commitments under the Convention has given Parties an incentive to explore a range of mechanisms which could help them meet any new emission reduction targets. The second national communications currently being received by the secretariat indicate that few Parties will succeed in returning their greenhouse gas emissions to 1990 levels by 2000. In this context, it is not difficult to understand why emissions trading, along with other mechanisms for flexibility, have attracted so much interest from many Parties.

3. Emissions trading and the AGBM: Status of discussions

- Although no formal analysis of emissions trading has taken place in the AGBM, the issue has been addressed by other bodies which provide input into the Group's work, notably Working Group III of the Intergovernmental Panel on Climate

Change (IPCC) and the Annex I Experts Group to the UNFCCC, which issued a report on "International greenhouse gas emissions trading" in conjunction with AGBM 6 (31 July - 7 August 1997).

• Several developed countries, known as Annex I Parties, have put forward proposals or made comments on the treatment of emissions trading in the new instrument. Although the lack of real discussion which has taken place to date makes it difficult to accurately gauge the positions of Parties on this matter, the following broad groupings can be identified:

- For some Annex I Parties (these could be termed the "active supporters"), notably the US, emissions trading forms an integral part of their proposals as an important mechanism to ensure flexibility. At the second Conference of the Parties to the UNFCCC (Geneva, July 1996) for example, the US asserted that "international emissions trading must be part of any future regime" (Statement on the Geneva Ministerial Declaration, FCCC/CP/1996/15, Annex IV) ;

- Other Annex I Parties (which could be described as "in-principle supporters"), the EU in particular, see potential in the concept of emissions trading, but only as part of a wider portfolio of domestic action to tackle climate change.

- A third group of Parties, especially developing countries which are known as "non-Annex I Parties" under the Convention, have tended to express serious misgivings at the principle of emissions trading.

• These contrasting attitudes to emissions trading mirror wider differences in approach among Parties to the broader issues under discussion in the AGBM.

• The active supporters of emissions trading are also those Parties for whom the need to secure flexibility in the new instrument is paramount. This emphasis stems from a concern to ensure that emission reduction commitments can be met at the lowest possible cost. These Parties have tended to reject the inclusion of mandatory policies and measures in the new instrument in favour of market-based and voluntary mechanisms.

• The active supporters argue that emissions trading would enable emission reduction commitments to be met as efficiently and cost-effectively as possible and could, in addition, provide a more powerful incentive for private sector companies to reduce their emissions in order to contribute to meeting national targets.

- The in-principle supporters, Parties who are more cautious in their support for emissions trading such as the EU, tend to be those who prioritize the establishment of a "level playing field" in emissions abatement, in order to avoid distortions in trade and competitiveness. Based on this approach, this group of Parties has emphasised the need to elaborate common policies and measures in the new instrument, including some which would be mandatory and/or coordinated, and has expressed concern that emissions trading should not be seen as a substitute for effective domestic action to reduce emissions. These Parties have also stressed that mechanisms for flexibility such as emissions trading should be accompanied by a strong emission reduction target. At AGBM 7 (22 - 31 October 1997), the EU put forward new text setting out certain criteria to be fulfilled before emissions trading could be permitted. These criteria include a requirement that trading be supplemental to domestic policies and measures to mitigate climate change and that trading in emission reductions achieved before the start of the trading system not be allowed.

- Developing countries have expressed suspicion at the concept of emissions trading, raising questions regarding its equity implications. This concern may be related to wider misgivings on the part of many developing countries over the different mechanisms for flexibility currently on the table in the AGBM. This group of countries stresses that Annex I Parties should not be able to buy their way out of making the changes to their domestic production, consumption and investment patterns which are needed to reduce emissions and to provide incentives for technological innovation.

- From a different perspective, the US and New Zealand have argued that the possibility of participating in an emissions trading system might provide an incentive for certain non-Annex I Parties to consider voluntarily entering into emission limitation commitments.

- Questions have been raised regarding the practical feasibility of establishing an emissions trading system before Kyoto. If agreement were to be reached on this issue, a possible way forwards may be to endorse emissions trading in principle, leaving the details of its operation to be worked out in subsequent deliberations post-Kyoto.

- The need to ensure cost-effectiveness in climate change mitigation has been a recurrent theme in the AGBM negotiations. Emissions trading is clearly a response to that need. Another response is that of joint implementation (JI), whereby one Party would be able to implement emission abatement projects in the territory of another Party and receive credit for any emission reductions achieved. In contrast to the varying levels of enthusiasm expressed for emissions trading, almost all

Annex I Parties have now lent support for provisions on JI to be elaborated in the new instrument, although there is debate over whether JI should extend to all Parties, or just to Parties with commitments. Consideration will need to be given to how suggested measures on emissions trading would relate to those put forward on JI in the new instrument.

4. Current proposals on emissions trading

• Only the US and New Zealand have submitted full proposals in legal language on a possible emissions trading system for inclusion in the new instrument, and the section on emissions trading in the negotiating text is one of the shortest. The two proposals are similar in a number of key respects:

- Both consist of a trading system based on so-called "emission budgets", allocated for an (as yet unspecified) period of time. A Party would be permitted to sell part of its emissions allowance to another Party, thus reducing its budget, or buy emissions, thereby increasing its allowance;

- Trading would only take place between Parties with legally-binding emissions reduction commitments, and would be voluntary;

- Greenhouse gases included in such a system would be calculated according to carbon equivalent emissions, and only greenhouse gases for which there is sufficient knowledge should be included. Such greenhouse gases would be listed in an annex to the new instrument.

- Trading would not be limited to State Parties, but would also be allowed between "domestic entities" such as private firms, NGOs, and even individuals. The State Party, however, would retain full responsibility for meeting its emission reduction commitments;

- Strict monitoring, verification and compliance provisions would be part of the system. Only countries with an accurate domestic emissions measurement system would be allowed to trade.

5. Future prospects for emissions trading

• The elaboration of emissions trading within the framework of the Convention is still at an embryonic stage of development. In order to be more widely accepted, emissions trading will require analysis and assessment to take place as part of the Convention process, so as to address the many concerns of Parties and garner broader support.

- In addition, the practical complexities involved mean that the establishment of a fully-fledged inter-governmental emissions trading system by Kyoto is probably unfeasible. The most likely way forwards is that the new agreement would endorse emissions trading in principle, with the details to be worked out after COP 3.

- It is certain, however, that the importance attached to emissions trading by a number of Parties ensures that the issue will only grow in importance, up to Kyoto and beyond.

THE UNITED STATES PROPOSAL FOR AN INTERNATIONAL CO_2 EMISSIONS TRADING SYSTEM*

JAMES WOLFE
U.S. Embassy Berlin
Neustädtische Kirchstrasse 4-5
D-10117 Berlin, Germany

* Reflects the position of July 25, 1997

1. Introduction

This paper will represent the U.S. government approach to climate change, and specifically, the U.S. draft protocol to guide green-house gas emissions reduction efforts in the post-2000 period. Before proceeding to the particulars of the U.S. proposal, however, this proposal will be placed into its proper political context. The entire U.S. proposal for a climate change protocol covers a mere thirteen sheets of paper, but is the balanced product of intense inter-Agency negotiation, of consultations with industry, environmental experts, non-governmental organizations and elected officials. These entities represent a wide range of views on climate change, from those who question the very existence of global warming, to those who favor radical reduction programs to address the problem.

The U.S. proposal for a climate change protocol must also be seen against the background of a major environmental initiative launched last year to more fully integrate environmental considerations into U.S. foreign policy. At a speech at Stanford University last April, former Secretary of State Christopher announced that the State Department would put global environmental issues where they belong among the key national security concerns that the world faces. More recently, Secretary of State Albright testified before the U.S. Congress that she will carry on this landmark initiative, stating that environmental threats may not be as dramatic as those posed by nuclear missiles or terrorism, but ignoring them would bear enormous costs - to our health, our livelihoods, and our quality of life. This administration takes environmental issues seriously, and it places climate change at the forefront of those environmental challenges facing us today.

There is a considerable range of opinions as to the exact cost of reducing global emissions; some studies indicate that reducing greenhouse gas emissions 20% below 1990 levels by the year 2010, as has been proposed by some signatories to the

J. Hacker and A. Pelchen (eds.), Goals and Economic Instruments for the Achievement of Global Warming Mitigation in Europe, 385–393.

Framework Convention on Climate Change, would reduce the gross domestic product of the United States by one to two percent, and cost nearly $100,000 million per year. Other studies suggest that the costs could be even higher, and would require a $280 per ton carbon tax or its equivalent. I would expect that Germany, with its own intense budget debate and efforts to create growth and employment, would share our view that governments must find the most efficient and cost-effective mechanisms possible to achieve the necessary reductions. We believe that emissions trading is such a mechanism. It addresses the seriousness of the environmental issue at hand, while recognizing the costs of regulation, and the potential for ill-designed regulation to adversely effect economies.

2. The U.S. Climate Change Proposal

There is relatively little time remaining before the Third Conference of the Parties to the Framework Convention on Climate Change in Kyoto, and there is much work to be done. Our guiding principles in formulating our proposal have thus been pared to the essentials: within the umbrella framework of binding targets, and with a view to long-term efforts under the convention, we hold it imperative that any climate change protocol be based on goals that are both realistic and achievable. We believe it important that any protocol contain a mechanism for implementation at the national level. Finally, varying circumstances will affect the ability of individual states to reduce emissions; we thus propound allowing party states maximum flexibility to meet emissions reductions goals, using a variety of tools and mechanisms.

3. Definition of Terms

The philosophy of the U.S. emissions trading proposal, it should be noted, is consistent with our preference for seeking market-based solutions to environmental problems. The U.S. climate change proposal builds upon the successful American experience in sulfur dioxide trading and adopts some features of that system.

Building on the concept of cumulative and averaged emissions, the U.S. proposal would allocate an "emission budget" to each developed country party for a given period of time. An emissions budget is the total amount of greenhouse gases that can be emitted over a period of several years. The budget would be calculated in the same manner for all Annex I parties, the so-called developed countries, including Russia and Eastern Europe. We have named this group Annex A in our proposal.

We have proposed multiple emissions budget periods, including a second period in which emissions are equal to or less than the first period. This concept will assure continued progress toward achievement of the Convention's objective of reducing the concentration of greenhouse gases in the atmosphere. It will further facilitate public and private sector investment in the new technologies that will be needed to address this problem effectively over time.

We generally favor the idea of multi-year budget periods, insofar as they increase party states' flexibility. Some party states, for example, might choose to use many emissions units early on, in the hope that towards the end of the budget period, new technologies would make it easier and cheaper to reduce emissions. Others, in contrast, might economize on the use of their budgets in the early years to provide for rapid industrial growth and increased emissions in later years. A multi-year budget period also would smooth year to year variability in economic and weather cycles.

The U.S. has made no specific recommendation as to how many years should be included in an emissions budget period, nor have we suggested the level at which budgets should be set. We are now actively engaged in an intensive analytic effort to assess what budget levels would be appropriate. In this context, we don't believe that it is reasonable to set a political target without a concept of how such a target might be met, or what costs can reasonably be expected to be associated with such a target. We have, however, identified key principles of emissions budgets: namely that they be legally binding; medium-term; aim for realistic and achievable levels of reduction; and allow individual states maximum flexibility in their implementation. Annex A emissions budgets would use 1990 as the base year.

In spite of their best intentions, all Parties are dealing with large variables in controlling their greenhouse gas emissions, for example, rates of economic growth, weather conditions, and world fuel prices. Our proposal would allow Parties considerable flexibility in achieving their binding reduction targets. If a Party state does not use its entire emissions budget during the specified period, it would be allowed to "bank" these emissions for future use. This provision of our instrument would allow a Party to take more aggressive actions and reduce emissions beyond the level required during one budget period, and save those reductions for use at a future time. In this way, our instrument both provides an incentive to take early reduction actions, and offers each Party the opportunity to maximize the cost-effectiveness of its own reduction program.

Similarly, we also believe it appropriate to allow parties to "borrow" a limited amount of emissions (with a penalty) from a subsequent period. Borrowing also makes it possible for a Party to plan its emissions trajectory beyond the established budget period. For example, if a country has an old, inefficient and greenhouse gas intensive power plant which is scheduled to be replaced by a new and more efficient plant the year after the budget period ends, it can borrow against the expected reductions knowing that it would then be in full compliance. Each of us is familiar with the concept of borrowing from a bank to develop new investment projects.

Restrictions analogous to those from the world of finance may also be applied to emissions borrowing - with penalties for late payment, and a cost to the borrower for the transaction. The benefits in reduced costs, and smoother transitions to a new emissions reduction regime make the concept well worth inclusion. We would recommend, however, that borrowing be limited to a very small percentage of a country's emissions budget.

A Party's total emissions budget thus equals: the party's initial allocation; plus any amount the country banked; plus any amounts borrowed, including any amounts acquired through international emissions trading or joint implementation; minus any amounts transferred to other parties through international emissions trading.

4. ANNEX B - Developing Country participation

In 1997, on the occasion of Earth Day, the Department of State presented its first annual report on the Environment and Foreign Policy to the Congress. One section on climate change stated:

> "As the world's largest economy and emitter of greenhouse gases, the United States has a special responsibility to take meaningful action to attack the causes of climate change and mitigate its effects. Acting alone, however, will not solve the problem. Over three quarters of global emissions come from outside the United States, and as developing countries such as China and India continue to grow economically, their emissions will become an increasingly large portion of the problem."

In keeping with this approach, we have thus proposed a new category to encourage rapidly developing countries to voluntarily adopt emissions budgets. We called this group Annex B, and propose that they would have a different budget than that assigned to Annex A countries. While membership in Annex B would be voluntary, we believe that the benefits derived from membership - including the opportunity to participate in emissions trading - will entice many to join. We expect to have further suggestions about emissions budgets and budget periods for Annex B parties, although we envision that these could involve percentages, budget periods and base years different from those applicable to Annex A parties.

5. ANNEX C - Relevant Greenhouse Gases

The U.S. is convinced that only those greenhouse gases that are quantifiable and verifiable should be included in emissions budgets; a list of these would be included in Annex C of our proposal. It will include all greenhouse gases not covered by the Montreal Protocol, but would exclude gases for which there is insufficient knowledge of the global warming potential. Annex C will thus be selective; it might include methane from pipelines - but might not include methane from rice fields; it might include CO_2 from flaring or natural gas at the well head, but might not include CO_2 as embedded carbon in wood products. We would like to note, however, that our proposal does anticipate a provision for controlling of greenhouse gases not included in Annex C. We have thus included an obligation to take appropriate steps (which we are now working to better define) to control such emissions.

6. Emissions Trading

Emissions budgets would be used to facilitate "emissions trading". Emissions trading, as we propose it, would be allowed only between parties that each have budgets and that are in compliance with their measurement and reporting obligations under the agreement. If a party has successfully reduced emissions below its budget, it will have excess "credits" which it can choose to sell. Other Parties who have exceeded their budgets and who might not have good prospects for borrowing can choose to purchase emissions credits from another Party or from some entity that is authorized to trade in emissions.

Our submission proposes that participating countries may authorize any domestic entity, for example, government agencies, private firms, non-governmental organizations, or individuals, to participate in actions leading to the transfer and receipt of tons of carbon equivalent. However, only countries with emissions budgets would be able to register such trades.

While the private sector may engage in trading (and we expect most trades to take place through private sector activity), the parties themselves retain full responsibility for the emissions traded. Compliance with budget obligations thus remains with the government. But the potential benefits of trading are enormous. First, trading is efficient, and will substantially reduce the cost of compliance. Rather than having rigid enforcement measures and targets proscribed for them, Party states can come up with a variety of tools and mechanisms for encouraging industry to meet reductions targets. Further, emissions trading will, in the long term, equalize the incremental cost to all Annex A Parties of the next unit of greenhouse gas emissions reductions.

We would expect emissions trading to have a direct impact on fossil fuel consumption, as well. For example, importers of fossil fuels will pass the costs of buying emissions credits to industrial consumers of their products. This will lead to behavior changes on the part of industry in order to use less fuel - emissions will be reduced, as a result.

7. How this might work on an international basis

The U.S. does not wish to propose the creation of elaborate new structures to facilitate emissions trading. Because our proposal contains substantial detail on reporting and monitoring obligations for Annex A and B parties (required to insure compliance with the budget even in the absence of trading), there need be no additional complex scheme to monitor trades. It becomes largely an accounting exercise. Only parties with emissions budgets may buy or sell "tonnes allowed". We envision that parties for whom the cost of reducing emissions is high will have incentives to buy emissions credits. Conversely, parties for whom the cost of emissions reductions is low, have an incentive to sell their excess credits. Each year, parties would report purchases or sales that year, adding to or subtracting from their budgets. The private sector can facilitate trades.

8. Conclusion

The topic of this paper has been the emissions trading element of the U.S. proposal, but this concept really can't be separated from the proposal as a whole. In our view, emissions trading is distinct from, but inextricably linked to other key elements of our proposal, joint implementation, national accountability, and long-term efforts under the Convention, etc. For this reason, the Fact Sheet on U.S. Climate Change Proposal is attached.

FACT SHEET ON U.S. CLIMATE CHANGE PROPOSAL

June 1997

In July 1996, the United States outlined a broad framework for negotiation of next steps under the Framework Convention on Climate Change. In December 1996, the U.S. further elaborated its ideas by describing the key elements that should be discussed for inclusion in a protocol to guide greenhouse gas emissions reduction efforts in the post-2000 period. The current U.S. proposal represents further elaboration of the proposed framework. In January 1997, the United States presented to the Convention's Secretariat draft text for consideration in international negotiations that will occur throughout 1997. Several technical elements left undefined in the January submission (principally those relating to compliance issues, and to which gases should be included in the agreement) were further developed and submitted in early June 1997. The revised U.S. draft protocol proposal containing these elements will form the basis for the U.S. position in the next session of the Ad Hoc Group on the Berlin Mandate (schedule to meet in late July). Consistent with previous U.S. statements and proposals, the U.S. submission addresses the following key topics:

EMISSIONS TARGETS

- A new concept is set forth to guide the establishment of developed country emissions targets. Building on the concept of cumulative and averaged emissions, the U.S. proposes creation of an "emissions budget". Multiple emissions budget periods are proposed, including a second period in which emissions are equal to or less than the first period, thus assuring continued progress toward achievement of the Convention's objective.

    - For a given period, each developed country Party would be allocated an emissions budget.
    - The submission does not identify either the size of the budgets or the duration of the periods.
    - Parties would be allowed to "bank" for future use emissions not used during a given period; to "borrow" a limited amount of emissions (with a penalty) from a subsequent period; and to trade emissions.
    - A new category is proposed to encourage rapidly developing countries to voluntarily adopt emissions budgets.

REPORTING AND COMPLIANCE

- The proposal establishes procedures to ensure adequate reporting, measurement, review and compliance.

    - Countries would have to set national systems for measuring emissions accurately, achieving compliance and ensuring enforcement.
    - Annual reports on measurement, compliance and enforcement efforts for the relevant budget period would be required and made available to the public.
    - Implementation efforts would be reviewed under the Convention by expert teams and discussed at appropriate meetings of the Parties.
    - The Parties would be assigned responsibility for determining the consequences of non-compliance, for example: denial of emissions trading/ joint implementation rights; loss of voting and other decision-making rights.

ADVANCING DEVELOPING COUNTRY EFFORTS

- The proposal advances implementation of developing country commitments under the Convention by:

    - Requiring developing countries to identify and adopt "no-regrets" measures to mitigate net greenhouse gas emissions.
    - Requiring developing countries to prepare annual emissions inventories and report on steps taken to reduce emissions.
    - Establishing a process for reviewing developing country reports and improving emissions reduction strategies.

EMISSIONS TRADING AND JOINT IMPLEMENTATION

- The U.S. submission provides for full emissions trading among countries with emissions budgets (provided they are in compliance with all obligations under the agreement); it also provides for "joint implementation," through which countries without emissions budgets could create and transfer emissions reduction credits achieved by qualified projects.

LONG-TERM EFFORTS UNDER THE CONVENTION

- The proposal calls for periodic review of the agreement as scientific knowledge and information grow; it also requires the Parties to seek establishment of a long-term goal for atmospheric greenhouse gas concentrations.
- The proposal calls for establishment of a date certain for negotiation of emissions obligations for all Parties, and calls for development of graduation mechanisms to strengthen the obligations of developing nations.

SECTION IV

TRADEABLE EMISSION PERMITS -

EMISSION QUOTA TRADE - A SIMULATION EXPERIMENT

RESULTS OF THE EMISSION QUOTA TRADE SIMULATION EXPERIMENT BY THE PARTICIPANTS

Dr. ARTHUR PELCHEN
UMB UmweltManagementBeratung Hacker GmbH
Selchowstr. 1, D-14199 Berlin, Germany

1. Introduction

During the course the participants carried out an emission quota trade simulation experiment along the lines of an experiment that was prepared on behalf of the Nordic Council of Ministers' Ad Hoc Group on Energy Related Climate Issues by Professor Peter Bohm and Björn Carlén at Stockholm University last year among Denmark, Finland, Norway, and Sweden (Bohm, 1997).

Special thanks to Morton Søberg, who 'translated' the original experiment for the purpose of this course, gave an introductional explanation, analysed the trading data and presented the results.

Within the course the aim of this simulation experiment was:

- to balance the more theoretical lecture in the course by the participants' own activities,
- to promote the discussion of the subject among the participants,
- to deepen the understanding of the trading mechanism,
- to give an example for differences in the marginal abatement costs of different countries and
- to show the potential gains from a system of tradeable emission permits.

2. Experimental design and basic assumptions

The participants were divided in four groups each representing one of the four Nordic countries. The experiment is based on the assumption:

- that all four countries have committed themselves to a stabilisation of their individual CO_2 emissions on 1990 levels by the year 2000 and

J. Hacker and A. Pelchen (eds.), Goals and Economic Instruments for the Achievement of Global Warming Mitigation in Europe, 397–406.

- that individual countries are allowed to exceed their targets, if this is compensated by an additional reduction in another Nordic country and the total emissions within the Nordic countries are stabilised.

The goal for the participants as representatives of the individual countries was to minimise the social abatement cost for their countries. According to the assumptions each individual country may meet its target entirely on its own or partly by purchasing contracts of emissions reductions from the other participating countries. The latter option is attractive, if it is less costly than making the same volume of reductions at home. A country selling (profitable) emissions reductions contracts would have to make emissions reductions in the year 2000 equal to its own emissions target *plus* contracted additional emissions reductions commitments.

The emissions according to the business-as-usual (BAU) scenarios, the emissions according to the stabilisation target based on the 1990 emission level and the necessary emission reduction are given for each Nordic country in TABLE 1.

TABLE 1. Business-as-usual scenarios (BAUs), targets for the year 2000 and the resulting emission reduction commitment (CO_2)

Country	BAU (Mton)	Target (Mton)	Emission reduction (Mton)
Denmark	53.0	52.1	1.7
Finland	60.0	54.0	6.0
Norway	41.0	35.6	5.4
Sweden	62.9	61.3	1.6
Total	217.7	203.0	14.7

3. CO_2 abatement costs

The basis for mutually profitable emission reduction trades between two countries is that their marginal emissions abatement costs differ at their respective target levels for emissions in the year 2000. Since all participating countries have some version of a CO_2 tax system in operation now and most likely also for the year 2000, the basic instrument for adjusting the emissions in each of the four countries can be taken to be changes in the CO_2 tax required for reaching the specific emissions levels for that year. Hence the technical abatement costs refer to the costs associated with increasing the CO_2 tax rate in order to implement CO_2 emission reductions. The estimated marginal technical abatement cost functions are assumed to be known publicly for all four countries. They are given in FIGURES 1 - 4.

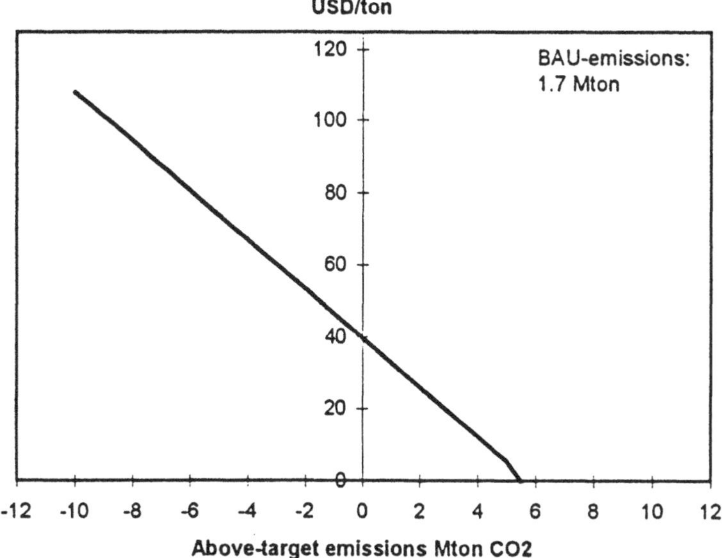

FIGURE 1. Marginal technical abatement costs for Denmark (centred around the year 2000 emission target)

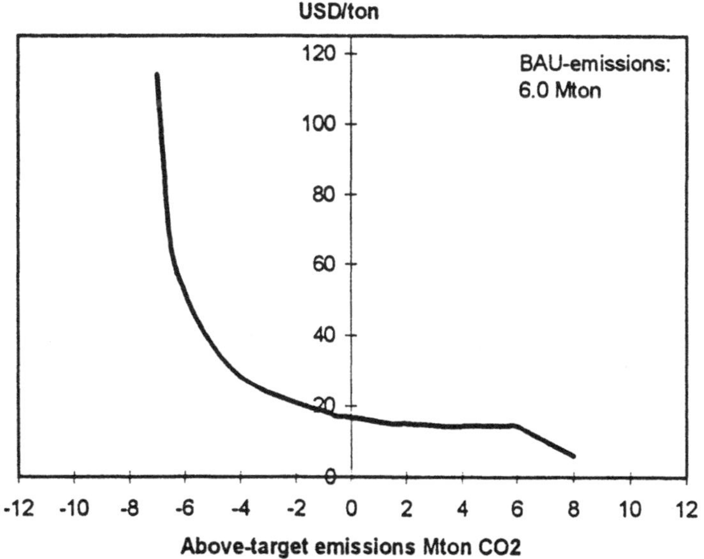

FIGURE 2. Marginal technical abatement costs for Finland (centred around the year 2000 emission target)

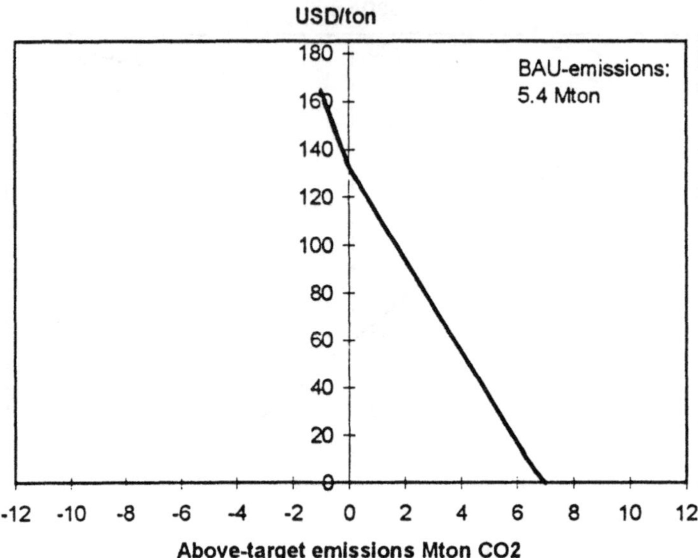

FIGURE 3. Marginal technical abatement costs for Norway (centred around the year 2000 emission target)

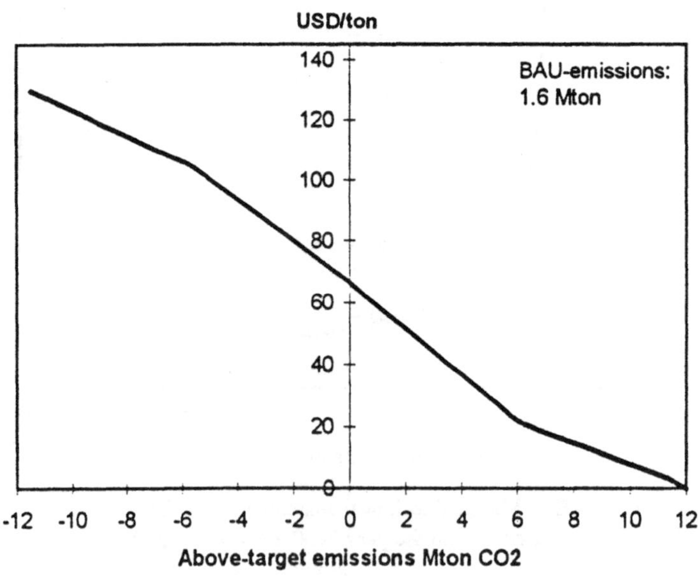

FIGURE 4. Marginal technical abatement costs for Sweden (centred around the year 2000 emission target)

Note that the cost curves are centred around the year 2000 emissions targets for the four countries. Thus, in the case of Denmark, zero emissions above-target corresponds to emissions of 52 Mton in the year 2000.

The amount of emissions reductions required in the absence of Joint Implementation is the difference between the target level and the business-as-usual level. Denmark's reduction thus would be 1.7 Mton. Hence, the aggregate *technical* abatement costs incurred were Denmark to fulfil its obligations unilaterally is equal to the area between 1.7 and 0 under the marginal technical abatement costs curve. As a matter of course, the same kind of reasoning applies to the other countries as well. Thus, the stated BAU-emissions level in each figure is equal to the country's national CO_2 emissions reductions target. Given this framework, joint implementation implies that a buyer country of emissions reductions is shown by the fact that its emissions exceed its target level. And vice versa: a seller country will be characterised by having located its emissions level below the target level.

However, the relevant costs are not likely to be the technical abatement costs, but rather the *social* abatement costs which include considerations of real-world political constraints concerning, e.g., employment and distribution effects. These social costs would probably be known much less perfectly outside the individual country incurring the costs. Hence these are viewed as private information during the course of this simulation experiment. The marginal social abatement cost function for each of the Nordic countries were distributed only to their associated negotiating teams.

The importance of the technical costs is that the negotiating teams can use this information to guess what the politically relevant and hence, trade relevant, cost data of the other countries can be, since the marginal social abatement costs are typically higher than the marginal technical abatement costs. Actual trade negotiations should therefore be based on estimates about the costs relevant for potential trade partners.

4. Negotiation rules

The negotiations were restricted to two rounds of negotiation on different days, one of 45 minutes and one of 90 minutes. Between the negotiations no trading was allowed. The results of the experiment were presented on the day after the final round of trading. The negotiation rules, which define the basic terms and the wording to be used in the negotiations are given in FIGURE 5.

The basis for communication between the groups was an e-mail system with one personal computer for each group and an additional computer for the

- Messages must be of the following types, where country names are used within quotation marks:

  Bid: „..............." wants to purchase ..... million metric tons of CO_2 emission reductions from „..............." for USD ....

  Ask: „..............." offers ..... million metric tons of CO_2 emission reductions to „..............." for USD ....

  Acceptance: „..............." accepts „...............“=s bid/ask and sells/buys ...... million metric tons of CO_2 emission for USD ....

- Upon acceptance a trade constitutes a preliminary contract. However, the negotiators may choose to offer/accept to bind and annul such a contract as well as a binding contract by means of the following messages:

  Offer to bind a preliminary contract: „..............." wants to bind the contract where we sell/buy ..... million metric tons of CO_2 emission reductions to/from „..............." for USD .... If you agree, both parties are mutually bound by this contract, at a penalty (as described below).

  Accepting to bind a preliminary contract: „..............." accepts the offer to bind the contract where we buy/sell ..... million metric tons of CO_2 emission reductions from/to „..............." for USD ....

  Annulling a binding contract: „..............." annuls the contract with „..............." where we agreed to buy/sell ..... million metric tons of CO_2 emission reductions for USD....... We acknowledge that this means that we owe you a penalty of USD........... (see below).

  Annulling a preliminary contract: „..............." annuls the contract with „..............." where we agreed to buy/sell ..... million metric tons of CO_2 emission reductions for USD ....

- Preliminary contracts that have not been annulled at the end of the last round of negotiations will then be binding. A preliminary contract can be annulled at no cost.

- A binding contract can be annulled at the cost of a penalty of 15 percent of the contract value during the first round of trading and 30 percent during the last round of trading. Penalties are debited to the annulling party and credited to the party affected.

- Disagreements between the parties are decided on by the Secretariat and cannot be appealed.

FIGURE 5. Negotiation rules

'secretariat' of the experiment, which received a copy of every message sent between the negotiation teams for later analysis.

5. Incentives for good performance of the negotiation teams

The driving force behind emissions trade negotiations, had they been real, would be for each country's negotiators to try to minimise the country's net costs of meeting its international CO_2 emissions commitments. When the trade negotiations are hypothetical, as is the case here, a feasible incentive mechanism which as much as possible mimics this driving force needs to be instituted. In the Nordic experiment this mechanism consisted of the countries depositing, prior to the start of the negotiations, their estimated social emissions reductions cost functions, i.e., the relationship that guides the negotiators in each country, to an international evaluations team. This team, which consisted of professor Scott Barrett, London Business School, Dr. Jean-Charles Hourcade, CIRED-CNRS, Paris, and professor Robert Stavins, Harvard University, evaluated the emissions trade agreements reached against this background and published its findings of how relatively successful the participating countries had been in their negotiations (Barrett et al., 1996).

A similar mechanism was applied to the simulation experiment in this course in that the cost savings achieved by each negotiation team have been made official as well as discussed during the presentation and assessment after the end of the experiment.

6. The course of the experiment

Besides some minor technical problems with the computer network, the general course of the experiment was satisfactory. Nevertheless a few points are worth mentioning:

- The fact that the expressions 'Joint Implementation' and 'trading with emission permits between states' (and/or private entities) were introduced and used within the Advanced Study Course with a slightly different meaning (see the relevant papers in this volume), unfortunately led to some confusion among the participants. This only became clear in the final discussion on the results of the experiment.
- At the same time the incentives to take the experiment fully serious were obviously not strong enough, which showed itself in some of the e-mails not adhering to the formulated instructions and stating contents far away from the necessities of the experiment. Some participants rather used the time to compensate for the tiring lectures and to have some fun sending

nonsense e-mails. A fact that certainly was not observed in the original Nordic experiment with the government officials of the four countries.

- This also led to some 'asks' and 'bits', especially at the beginning of the negotiations, that were not rationally explainable from the marginal social abatement cost curves. Some trades that were not negotiated to an end, because one of the teams stuck to irrational prices (see also '7. Results').
- In addition the discussion in some of the negotiating teams was at least part of the time dominated by individual participants, who possibly were not the most qualified to lead the decision-making-process. This also yielded decisions that were not always carried by the team. This was even true in cases where either the majority of the negotiation team did not want to sell at a certain given price, because it was irrationally low. Or vice versa, if the majority wished to reduce the price for an 'ask', because it was still above the marginal social abatement cost curve.
- Many of the trade agreements were only finalised during the last few minutes of the experiment with some confusion.

Since it was not necessarily the aim of the simulation experiment to yield a result at the perfect competition price realising the (theoretical) maximum net gains, the overall aims as stated in the introduction were achieved in spite of these points.

7. Results

During the experiment three trade agreements were reached. The details are given in TABLE 2.

TABLE 2. Details of trade agreements

Seller	Volume (Mton)	Price (USD/t)	Buyer
Denmark	3.0	65	Norway
Finland	0.5	60	Denmark
Finland	0.5	50	Sweden

It is interesting to see that Denmark bought emission permits from Finland and managed to sell them on to Norway at a higher price. This yielded Denmark a net gain in excess of the efficient net gains, but reduces the efficiency gains for the group of Nordic countries (see also TABLE 5). On the other hand Finland could have sold much more if they had not stuck to their high prices. Finland therefore lost a significant share of the theoretical efficiency gains. Finally Sweden could have offered a higher price, since their marginal social abatement costs are higher even for the BAU scenario.

TABLE 3 compares the average prices achieved in this experiment with those from the Nordic experiment and the optimum of the perfect competition prices.

TABLE 3. Comparison of prices

Weighted average price	62.50 USD/ton
Perfect competition price	50.31 USD/ton
Nordic experiment average	52.27 USD/ton

It is obvious from TABLE 3 that the average price achieved in this experiment is higher than the perfect competition price. This is due to the fact that both Finland and Denmark tried to bargain for a price higher than would have been rational, i. e. higher than their social abatement costs at this level of reduction.

TABLE 4 presents a comparison of the situation without trading, i. e. with unilateral reductions, with the results achieved by the negotiating teams in this experiment and the theoretical net gains.

TABLE 4. Main results of the simulation experiment

	Unilateral reduction		Trade	Ex-post JI		
Country	Em.red.	Cost	Export/Import (-)	Em.red.	Cost	Net gain
Denmark	1.7	61	2.5 (1.19)	4.2	189	37 (5.1)
Finland	6.0	94	1.0 (5.76)	7.0	114	35 (132.2)
Norway	5.4	456	-3.0 (-4.40)	2.4	127	134 (194.5)
Sweden	1.6	102	-0.5 (-2.55)	1.1	68	9 (25.1)
Total	14.7	713	Exp=Imp=3.5	14.7	498	215

(Note: Fully efficient numbers in parentheses.)

According to TABLE 4 Joint Implementation in this form reduces aggregate abatement costs by 30%. But only 60% of the maximum net gains were realised by the negotiation teams in this experiment. In physical terms 3.45 Mton (or 49%) of the efficient trade is not carried out.

This leads to the final assessment of the performance of different groups, which is shown in TABLE 5.

As mentioned above Denmark gained from buying permits relatively cheap and selling them on at a higher price leading to higher net gains. But it has to be kept in mind that this is at the expense of the efficiency achieved by all four

TABLE 5. Assessment of the performance of the evaluation teams

	Net gains as % of unilateral abatement costs	Net gains as % of efficient net gains
Denmark	60	725
Finland	7	26
Norway	29	68
Sweden	8	35

Nordic countries taken together. Hence ignoring Denmark for this reason the best performance was delivered by Norway, that realised more than two thirds of the efficient net gains. Least efficient was the team from Finland with less than a third of the possible gains.

8. Conclusion

Although the results of the experiment were not as good as in the original Nordic experiment it was shown that trading between the countries reduces the cost of global warming mitigation even if carried out between countries with fairly similar economic conditions. The gains will be even greater if carried out between countries with marginal abatement cost curves that differ more than in this case. At the same time more flexibility can be added to a future system of global warming mitigation if the trading is performed by and between private entities (companies etc.) and not only between government officials. Taken together the experiment fulfilled the expectations and aims as stated in the introduction.

Nevertheless there are some possible improvements, if the experiment should be carried out again:

- Clarify the meaning of the important terms (mainly Joint Implementation and trading) and make sure that they are used in a uniform way throughout the course.
- Offer a reward for the best performing negotiation team to insure that the experiment is taken appropriately serious.

References

Bohm, P. (1997) *Joint Implementation as emission quota trade: An experiment among four Nordic countries*, Nordic Council of Ministers, NORD 1997:4, Nordic House of Publishing, Copenhagen

Barrett, S., Hourcade, J.-S. and Stavins, R. (1996) Evaluation report, in: P. Bohm (ed.), *Joint Implementation as emission quota trade: An experiment among four Nordic countries*, Nordic Council of Ministers, NORD 1997:4, Nordic House of Publishing, Copenhagen

SECTION V

COMPARISON OF INSTRUMENTS AND SUMMARY

ASSESSING THE DIFFERENT INSTRUMENTS IN CLIMATE CHANGE MITIGATION FROM THE PERSPECTIVE OF ECONOMICS*

Prof. Dr. ALFRED ENDRES
Distance Teaching University
Profilstr. 8, D-58084 Hagen, Germany

* *The paper benefited from the discussions within the project on "Kooperative Lösungen für internationale Umweltprobleme: Eine ökonomische Analyse unter besonderer Berücksichtigung des Einflusses gesellschaftlicher Interessengruppen" sponsored by the Volkswagen-Foundation, Germany (II/69982).*

1. Introduction

In this conference international experts from scientific institutions, governments and NGOs have discussed the serious problems posed by greenhouse gas emissions and alternative means for mitigation.

The role of the paper at hand is to summarize, point out certain highlights and draw a few policy conclusions. The paper also attempts to put the many details of information provided during the conference into a unifying economic perspective.

In order to move towards this goal I am going to proceed as follows:
First, I will briefly review the traditional approach of economics towards the evaluation of environmental policy instruments.[1] The economists among the readers may skip this paragraph. Its role for this paper is to provide a common background for an interdisciplinary readership.

Among the main criteria in the traditional economic assessment of environmental policy instruments are

- the efficiency of an environmental policy instrument, i.e., its ability to attain a predetermined emission level at minimum cost
- the accuracy of an instrument, i.e., its ability to arrive at the target without cumbersome trial and error processes

---

[1] e.g., A. Endres (1994).

J. Hacker and A. Pelchen (eds.), Goals and Economic Instruments for the Achievement of Global Warming Mitigation in Europe, 409–425.

- the political acceptability which relates to an instrument's appeal to interest groups influencing the political decision making process.

After these foundations I will put the criteria of instrumental assessment in the context of global warming. Not all of the criteria have the same features in the international setting that they have for traditional regional environmental analyses. These qualifications may lead to suggestions regarding the design of environmental policy instruments used to mitigate global warming. I am going to wrap up with a summary and some policy conclusions.

2. Traditional Assessment of Environmental Policy Instruments...

2.1. EFFICIENCY

In the literature, effluent charges are considered to be efficient instruments. Reducing emissions to the point at which marginal abatement costs equal the effluent charge rate, polluters automatically meet the efficiency condition of equal marginal abatement costs across all sources. Transferable discharge permits (TDPs) are equally efficient, substituting the permit price for the tax rate. In contrast, command and control regulations forcing the polluters to maintain certain emission standards are not efficient, in general. Regulators do not have sufficient information to differentiate the standards for individual sources according to their marginal abatement costs.

2.2. ACCURACY

In regard to its ability to attain with precision a predetermined pollution target level, the effluent charge suffers from the fact that it acts as a „price" for discharging to which each firm is free to adjust its discharge quantity. The predetermined emission target will therefore be achieved only after a process of trial and error.[2] In contrast, in the case of transferable discharge permits the level of pollution is under direct control of the environmental agency through the number of permits auctioned off (or given away free). So permits exhibit maximum accuracy. This also holds true for a command and control policy that specifies absolute emission ceilings for the polluting sources.[3]

---

[2] It is assumed that the agency setting the tax is unable to accurately estimate aggregate marginal abatement cost.

[3] If the regulation sets forth terms of emission *concentration* the accuracy of the command and control approach is deficient.

2.3. POLITICAL ACCEPTABILITY

In the recent years the use of market based instruments for environmental protection has considerably increased. The OECD (1994a), (1995) and others[4] have presented surveys and analyses that are quite impressive. However, it is still true that command and control policies are the dominating approach to environmental policy by far. Moreover, where economic incentives are used they are very often designed in a manner contradicting environmental economic wisdom. Furthermore, they are often combined with command and control elements, also considerably attenuating their efficiency and their ecological effectiveness.[5]

So for environmental economists, claiming that market based instruments are superior for the economy as well as for the environment, there is an obvious need to explain why they are not used more often in political practice.

Taking a public choice view, political and economic reality is not to be explained according to some criterion of social optimality. To the contrary, the outcome of political decision making is reasoned to be the result of a struggle among different groups of society with diverging interests. Following this approach the prevailing policy mix in environmental policy (and other policies) must be explained by affinity to the interests of powerful groups in society.[6]

It has been argued in the literature that for many reasons incentive based instruments are neither very attractive to the polluting industries nor to the environmental bureaucracy. Moreover, they seem to be doubtful to most of the public at large for psychological reasons.

In addition to those reasons, elaborated very well in the literature mentioned above, the acceptability of transferable discharge permits might suffer from the „ecological transparency" of this policy. TDPs are the only system among all the alternative policy suggestions in which the government has to clearly specify its goals in terms of total discharge quantities for each pollutant tackled. Certainly, this is a no nonsense kind of an approach which is to be welcomed from an ecological point of view. However, governments are not so keen of giving away this kind of a precise target information for this provides the public and the political opponents with a very clear reference to judge the government's performance *ex post*. Moreover, within the transferable discharge permit system it would be quite visible if

---

[4] See, e.g. D. Cansier, R. Krumm (1996), P. Michaelis (1996).

[5] See D. Ewringmann et al. (1993) and E. Gawel (1993) discussing the recent amendments of the German water waste charge law (Abwasserabgabengesetz).

[6] The „classical" reference regarding work using the public choice approach to explain instrumental choice of practical environmental policy is J.M.Buchanan, G. Tullock (1975), Further references are, e.g,, A. Endres, M. Finus (1996), B.A. Forster (1993), B. S. Frey, F. Schneider (1997), F. Jaeger (1994), Ch. 12, P. Pashigian (1985), M. Pearson (1995), L. de Savornin Lohman (1994), B. Yandle (1989), K. Zimmermann (1996).

certain interest groups are privileged by exemptions. This is a general practice which can be much easier hidden from the general public if CAC-policy (command and control) or effluent charges are used. So if it is true that it is beneficial for politicians to act in an environment which is quite intransparent it does not come as a big surprise that an environmental policy instrument which is tough and clear is not so popular in political circles.

3. ...Applied to GHG-Reduction

3.1. EFFICIENCY IN THE GREENHOUSE

3.1.1. Introduction

Of course, the question of efficiency is particularly important in the context of reducing global warming. We are not dealing with a trifle problem where only a moderate amount of resources is at stake. To the contrary and quite obviously, deciding on which instruments to be used to fight global warming, we are at the crossroads in terms of the world wide use of energy resources.[7] A wrong decision not using the potential of efficient policy instruments will cost billions of ECUs.

Of course, the superiority of market based environmental policy compared to the command and control approach may be somewhat lower in practice than the environmental economic textbooks suggest. This is so because neither is the practical command and control approach always as clumsy as its textbook dummy nor do practical incentive based policy instruments reach the perfection of their counterparts in economic „Nirvana-models". Let me give three examples for the latter point: First, involved firms usually do not operate in a perfectly competitive setting and this is an efficiency problem in its own right interacting with the efficiency properties of incentive based instruments (IBIs).[8] Second, the bilateral and sequential approach often taken in practice (e.g. by joint implementation) cannot capture gains from trade as perfectly as a fully competitive market could.[9] Third, incompleteness of information prevents that the potential of efficiency gains is perfectly exploited.[10] Today, enlighted environmental economists are aware of these qualifications, and others.

Still, there is no doubt that market based incentive instruments are significantly

---

[7] There is a very nice introduction into the economics of global warming in J. Blank, W. Ströbele (1994). See also G. Boero (1995), R. Perman (1994).

[8] See A. Endres (1997b) For a general review of environmental policy under imperfect competition. Specifics of the design of carbon taxes in countries with non-competitive markets are discussed in R. Golombek, J. Braten (1994).

[9] See, e.g., G. Klaassen (1996).

[10] See, e.g., C. Hagem (1996)

superior to the command and control approach in terms of efficiency. There is overwhelming theoretical and impirical evidence for this hypothesis and I do not think that this claim is seriously controversal in the scientific[11] or in the political arena.

So the interesting question is not whether to use IBIs or CAC-policy.[12] The question is how market based incentives can be designed to meet the high expectations regarding their efficiency. Another question is which ones of the many forms of market based instruments are suited best to meet the demands of global warming mitigation.[13]

Concentrating on the first question, for the time being, it is useful to remember the reasons for IBIs' high potential for efficiency: The starting point is that different polluters incur different costs when obliged to fight the same kind of environmental pollution. Using IBIs puts a self-selection process into motion differentiating equilibrium pollution abatement efforts of different polluters according to their reduction costs. In order to achieve maximum efficiency this decentralized process must be allowed to function *comprehensively* and *smoothly*.

3.1.2 Comprehensiveness

Applying the principle of comprehensiveness to using IBIs for the mitigation of global warming it first must be noted that global warming is caused by CO_2 emissions predominantly but not exclusively. So a comprehensive market based approach to mitigate global warming requires that all pollutants (as many pollutants as possible) contributing to global warming are integrated in the market process.[14] That is a tax must be on CO_2 equivalents and not on CO_2. In case permits are used the trades must be in units of CO_2 equivalents and not in terms of CO_2 alone. Another element of the „comprehensiveness for efficiency"-program is that as many countries as possible should participate. It is particularly productive to have countries participating which are heterogeneous in terms of their industry structure because this increases the variety of reduction cost functions involved in the market process. Moreover, comprehensiveness requires that not only countries are involved in the market system but all other relevant institutions, especially individual

---

[11] See, e.g., the Statement on Climate Change, signed by more than 2000 economists including seven Nobel Laureats, published in the International Society for Ecological Economics' „Ecological Economics Bulletin", Vol. 2 (1997), pp. 16 - 18.

[12] In this paper, it is implied that GHG-policy is applied by more than one country. For unilateral activities see, e.g., L. Bovenberg (1993), W.F. Richter (1997).

[13] The most systematic procedure requires: First, for a given environmental problem design the most appropriate variant for each type of an environmental policy instrument. Second, compare these optimal variants and choose the optimum optimorum.

[14] Theoretical issues of environmental policy with multiple pollutants are discussed in A. Endres (1986), P. Michaelis (1992) and T. Zylicz' paper in E.C. v. Ireland (1994).

firms. This is particularly important for the design of transferable discharge permit systems. Here, the harmonization between international and national environmental policy is a prominent feature to serve the principle of comprehensiveness. An important part of the efficiency potential of the transferable permit approach would be foresaken if countries would allocate national emission reduction targets via an international permit market and then the national emission budget would be allocated among national polluters by the command and control approach. So in order to capture its efficiency potential a permit policy would have to be designed vertically consistent from the individual emitting source to the global transaction level.[15]

3.1.3 Smoothness

There is a second requirement to enable market based instruments to live up to the expectations regarding their efficiency: The design does not only have to be comprehensive it also has to ascertain the smoothness of market operations. Efficiency suffers if the selfselection process mentioned above is hampered by governmental regulations. This issue has plagued market incentive instruments wherever they have been used so far on a national or regional level and it has significantly diminished the practical performance of these systems. In order to arrive at a successful incentive based policy for global warming mitigation we really have to learn from these experiences. These regulations, strangleing market forces, may be environmentally or otherwise motivated. An example for the latter type is the regulation of energy markets. In this sector (real or alleged) reasons regarding a reliable provision of energy have led to various kinds of regulations. These cause a significant threat to the use of IBIs undermining their conditions of operation. The idea of IBIs is to confront the users of scarce resources with prices „which tell the truth" and then let them make their own choices regarding what to consume, what to produce and how to produce it. If prices reflect the opportunity costs of resources, profit maximizing and utility maximizing choices result in efficient allocations. However, regulations constrain the individual choice of production processes and inputs. Subsidies distort resource prices. There are many examples for that as in the case of regulations requiring the input of a certain amount of home coal or subsidies for this kind of coal. Most recently, adverse effects of regulation in the US energy market on the effectiveness of the transferable discharge permit system within the US acid rain program have been demonstrated.[16] Under the current energy regulation firms prefer to comply to environmental standards by input substitution or by

---

[15] In the general controversy about TDPs, the argument that „thin markets" might pose an important threat to the functioning of the system played a crucial role. It is important to note that in a global permit system which is comprehensive in the sense defined above nobody has to worry about thin markets.

[16] See the excellent analysis of R. Schwarze (1997). Related issues are dealt with in M.B. Kornshaw, J. B. Kruse (1996).

installation of scrubbers compared to using emission permits, even if the former strategies are more expensive than the latter. The reason is that the firms are not able to appropriate any gains from using least cost permit strategies since the system uses cost based rate regulations. Firms selling permits must hand over gains from trade to their customers via reduced rates. Moreover, investments in scrubbers and substitution processes count as capital costs with the consequence that a „fair rate of return" is incorporated in the rates approved by the regulatory agency. Opposed to that, expenditure for permits counts as variable cost. For this, no return is allowed to be calculated in the rates.

It follows from this and other experience that in a system where individual choice is constrained and input prices are heavily distorted environmental policy cannot be expected to be efficient even if IBIs are used. So hoping that our readers are not tired of all these double dividend hypotheses around, we venture to present another one:

Deregulation not only increases consumer welfare but also enables IBIs in environmental policy to unfold their potential to provide environmental protection at lower cost and to stimulate environmentally friendly technical progress.

Of course, what has been said above also applies to regulation in the environmental sector itself.[17] There is a long story of command and control environmental regulations inhibiting IBIs which have been added to these regulations to work properly. Controlled emissions trading in the seventies and eighties in the USA is an example[18] as well as the German water effluent charge act. The latter has been constrained heavily right from the start and has been recently strangled by additional regulations.[19]

3.2. GLOBAL POLLUTION AND STOCK EFFECTS: IMPLICATIONS FOR ECOLOGICAL ACCURACY

3.2.1. Regional Ecological Incidence

It has been argued above, that it is an important advantage in terms of efficiency that IBIs allocate emission reduction loads among the polluters according to their ability to reduce. For this advantage there is a price to pay in terms of the ecological incidence of pollution:

Since the polluters are distributed in space the regional profile of equilibrium pollution is a result of the market process. It cannot be easily foreseen and it might

---

[17] See H. Zimmermann (1993) on conceptual issues of combining charges and direct regulations.

[18] See, e.g., R.W. Hahn, G.L. Hester (1989), T.H. Tietenberg (1990).

[19] See, e.g., D. Ewringmann et al. (1993), E. Gawel (1993). More recently, CAC environmental regulation has severely constrained the market incentives of the US acid rain program. See R. Schwarze (1997).

imply local loads of emissions which are intolerable.[20] This might considerably weaken the political acceptability of IBIs, an issue that will be taken up later in this paper.[21]

It is a very important point in the context we are discussing today that the problems of regional equilibrium distribution of pollution using incentive based instruments are not relevant in the case of global warming. The effect the emissions of greenhouse gases have on global warming do not depend on their place of origin.[22] So it should be noted that certain problems which are a matter of concern regarding the application of incentive based instruments in the general discussion can be ignored in the context of greenhouse gas mitigation. So in this respect the conditions for the application of IBIs are particularly favourable in the case of global warming.

3.2.2 Controlling Aggregate Emissions

Another issue regarding the ecological incidence of an environmental policy application (beyond regional distribution) is how precisely the level of *aggregate* emissions can be controlled. As I mentioned earlier transferable discharge permits have important advantages in this respect compared to effluent charges.

It might be argued that this property is not so relevant in the context of greenhouse gas mitigation. The reason for this position is that the detrimental effects of greeenhouse gases on temperature are not generated by the annual flow of pollutants. It is rather the accumulated stock of these pollutants which does the damage. Accordingly, it is not so important to precisely control the annual discharge quantity but to make sure that the stock of pollution does not exceed a certain level in the long run. Following this reasoning a drawback of effluent charges compared to transferrable discharge permits, thought to be important in the general discussion, seems to vanish in the context of global warming.

Even though certainly plausible, this argument loses some of its appeal at second glance. The reason for the lacking ecological accuracy of effluent charges is not only that marginal abatement costs are unknown to the policy maker setting the tax

---

[20] There are severeal sophisticated ideas for permit design to cope with this problem. However, this modified variants of the permit approach suffer from lower efficiency and higher transaction costs compared to their textbook counterpart. (See, e.g., S.E. Atkinson's paper in G. Klaassen, F.R. Førsund (1994) and G. Klaassen (1996)).

[21] Moreover, if the damage function relating pollution concentrations in the environmental media to monetized environmental damages is non linear, total damage does not only depend on total pollution but also on the regional distribution of pollutants. So a market based approach leading to hotspot problems may well incur higher total damages compared to a command and control policy spreading the same amount of aggregate pollution more evenly.

[22] Let me note in passing that it has to be taken into consideration whether a certain greenhouse gas has other unwarranted environmental impacts than contributing to global warming. These may or may not have a regional component.

rate. Another reason is that the emissions discharged in the tax adjusted equilibrium increase in a process of inflation and economic growth. Even though inflation is not among the most pressing macroeconomic problems today, this might well change some time in the future. Most certainly, there is an enormous growth potential in the world economy particularly due to the development of countries which are now on the verge of industrialization. If global warming mitigation would be achieved by effluent charges there would be questions regarding the *persistence* of this policy's efficacy. Most probably the economic leverage effect of this policy would be undermined by the macroeconomic tendencies mentioned. I would consider it very optimistic regarding the flexibility of international environmental policy to expect that this erosion would be compensated by adjusting the tax rates once negotiated.

3.2.3. Intertemporal Ecological Incidence

Above, the ecological incidence of applying environmental policy instruments to mitigate global warming has been discussed with respect to total emissions and to the regional distribution of these emissions. However, there is a third dimension of ecological incidence that is often overlooked in environmental economic assessments. I am referring to the distribution of greenhouse gas pollution over time.

Suppose it is the goal of environmental policy to stabilize the accumulated stock of greenhouse gas pollutants at a certain level within a certain transition period (and possibly to reduce this stock after the transitional phase). Then, *aggregate* greenhouse gas pollution emitted during this transitional face is not allowed to exceed a certain amount compatible with the stock oriented goal. Since the policy is aiming at a certain result in terms of stocks and not in terms of the flows of the individual years, the distribution of the annual emissions during the transitional phase is ecologically irrelevant, as long as the constraint defined in terms of aggregate emission is met. If this is so the distribution of annual emissions in the transitional period can be decided according to economic criteria. Choosing the efficient trajectory of emission reductions over time is a source of welfare gains just as important as the efficient distribution of pollution reduction across different polluting sources. It is the instrument of transferable discharge permits which has the biggest potential to find this efficient trajectory and make sure that it is followed closely by the practical adjustment process. In this statement it is presupposed that the discharge permit policy is designed allowing for emission banking and emission borrowing. Within this system each firm can „reshuffle“ individual reduction loads according to its own technological and economic needs and possibilities. Moreover, there are additional efficiency gaines from intertemporal permit trade among different firms. Firms which are able to reduce emissions fast have an incentive to do so selling permits to firms which are less flexible over time.

Neither effluent charges nor the command and control approach allow for this flexibility to gain from intertemporal reassignment of reduction loads as the transferable discharge permit approach including banking and borrowing does.[23]

3.3. AN INTERNATIONAL PERSPECTIVE OF POLITICAL ACCEPTABILITY

3.3.1 Global Optimality versus Individual Rationality

In section 2.3, above, several problems for market based environmental policy approaches to be accepted in society and in the political arena have been mentioned. Certainly, these problems also apply in an international setting. Even worse, the international nature of the global warming problem causes additional problems.
In the national context environmental economics assumes that there is a central agency - the government - willing and able to design and enforce environmental policy, even if the interests of some groups of society are violated. In the light of a public choice perspective this hypothesis is somewhat doubtful, even in the national context. In an international setting, however, it is completely inappropriate. Since there is no world government, international environmental policy has to be agreed upon among volontarily participating sovereign countries. This has important implications for the focus of environmental economic analyses as well as for the selection and design of environmental policy instruments.[24]

If we believe in the government striving for the common good it make sense for economics to focus on the characterization of pollution levels and the design of environmental policy instruments which are optimal in the sense of social welfare maximization. In the international setting with its lack of a central institution striving for the common good, however, social (global) optimality is not a criterion which is very relevant for practical decision making. Instead, the question to be asked is how environmental policy goals can be specified and how environmental policy instruments can be designed in order to make each individual country benefit from their endorsement and application. Therefore, in the international setting, the primary focus of environmental economic analysis is the *individual rationality* of environmental policy for each individual country instead of social optimality for the international community as a whole. Applied to the choice of environmental policy instruments and their design this means that each country must achieve a positive net benefit from the application of this instrument.

In order to make the fundamental difference between the two concepts clearer it may be helpful to have a look at the formal conditions for social optimality on the

---

[23] To reach the optimal trajectory by effluent charges the governmental agency setting the time path of tax rates has to have complete information on the technical and economic flexibility of the firms over time. For intertemporal efficiency in the TDP-system it is sufficient for the individual firm to know its own flexibility.

[24] See, e.g., A. Endres (1996), (1997a).

one hand and individual rationality on the other hand. Interpreting the reduction of greenhouse gases as a pure public good it is wellknown that the condition for social optimality is that the sum of country specific marginal benefits from pollution reduction is equal to each individual country's avoidance cost, i.e., $MAC_i = \sum MB_i$ where i is the index of one of the total number of n countries. Opposed to that, the condition for individual rationality is that the difference between benefits and cost for each country i in the situation applying the environmental policy instrument (situation 1) is bigger than the net benefit in the situation without the environmental policy instrument (situation 0), i.e., $B_i(1) - C_i(1) > B_i(0) - C_i(0)$.

The readers might realize that I understand the issue of political acceptability not so much as a question of justice but as a question of distribution. I concede that these two concepts are related but I maintain that they are not identical. Justice is concerned with the conformity of practical institutions or results of allocation processes with exogenous ethical norms. Distribution is concerned with the question how to design institutions and attribute the results of allocation processes in a way to make every participant happy (happier).

Interpreting political acceptability as a matter of distribution leads to the statement that a necessary condition for political means to be acceptable is that any country is better of applying this means than it has been before. Applied to the question of mitigating the greenhouse effect it follows from this approach that it is necessary for any effective and efficient greenhouse policy to include a system of transfers between countries. To understand that we must remember that for an efficient solution countries have to abate the more emissions the lower their avoidance costs are. Without a system of international transfers this would only be a feasible result of an international agreement in the special case where the countries with low emission avoidance costs are also the ones with the highest benefit from emission reduction. Of course this does not have to be so in practice. It is possible that a country with low avoidance cost has relatively little interest in the reduction of greenhouse gases. On the other hand, there may be countries that are highly interested in reducing global warming but do not have favourable technical or economical conditions to do that. In this situation tayloring the abatement efforts according to abatement costs can only be achieved if low cost /low benefit countries are compensated for their efforts by high cost/high benefit countries.

It is obvious that a system of transferable discharge permits is quite well equipped to meet the requirement that greenhouse policy must include transfer payments. Transfers are a built-in property of this instrument. Of course, it is not impossible to include transfers also in a system of effluent charges or even command and control. E.g., the proceeds from an energy tax might be paid into a fund from which transfers could be financed. However, there are serious and obvious doubts as to whether large public funds would be managed efficiently in terms of the reduction of greenhouse gases. It seems to be more promising to leave the decision on who compensates whom for what kinds of emissions reduction to the market instead of to public decision.

3.3.2 Starting Point Distribution of Property Rights

Returning to the transferable discharge permit policy, it must be noted that the answer to the question of who sells and buys how many permits crucially depends upon how the permits are initially distributed among the countries. It is obvious that this issue of initial permit endowment is essential for the political acceptability of the permit system. There are a whole lot of alternative possibilities which have been discussed in the paper of Dr. Collier and therefore will not be repeated here.

I just want to make the critical (and self-critical) remark that traditional economic theory has not been much of a help to deal with this important issue. The reason for this deficit is that traditional environmental economics has been primarily concerned with questions of efficiency and - at least in theory - the initial distribution of property rights is irrelevant regarding the result of the allocation process. So believe in the Coase-theorem and in the benevolance of the government (which is supposed to solve distributional issues separately from the allocation problem) has led the mainstream of the economic profession to ignore a problem which is crucial for the solution of international environmental problems. Probably the unsolved distributional problem of the initial allocation of transferable discharge permits is the most important obstacle to the introduction of this policy at all. Of course, I do not deny the importance and explanatory power of neo-classically based environmental economics, but in this important issue it has to be supplemented by evolutionary and institutional economics. Maybe, we can get important insights from this kind of theorizing on the evolution of property rights. Maybe we can also learn from the analysis of the evolution of the national states and of international institutions outside of the environmental field. I think that in these issues, new dimensions of environmental economic research programs can be found.

So due to this lack of theoretical foundations regarding distributional issues I cannot come up with a ready made proposal based on economic theory regarding the best initial distribution of permits among countries. However, the concept of individual rationality mentioned earlier may well provide an important contribution. The core of the idea using the individual rationality priciple would be to allocate initial permit endowments among the countries according to the countries' specific net benefits of pollution reduction. The higher the net benefit of pollution reduction is for a specific country the higher is the individual emissions reduction load acceptable for this country. Of course it would not be efficient to actually enforce country specific emission reductions according to this initial allocation of permits, since a distribution of reduction efforts according to pollution reduction net benefit is not identical to the allocation according to marginal abatement costs. However, this deficit would be eliminated by permit trade between the countries. Permit trade subsequent to initial allocation would further efficiency and not violate individual rationality because by voluntary trade no participant would be made worse off. Of course, a requirement for the realization of this idea which is hard to meet in practice is that country specific marginal avoidance costs and marginal avoidance benefits have to be quantified.[25] Still, the idea of individual rationality in initial permit allocation, sketched here, might revitalize the discussion which seems to have arrived at some kind of a deadlock to date.

3.3.3 Stability of International Agreements

I am afraid that adding the complicated issue of individual rationality to the list of topics which are discussed in the national context under the headline of political acceptability is not enough to characterize specific acceptability problems in the international setting. A final problem has to be mentioned:

Even if we assume that it is possible to find a policy which improves the situation of any individual country, unfortunately this does not imply that the situation is stable. This is so because it might be attractive for each individual participant in an international agreement to withdraw in order to enjoy the benefits of a free rider. It is wellknown in the literature that international agreements on environmental and other issues are plagued with the temptations of the prisoners' dilemma even if each participating country benefits from the arrangement in the first place. I cannot elaborate this problem further[26] but I would like to mention that international greenhouse policy must also include sanctions against breaches of contract. This is true irrespective of which instrumental approach is persued.

3.3.4 The Role of National IBI-Experience

I raised a couple of issues complicating the political acceptability question in the international context compared to the national context. However, I can also contribute to consolation. Contrary to the national context, at least as far as Germany is concerned, there is one important player in the game, endorsing a system of emissions trading for CO_2. As J. Wolfe has elaborated in his paper, the United States have made a detailed proposal for such a system. This proposal is based on the experience of the US using national programs of emission trading. I think that the international community can benefit a lot from that experience. This reduces an important argument against the use of tranferable discharge permits, i.e. that it is a system unknown and therefore hard to assess for the parties involved. Of course the

---

[25] A selection of journal articles dealing with this issue is reprinted in T. Tietenberg (1997). (This sampler also features the most influential articles on carbon taxes and TDPs for the control of global warming.) A demand revealing mechanism designed to reveal information on abatement cost and benefits is suggested by J. Falkinger, F. Hackl, G.J. Pruckner (1997). General methodological aspects of monetisation are dealt with in A.Endres, K. Holm-Müller (1998).

[26] See, e.g., A.Endres (1997a).

weight the United States carry in international negotiations is a light of hope for many frustrated followers of the transferabe discharge permit idea in Europe.

4. Conclusion

Based on the analysis presented above, I come out with considerable sympathy for a transferable discharge permit system as the means to fight global warming. I do not want to be dogmatic about it: Closer analysis might turn out reasons for a mixed policy system where some sources of GHG pollution have to be regulated by other variants of environmental policy.[27] Still, from the economic point of view TDPs should play the dominant role in greenhouse policy.[28]

Let me sum up the reasons for this assessment:

- Efficiency is a must in greenhouse policy. Neither industrialized nor developing countries can afford (are willing to afford) to waste resources. I am afraid that if we cannot manage to slow the emission of GHGs efficiently, we will not get any significant reduction at all, incuring the risk of disasterous consequences for future generations.
- Among the efficient policy instruments transferable discharge permits stand out since
    - they also allow for resource saving adjustments over time,
    - they have a built-in compensation mechanism to advance countries' individual rationality for systems participation,
    - they do not generate huge funds of public money which are threatened to be mismanaged and
    - they are able to provide a reduction in GHGs which is persistant and therefore able to achieve the intended effect on GHG-stocks.

So my recommendation is that the European Union should push for the introduction of an international transferable discharge permit system in the political arena.

The time seems to be just about right to do so because the US (and New Zealand) have submitted high quality transferable discharge permit proposals, as we have learnt from the papers of Joanna Depledge and James Wolfe. The papers discussing theory and practice of TDPs at this conference have convinced us that

---

[27] A possible reason is that for some non-CO_2-GHGs monitoring is difficult. (See D.G. Victor (1991).)

[28] In this conference, many contributors have dealt with practical issues of GHG-permit design including issues of certification, monitoring and enforcement. Additional references are J. Heister, P. Michaelis (1991), P. Koutstaal, A.Nentjes (1995) and T. Tietenberg's paper in E.C. v. Ireland (1994).

OECD, IEA and other institutions can provide ample know-how to help putting these proposals into practice. So the European Union should join forces with these countries and institutions.[29] Maybe with a major joint effort it is possible to fulfill an important and difficult task: To mobilize the tremendous powers of the market system to secure the living conditions for the future generations on this planet.

References

Barrett, S. (1992) International Environmental Agreements as Games, in R. Pethig, *Conflicts and Cooperation in Managing Environmental Resources*, Springer, Berlin, Heidelberg, New York, pp. 11-37.

Blank, J. and Ströbele, W. (1994) Das CO_2-Problem aus ökonomischer Sicht, *WiSt*, **23**, 552-557.

Boero, G. (1995) Global Warming: Some Economic Aspects, *Scottish Journal of Political Economy*, **42** , 99-112.

Bovenberg, A.L. (1993) Policy Instruments for Curbing CO_2 Emissions: The Case of The Netherlands, *Environmental and Resource Economics*, **3**, 233-244.

Buchanan, J. and Tullock, G. (1975) Polluters' Profits and Political Response, *American Economic Review*, **65** , 139-147.

Cansier, D. and Krumm, R. *(1996) Empirische Analyse zur Besteuerung von Luftschadstoffen*, Tübinger Diskussionsbeiträge, No. 68, Tübingen.

Capros, P., C. Konrad et al., (1996) Double Dividend Analysis: First Results of a General Equilibrium Model (GEM-E3) Linking the EU-12 Countries, in Carraro, C. and Siniscalco, D. (Eds.) *Environmental Fiscal Reform and Unemployment*, Kluwer Academic Publishers, Dordrecht, pp. 193-227.

Carraro, C. and Siniscalco, D. (Eds.) (1993) *The European Carbon Tax*, Kluwer Academic Publishers, Dordrecht.

Carraro, C. and Siniscalco, D. (Eds.) (1996) *Environmental Fiscal Reform and Unemployment*, Kluwer Academic Publishers, Dordrecht.

Cesar, H. and Zeeuw, A. d. (1996) Issue Linkage in Global Environmental Problems, in A. Xepapadeas, *Economic Policy for Environment and Natural Resources*, E. Elgar, Cheltenham, pp. 158-173.

Ebert, U. (1991) Pigouvian Tax and Market Structure, *Finanzarchiv*, **49**, 154-166.

Emons, W. (1994) The Provision of Environmental Protection Measures under Incomplete Information, *International Review of Law and Economics*, **14** , 479-491.

Endres, A. (1986) Charges, Permits and Pollutant Interactions, *Eastern Economic Journal*, **12**, 327-336.

Endres, A. (1994) *Umweltökonomie*, Wissenschaftliche Buchgesellschaft, Darmstadt.

Endres, A.(1996) Designing a Greenhouse Treaty: Some Economic Problems, in: R. van den Bergh, E. Eide (Eds.), *The Law and Economics of the Environment*, Juridisk Forlag, Oslo, pp. 201-224.

Endres, A.(1997a) Negotiating a Climate Convention, The Role of Prices and Quantities, *International Review of Law and Economics*, **17**, 147-156.

---

[29] Theoretical issues of coalition formation in international environmental policy are dealt with in M.Hoel, K. Schneider (1997) and T. Sandler, K. Sargent (1995). See also S. Kverndokk's paper in E.C. v. Ireland (1994).

Endres, A. (1997b) Incentive-Based Instruments in Environmental Policy: Conceptual Aspects and Recent Developments, *Konjunkturpolitik*, forthcoming.

Endres, A.and Finus, M. (1996) Zur Neuen Politischen Ökonomie der Umweltgesetzgebung - Umweltschutzinstrumente im politischen Prozess, *Zeitschrift für Angewandte Umweltforschung*, **9**, 88-103.

Endres, A. and Holm-Müller, K. (1998) *Die Bewertung von Umweltschäden - Theorie und Praxis sozioökonomischer Verfahren*, Kohlhammer, Stuttgart.

Ewringmann, D. et al. (1993) Abschied von der Abwasserabgabe, *Zeitschrift für Angewandte Umweltforschung*, **6**, 153-171.

Falkinger, J., Hackl, F. and Pruckner, G.J. (1996) A Fair Mechanism for Efficient Reduction of Global CO_2-Emissions, *Finanzarchiv*, **53**, 308-331.

Forster, B. A. (1993) *The Acid Rain Debate: Science and Special Interests in Policy Formation*, Iowa State University Press, Ames.

Frey, B.S. and Oberholzer-Gee, F. (1996) Zum Konflikt zwischen intrinsischer Motivation und umweltpolitischer Instrumentenwahl, in H. Siebert (ed.), *Elemente einer rationalen Umweltpolitik*, Mohr, Tübingen, pp. 207-238.

Frey, B.S. and Schneider, F. (1997) Warum wird die Umweltökonomik kaum angewendet? *Zeitschrift für Umweltpolitik und Umweltrecht*, **20**, 153-170.

Gawel, E. (1993) Novellierung des Abwasserabgabengesetzes - Ein umweltpolitisches Lehrstück, *Zeitschrift für Umweltrecht*, **5**, 159-164.

Golombek, R. and Braten, J. (1994) Incomplete International Climate Agreements, *Energy Journal*, *15*, 141-165.

Hagem, C. (1996) Joint Implementation Under Asymmetric Information and Strategic Behavior, *Environmental and Resource Economics*, **8**, 431-447.

Hahn, R.W. and Hester, G.L. (1989) Where Did All the Markets Go? An Analysis of EPA's Emission Trading Program, *Yale Journal of Regulation*, **6**, 109-153.

Hamm, R. and Hillebrand, B. (1992) Elektrizitäts- und regionalwirtschaftliche Konsequenzen einer Kohlendioxid- und Abfallabgabe, Untersuchungen des Rheinisch-Westfälischen Instituts für Wirtschaftsforschung , **5**, RWI, Essen.

Hansjürgens, B.(1992) *Umweltabgaben im Steuersystem*, Nomos, Baden-Baden.

Heister, J. and Michaelis, P. (1991) *Umweltpolitik mit handelbaren Emissionsrechten*, Mohr, Tübingen.

Herber, B.P. and Raga, J. T. (1995) An International Carbon Tax to Combat Global Warming, *The American Journal of Economics and Sociology*, **54**, 257-267.

Hoel, M. and Schneider, K. (1997) Incentives to Participate in an International Environmental Agreement, *Environmental and Resource Economics*, **9**, 153-170.

Ireland, E.C. van, (ed.) (1994) *International Environmental Economics*, Elsevier, Amsterdam.

Jaeger, F. (1994) *Natur und Wirtschaft*, Rüegger, Chur.

Jochimsen, M. and Kirchgässner G. (eds.) (1995) *Schweizerische Umweltpolitik im international Kontext*, Birkhäuser, Basel.

Klaassen, G., Emission Trading for Air Quality Standards: Opening Pandora's Box? Paper presented at the Seventh Annual Conference of the European Association of Environmental and Resource Economists (EAERE) in Lisbon, June 27-29, 1996.

Koutstaal, P. and Nentjes, A. (1995) Tradeable Carbon Permits in Europe: Feasibility and Comparison with Taxes, *Journal of Common Market Studies*, **33**, 219-233.

Kuik, O., Peters, P. and Schrijer, *N. (1994) Joint Implementation to Curb Climate Change*, Kluwer Academic Publishers, Dordrecht.

Michaelis, P. (1992) Global Warming: Efficient Policies in the Case of Multiple Pollutants, *Environmental and Resource Economics*, **2**, 61-77.

Michaelis, P.(1996) *Ökonomische Instrumente in der Umweltpolitik*, Physica, Heidelberg.

OECD (1994) *Managing the Environment: The Role of Economic Instruments*, OECD, Paris.

OECD (1995) *Environmental Taxes in OECD Countries*, OECD, Paris.

Opschoor, J.B. and Turner, R.K. (Eds.) (1994) *Economic Incentives and Environmental Policies*, Kluwer Academic Publishers, Dordrecht.

Pashigian, P. (1985) Environmental Regulation: Whose Self-Interests are Being Protected,*Economic Inquiry*, **23**, 551-584.

Pearson, M. (1995) The Political Economy of Implementing Environmental Taxes, *International Tax and Public Finance*, **2**, 357-373.

Perman, R. (1994) The Economics of the Greenhouse Effect, *Journal of Economic Surveys*, **8**, 99-132.

Pethig, R. (Ed.) (1992) *Conflicts and Cooperation in Managing Environmental Resources*, Springer, Berlin, Heidelberg, New York.

Richter, W.F. (1997) Über die Ineffizienz einer nationalen Energiesteuer, *WiSt*, **26**, 124-130.

Sandler, T. and Sargent, K. (1995) Management of Transnational Commons, *Land Economics*, **71**, 145-162.

Savornin Lohman, L. de (1994) Economic Incentives in Environmental Policy: Why are they White Ravens? in J.B. Opschoor, R.K. Turner, *Economic Incentives and Environmental Policies*, Kluwer Academic Publishers, Dordrecht, 55-67.

Schmutzler, A. (1996) Pollution Control with Imperfectly Observable Emissions, *Environmental and Resource Economics*, **7**, 251-262.

Schwarze, R.(1997) SO_2 im Sonderangebot? Zur Entwicklung des US-Marktes für SO_2-Lizenzen und den Perspektiven von Zertifikatmodellen in der Luftreinheitpolitik, *Zeitschrift für Ange wandte Umweltforschung*, **10**, forthcoming.

Smith, S. (1995) The Role of the European Union in Environmental Taxation, *International Tax and Public Finance*, **2**, 375-387.

Tahvonen, O.(1995) International CO_2 Taxation and the Dynamics of Fossil Fuel Markets, *International Tax and Public Finance*, **2**, 261-278.

Tietenberg, T.H.(1990) Economic Instruments for Environmental Regulation, *Oxford Review of Economic Policy*, **6**, 17-33.

Tietenberg, T.H. (Ed.) (1997) *The Economics of Global Warming*, E. Elgar, Cheltenham.

Ulph, A. and Ulph, D. (1994) The Optimal Time Path of a Carbon Tax, *Oxford Economic Papers*, **46**, 857-868.

Victor, D.G. (1991) Limits of Market-Based Strategies for Slowing Global Warming: The Case of Tradeable Permits, *Policy Sciences*, **24**,199-222.

Zimmermann, H. (ed.) (1993) *Umweltabgaben*, Economica, Bonn (Economica).

Zimmermann, K. (1996) Zur Politischen Ökonomie von Umweltsteuern, *ORDO*, **47**, 169-194.

ASSESSMENT OF THE DIFFERENT ECONOMIC INSTRUMENTS BY THE PARTICIPANTS

Dr. ARTHUR PELCHEN
UMB UmweltManagementBeratung Hacker GmbH
Selchowstr. 1, D-14199 Berlin, Germany

Abstract. The evaluation of the different instruments at the end of the course showed that participants where about 80 % certain about the scientific evidence for an anthropogene enhancement of the greenhouse effect. About three quarters preferred a per capita based allocation of the necessary emission reduction. In the overall assessment the participants rated the instruments in average in the following ascending hierarchy: voluntary agreements, tradeable permits with state actors, Joint Implementation, tradeable permits with individual actors, energy tax and carbon tax. There were differences in this rating according to different educational backgrounds.

1. Introduction

At the end of the Advanced Study Course on 'Goals and instruments for the achievement of global warming mitigation in Europe' the participants were given a chance to assess the different instruments that they have been discussing during the course. The relevant instruments for the assessment were:

- Voluntary agreements on emission reduction [see this volume, pp. 159-204]
- Energy/carbon taxes [see this volume, pp. 205-251]
- Joint Implementation [see this volume, pp. 253-292]
- Tradeable permits (state actors) [see this volume, pp. 293-393]

J. Hacker and A. Pelchen (eds.), Goals and Economic Instruments for the Achievement of Global Warming Mitigation in Europe, 427–435.

2. Material and methods

For this purpose a questionnaire was distributed among the participants (see Annex I). In order to find out whether participants differed in their evaluation of global warming and the possible instruments for mitigation according to their educational backgrounds and nationalities, the latter were asked as basic information to subdivide the sample. Furthermore the participants were asked to state their certainty (on a scale between 0 = uncertain and 10 = certain) about the scientific evidence of an anthropogene enhancement of global warming and whether they rather preferred a flat rate or a per capita allocation of the necessary global emission reduction. In the main part of the evaluation participants were asked to give an overall assessment on all instruments by assigning a value between '-5' (bad instrument) and '+5' (good instrument). Additionally they were asked to give a detailed assessment (values between '-5' and '+5') according to the following criteria that had been discussed for this purpose in the course (see article by Pelchen in this volume, pp. 146-151):

- Speed of achieving the environmental target
- Accuracy of achieving the environmental target
- Economic efficiency
- Compatibility with market economy
  - Type of economic planning
  - Type of price fixing
  - Type of market
  - Type of ownership of means of production
  - Type of goal of businesses
- Compatibility with the polluter-pays-principle

Finally participants could state their own additional criteria for evaluation and again assign values between '-5' (bad instrument) and '+5' (good instrument).

For the question of allocation of emission reductions it was assumed that there might be a difference in the evaluation by participants from nations with high versus low per capita GHG emissions. Therefore the data were analysed for the complete sample and for the two groups of nations given in the questionnaire, based on above or below average GHG emissions compared to the European Union mean value. In the analysis of the assessment of the different instruments the data were evaluated for the complete sample and for each of the educational backgrounds, because it was assumed, that the individual rating depended on the educational background. Furthermore it was planned to compare the overall assessment stated by the participants with that calculated from the detailed assessment according to the methods outlined earlier in the course (see article by Pelchen in this volume, pp. 151-156).

3. Results and discussion

3.1. DATA BASIS

TABLE 1 shows the data basis and its distribution on educational backgrounds and groups of nationalities.

TABLE 1. Data basis of the assessment

Group	N	%
Total	30	100,0
Natural scientists etc.	9	30,0
Economists etc.	13	43,3
Political Sciences etc.	8	26,7
Group of countries with above average per capita emissions	18	60,0
Group of countries with below average per capita emissions	9	30,0
Other countries (not included in the assessment of the subgroups)	3	10,0

3.2. CERTAINTY OF EVIDENCE

The results of the question on the certainty of scientific evidence for an anthropogene enhancement of the greenhouse effect are given in FIGURE 1. Hence after the course the participants were 80,5% certain, that there is an anthropogene enhancement of the greenhouse effect.

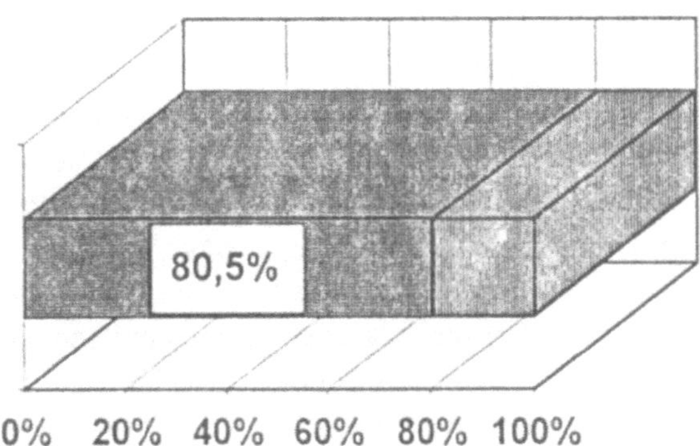

FIGURE 1. Certainty of evidence for an anthropogene enhancement of the greenhouse effect

3.3. FLAT RATE VERSUS PER CAPITA ALLOCATION

According to FIGURE 2, which presents the results of the question on the allocation of the necessary global emission reduction, 73% of all participants prefer a per capita based and 10% flat rate allocation. 17 % favour other methods of allocation. There was no significant difference on this question between the various groups of countries. This might be due to ignorance on the actual level of emissions of the participants country in relation to the EU mean value or to the fact, that a per capita based system is considered to be more just. Therefore only the results of the whole sample are presented here.

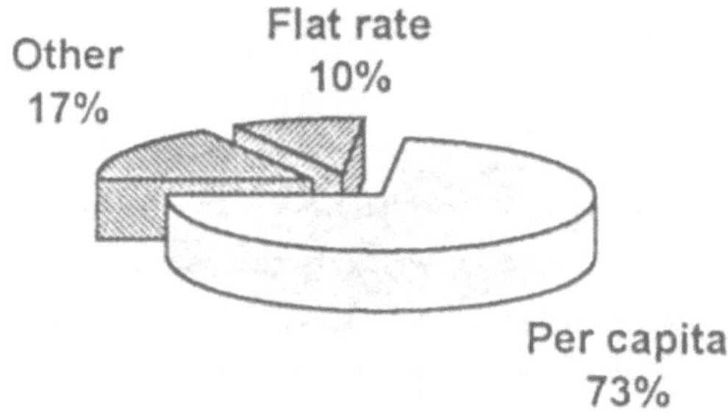

FIGURE 2. Preference of all participants on the allocation of the global emission reductions

3.4. OVERALL ASSESSMENT

FIGURE 3 shows the results of the overall assessment for the whole sample as well as for the subgroups according to the different educational backgrounds.

According to FIGURE 3 the average of all participants rated a carbon tax as the best instrument for the mitigation of global warming, followed by an energy tax. The other instruments were in descending order the tradeable permits with individual actors, JI, the tradeable permits with state actors and voluntary agreements. Comparing the different subgroups it is remarkable that the natural scientists rate all the instrument worse than the other subgroups indicating that they might prefer instruments that have not been presented in the course and the assessment. For the instruments voluntary agreements, carbon and energy tax as well as JI there is a clear order between the ratings of the natural scientists, the economist and the political scientists: the latter rate all these instruments better than the former. For tradeable permits this order changed: these instruments are rated better by the economists than by the political scientists. A possible explanation for this might on the one hand be in the deeper understanding of the instrument by the economists or on the other hand in the familiarity of the political scientists with the tax based instruments.

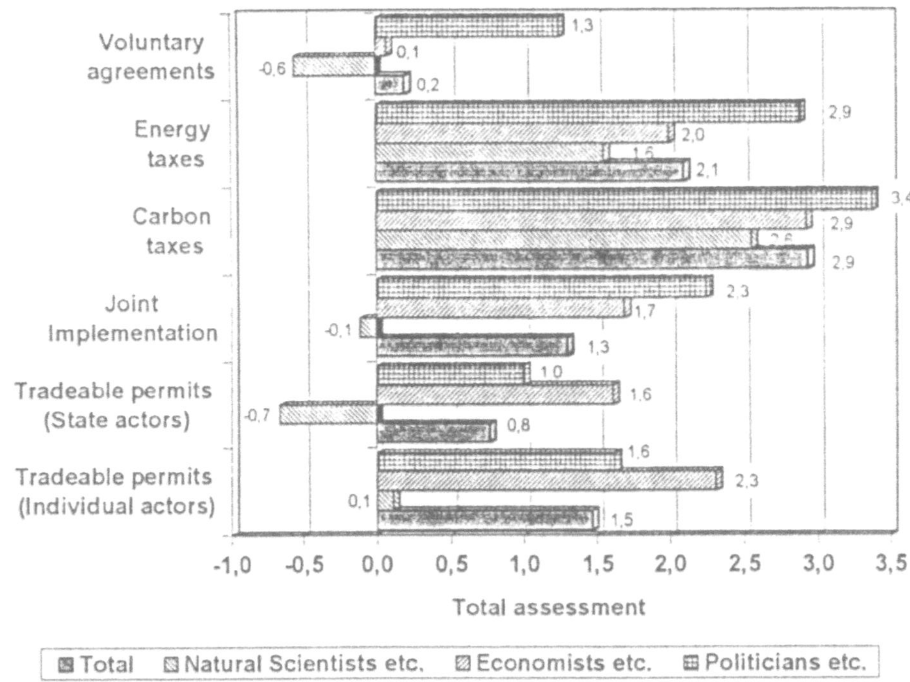

FIGURE 3. Average overall assessment of instruments by all participants and by subgroups according to educational background

3.5. DETAILED ASSESSMENT

Due to a lack of completely filled in questionnaires the detailed assessment as well as the comparison between the overall assessment stated by the participants and that calculated from the detailed information could not be carried out.

3.6. ADDITIONAL CRITERIA

The following additional criteria where stated by the participants:

- Social justice/distribution/equity/redistributive goals
- Revenue yield
- Political/systematical compatibility
- International competitiveness
- Possibility/stability of an international agreement

4. Conclusion

In relation to the aims of the Advanced Study Course the results of this assessment show, that at the end of the course all the participants acknowledge the problem of global warming. On the one hand it was obviously possible during the course to build a basic knowledge on the physics of global warming and its economic and social implications among all the educational backgrounds. The differences in the assessment of the economic instruments according to educational background on the other hand prove that much more in depth information and discussion is needed to overcome the prejudices between different schools of thought. In this light another Advanced Study Course concentrating on the economics of climate change might be a useful follow-up.

Annex I (Questionnaire, abbreviated)

EU Advanced Study Course Berlin, 20.-26.07.1997

'Goals and instruments for the achievement of global warming mitigation in Europe'

Evaluation of the political and economic instruments for the achievement of global warming mitigation in Europe

Introduction:

This evaluation is based on the method used for the 'First theoretical evaluation of political and economic instruments' earlier in the course. In order to facilitate the analysis of the data according to different subgroups please answer the following additional questions:

Educational focal point:	Natural Sciences	☐
	Economical Sciences	☐
	Political and administrative Sciences	☐
Nationality:	Luxembourg, Germany, Belgium, Netherlands, Finland, Denmark, Great Britain, Ireland	☐
	Austria, Greece, Italy, France, Sweden, Spain, Portugal	☐
	Other	☐

How certain is in your personal opinion the evidence for an anthropogene enhancement of the greenhouse effect on a scale from 1 (uncertain) to 10 (certain)? ☐

For the allocation of the necessary global reduction to individual countries do you rather prefer

a flat rate reduction for all countries or ☐
a differentiated approach on a per capita basis? ☐

Overall assessment:

Please assign values from +5 to -5 as overall assessment for the following political and economic instruments:

- Voluntary agreements on emission reductions
- Energy taxes
- Carbon taxes
- Joint Implementation
- Tradeable permits (state actors)
- Tradeable permits (individual actors)

Overall assessment

		+5	+4	+3	+2	+1	0	-1	-2	-3	-4	-5	
Voluntary agreements on emission reductions	good												bad
Energy taxes	good												bad
Carbon taxes	good												bad
Joint Implementation	good												bad
Tradeable permits (state actors)	good												bad
Tradeable permits (individual actors)	good												bad
		+5	+4	+3	+2	+1	0	-1	-2	-3	-4	-5	

Additional criteria of your choice:

If you wish to assess the given instruments against any other criteria of your choice, feel free to add two criteria in the tables below. Indicate the extreme positions of the criteria and fill in your values for the evaluation.

		+5	+4	+3	+2	+1	0	-1	-2	-3	-4	-5	
Voluntary agreements on emission reductions													
Energy taxes													
Carbon taxes													
Joint Implementation													
Tradeable permits (state actors)													
Tradeable permits (individual actors)													
		+5	+4	+3	+2	+1	0	-1	-2	-3	-4	-5	

Detailed assessment:

For the following more detailed evaluation you are asked to assign values from +5 to -5 for the 10 different criteria and their stated characteristics given in the tables below. (The criteria have been introduced to you in an earlier lecture of the course.)

Voluntary agreements on emission reductions

Energy taxes

Carbon taxes

Joint Implementation

Tradeable permits (state actors)

Tradeable permits (individual actors)

		+5	+4	+3	+2	+1	0	-1	-2	-3	-4	-5	
Type of economic planning	decentral												central
Type of price fixing	market based												admini-strative
Type of market	competition												monopoly
Type of ownership of the means of production	private												state
Type of motivation of economic subjects	economic												reason/ obligation
Type of goal of businesses	profit making												economic reason
Accuracy of achieving the environmental target	high												low
Speed of achieving the environmental target	fast												slow
Economic efficiency	high												low
Compatibility with the polluter-pays-principle	compatible												non-compatible
		+5	+4	+3	+2	+1	0	-1	-2	-3	-4	-5	

(The original questionnaire contained one individual table of this kind for each of the political and economical instruments mentioned above.)

SELECTED RESULTS OF THE WORKING GROUPS

Dr. A. PELCHEN
UMB UmweltManagementBeratung Hacker GmbH
Selchowstr. 1, D-14199 Berlin, Germany

1. Introduction

During the course two sessions of four working groups were carried out with the aim:

- to balance the more theoretical lectures by the participants' own activities,
- to promote the discussion of the subjects among the participants,
- to deepen the understanding of the main issues,
- to summarise important issues of the course and
- to evaluate the contents of the different sections of the course.

For the working groups the participants were divided in four groups, that were mixed according to nationalities and educational background. Some selected results from the working group discussions, that are thought to be interesting for the general reader, are presented in this paper.

2. Selected results

2.1. EVALUATION OF SESSION II, WORKING GROUP IV

2.1.1. *Tasks*

Introduction: Imagine you are the delegation of an industrialised country to the Conference of the Parties (CoP III) in Kyoto in December.

Task 1: Present a strategy for a reasoned allocation of the necessary global emission reduction, that would ideally suit the conditions of an industrialised country and that gives your country a maximum of benefits.

Task 2: What strategic sacrifices to your optimum strategy designed above would you be willing to make in order to find a compromise with the developing countries?

J. Hacker and A. Pelchen (eds.), Goals and Economic Instruments for the Achievement of Global Warming Mitigation in Europe, 437–439.

2.1.2. *Results*

The working group presented an allocation of the necessary emission reductions including their strategic sacrifices for industrialised and developing countries as well as countries in transition on the assumption that the global emissions are reduced by 10 % in 2010 and by 40 % by 2050 compared to 1990 levels. The proposal for the allocation is shown in Table 1 and Figure 1.

Table 1: Proposal for regional emission reductions

	1990	Δ to 1990	2010	Δ to 1990	2050
Industrialised countries	60	-25 %	45	-66%	20
Developing countries	20	+25 %	25	0%	20
Countries in transition	20	0 %	20	0%	20
Index	100	-10%	90	-40	60

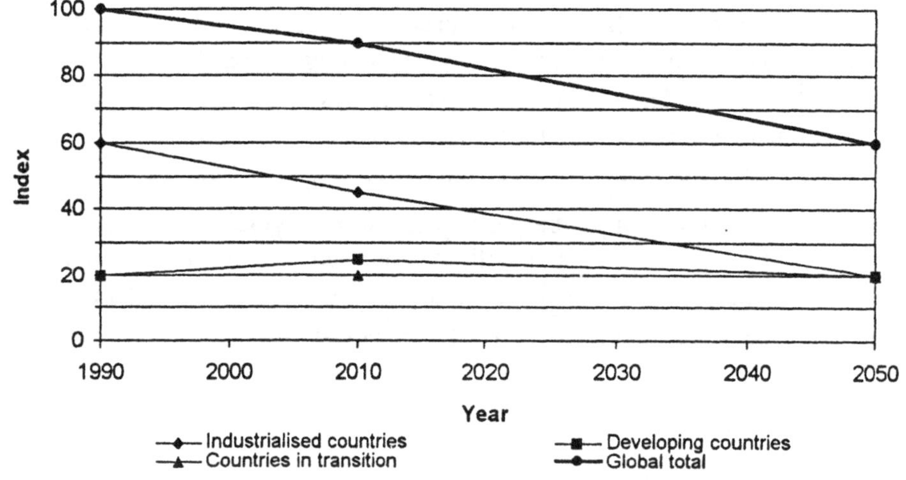

Figure 1: Proposal for regional emission reductions

The proposal was accompanied with the offer of financial and know-how transfers both towards developing countries and countries in transition.

There was a lot of discussion on this proposal in the plenum, because although it puts the main load of emission reductions on the industrialised countries, it still allows for fairly different per capita emissions in the individual groups of countries even in the year 2050, where all groups have an equal share in the total emissions. Nevertheless the developing countries will hold a much greater population at this time and therefore much lower per capita emissions compared to the industrialised countries.

2.2. EVALUATION OF SESSION IV, WORKING GROUP III AND IV

2.2.1. *Tasks*

Both working groups were asked to design a tradeable permit system under special consideration of:

- allocation of permits,
- who is allowed to trade,
- budget periods and banking/borrowing
- monitoring and enforcement.

2.1.2. *Results*

Concerning the results for the mechanism of allocation of permits working group III proposed:

- permits should first be issued to national governments on the international level according to the obligations decided upon in a future protocol, possibly by the Climate Secretariat,
- the national governments are then responsible for the national allocation to individual sources,
- grandfathering of permits for the fuel- or energy- intensive sectors,
- auctions of permits for the other sectors.

There was much discussion in the plenum on the second part of this proposal, because this gives preferential treatment to those industries that are the main polluters with regard to carbon dioxide and therefore violates the polluter-pays-principle.

Both working groups wanted to allow trading 'on all levels', i. e. between all possible partners like governments, institutions, private persons and industry.

The results concerning the borrowing of permits are summarised below:

- borrowing from later periods should only be allowed with discounting, i. e. to borrow 10 permits from the following year, 12 permits of that year would have to be sacrificed,
- borrowing should only allowed for the owner's own use and not for sale.

No precise suggestions were made for monitoring and enforcement.

3. Conclusion

It has to be emphasised, that all results were only expressing the opinion of the individual working groups and were not necessarily agreed on by all participants as was proven by the intense discussion in the plenum.

LIST OF AUTHORS

LIST OF AUTHORS

Christoph Bail
European Commission
Rue de la Loi 200
B-1049 Bruxelles
Belgium
Tel: +32-2-2 95 40 99
Fax: +32-2-2 96 95 57

Dr. Terry Barker
University of Cambridge
Sidgwick Avenue
Cambridge CB3 9DE
England
Tel: +44-1223-33 52 00
Fax: +44-1223-33 52 99
eMail: DAEADMIN@ECON.CAM.AC.UK

Richard Baron
International Energy Agency
9, rue de la Fédération
F-75739 Paris, Cedex 15
France
Tel: +33-1-40 57 67 24
Fax: +33-1-40 57 67 39
eMail: richard.baron@iae.org

Stephen Bill
European Commission
Rue de la Loi 200
B-1049 Bruxelles
Belgium
Tel: +32-2-2 95 78 83
Fax: +32-2-2 95 65 01

Karl Ludwig Brockmann
ZEW Center for European Economic Research
Postfach 10 34 43
D-68034 Mannheim
Germany
Tel: +49-621-12 35 214
Fax: +49-621-1 23 52 26

Dr. Ute Collier
Friends of the Earth Trust
26-28 Underwood Street
London N1 7JT
England
Tel: +44-171-5 66 16 71
Fax: +44-171-4 90 08 81
eMail: utec@foe.co.uk

Joanna Depledge
UN Climate Change Secretariat
Postfach 26 01 24
D-53153 Bonn
Germany
Tel: +49-228-8 15 16 02
Fax: +49-228-8 15 19 99
eMail: kschmidt@unfccc.de

Prof. Dr. Alfred Endres
Distance Teaching University
Postfach 940
D-58084 Hagen
Germany
Tel: +49-2331-98 44 50
Fax: +49-2331-98 44 49

Jürgen Hacker
UMB UmweltManagementBeratung
Hacker GmbH
Selchowstr. 1
D-14199 Berlin
Germany
Tel: +49-30-8977720
Fax: +49-30-8234597
eMail: umb-Hacker@t-online.de

Minh Ha Duong
Centre International de Recherche sur
l'Environnement et le Developpement
1, rue du 11 Novembre
F-92120 Montrouge
France
Tel: +33-1-46 12 18 50
Fax: +33-1-40 92 93 17
eMail: haduong@alize.msh-paris.fr

Prof. Dr. Klaus Hasselmann
Max-Planck-Institute for Meteorology
Bundesstr. 55
D-20146 Hamburg
Germany
Tel: +49-40-4 11 73-2 36
Fax: +49-40-4 11 73-2 50
eMail: klaus.hasselmann@dkrz.de

Prof. Catrinus J. Jepma
University of Groningen/Amsterdam
P.O.Box 800
NL-9700 AV Groningen
The Netherlands
Tel: +31-50-5 74 57 17
Fax: +31-50-5 74 57 17
eMail: c.j.jepma@eco.rug.nl

Frank Joshua
United Conference on Trade and
Development
Palais des Nations
CH-1211 Geneva
Switzerland
Tel: +41-22-9 17 58 34
Fax: +41-22-9 07 02 74
eMail: frank.joshua@unctad.org

Dr. Paul R. Koutstaal
ECN Netherlands Energy Research
Foundation
P.O. Box 1
NL-1755 ZG Petten
The Netherlands
Tel: +31-224-564516
Fax: +31-224-563338

Dr. Bo Lim
Organization for Economic Co-operation
and Development
Pollution Prevention and Control Division
2, rue André Pascal
F-75775 Paris, Cedex 16
France
Tel: +33-1-45 24 79 24
Fax: +33-1-45 24 78 76

Gordon McInnes
European Environmental Agency
Kongens Nytorv 6
DK-1050 Copenhagen
Denmark
Tel: +45-33-36 71 00
Fax: +45-33-36 71 99

Brian McLean
US Environmental Protection Agency
401 M Street, S.W. (6204J)
Washington D.C. 20460
USA
Tel: +1-202-2 33 91 50
Fax: +1-202-2 33 95 95

Dr. Klaus Müschen
Senator für Stadtentwicklung,
Umweltschutz und Technologie
Am Köllnischen Park 3
D-10179 Berlin
Germany
Tel: +49-30-24 71 10 60

Joaquim Oliveira Martins
Organisation for Economic Co-operation
and Development, Economics Department
2, rue André Pascal
F-75775 Paris, Cedex 16
France
Tel: +33-1-45 24 88 53
Fax: +33-1-45 24 91 75
eMail: Joaquim.Oliveira@oecd.org

Dr. Gerhard Petschel-Held
Potsdam Institute for Climate Impact
Research
Postfach 60 12 03
D-14412 Potsdam
Germany
Tel: +49-3 31-2 88 25 13
Fax: +49-3 31-2 88 26 00

Dr. Arthur Pelchen
UMB UmweltManagementBeratung
Hacker GmbH
Selchowstr. 1
D-14199 Berlin
Germany
Tel: +49-30-34500886
Fax: +49-30-34500887
eMail: Arthur.Pelchen@t-online.de

Dr. Klaus Rennings
ZEW Center for European Economic
Research
Postfach 10 34 43
D-68034 Mannheim
Germany
Tel: +49-621-12 35 207

Franzjosef Schafhausen
Federal Ministry for the Environment
Postfach 12 06 29
D-53048 Bonn
Germany
Tel: +49-30-2 85 50 23 50
Fax: +49-30-2 85 50 26 95

Dr. Franz Tattenbach
Oficina Costarricense de Implementación
Conjunta
P.O. Box 7170-1000
San José
Costa Rica
Tel: +5 06-2 20 00 36
Fax: +5 06-2 90 12 38

Dr. Ferenc Toth
Potsdam Institute for Climate Impact Research
Postfach 60 12 03
D-14412 Potsdam
Germany
Tel: +49-3 31-2 88 25 13
Fax: +49-3 31-2 88 26 00

Scott Vaughan
World Trade Organisation
Trade and Environment Division
Rue du Lausanne 154
CH-1211 Geneva 21
Switzerland
Tel: +41-22-7 39 50 91
Fax: +41-22-7 39 56 21

James Wolfe
Embassy of the United States of America
Neustädtische Kirchstr. 4-5
D-10117 Berlin
Germany
Tel: +49-30-2 38 51 74

LIST OF PARTICIPANTS

LIST OF PARTICIPANTS

Jochen Andretzky
Stephanusstr. 69
D-33098 Paderborn
Germany
Tel: +49-5251-76997
eMail: E85304@hrz.uni-paderborn.de

Clive Migosi Angwenyi
Kenya Marine & Fisheries Research Institute
P.O. Box 81651
Mombasa
Kenya
Tel: +254-11-4751515
Fax: +254-11-472215
eMail: kmfri@users.africaonline.co.ke

Ekaterina Atmatzidis
IZT Institut für Zukunftsstudien und Technologiebewertung
Schopenhauerstr. 26
D-14129 Berlin
Germany
Tel: +49-30-80308811
Fax: +49-30-80308888
eMail: 100726.2352@compuserve.com

Georg Bareth
Universität Hohenheim
Institut für landwirtschaftl. Betriebslehre
D-70593 Stuttgart
Germany
Tel: +49-711-4592544
Fax: +49-711-4593481
eMail: bareth@uni-hohenheim.de

Stefan Bayer
c/o Prof. Dr. D. Cansier
Universität Tübingen
Melanchthonstr. 30
D-72074 Tübingen
Germany
Tel: +49-7071-2974912
Fax: +49-7071-293926
eMail: stefan.bayer@uni-tuebingen.de

Silke Beck
Universität Bielefeld
Inst. f. Wissenschafts- und Technikforschung
Universitätsstr. 25
D-33615 Bielefeld
Germany
Tel: +49-521-10644342
Fax: +49-521-1066033
eMail: beck@iwt.uni-bielefeld.de

Jutta Borchers
Universität Bielefeld
Inst. f. Wissenschafts- und Technikforschung
Universitätsstr. 25
D-33615 Bielefeld
Germany
Tel: +49-521-1064342
Fax: +49-521-1066033
eMail: borchers@iwt.uni-bielefeld.de

Anjelika Borovikova
Mendelssohngasse 12/10
A-1220 Wien
Austria
Tel: +43-1-2098713
Fax: +43-1-2099571
eMail: A.Borovikova@iaea.org

Ulli Breitfuss
Douwesstr. 14
D-26721 Emden
Germany
Tel: +49-4921-35215
eMail: ulli@hermes.fho-emden.de

Georgina Broke
Durrell Institute of Conservation and Ecology
University of Kent
Canterbury/Kent CT2 7NJ
United Kingdom
Tel: +44-976-396311
eMail: gcb4@ukc.ac.uk

Catalina Cantero
C/Padre Jesus Ordonez 14, 2-C
E-28002 Madrid
Spain
Tel: +34-1-4111992
Fax: +34-1-3975161
eMail: ceceda@strogoff.adi.uam.es

Vicente Carabias
Technikum Winterthur
Fachstelle Ökologie
Postfach 805
CH-8401 Winterthur
Switzerland
Tel: +41-52-2677674
Fax: +41-52-2677473
eMail: crb@twi.ch

Richard A. Case
Department of Civil and Transportation Engineering
Napier University
Edinburgh
United Kingdom
Tel: +44-1751-431774
Fax: +44-1751-431774

Luis A. Collado
C/Lino 5, 2-2
E-28020 Madrid
Spain
Tel: +34-1-5534514
Fax: +34-1-3974328
eMail: ceceda@strogoff.adi.uam.es

Erwan Cotard
Cogan Europe
98, Rue Gulledelle
B-1200 Bruxelles
Belgium
Tel: +32-2-7728290
Fax: +32-2-7725044
eMail: cogen-europe@compuserve.com

Pablo Del Rio
C/Sirio 16, 6-Dcha
E-28007 Madrid
Spain
Tel: +34-1-4092705

Anita Engels
Universität Bielefeld
Inst. f. Wissenschafts- und Technikforschung
Postfach 100131
D-33501 Bielefeld
Germany
Tel: +49-521-1064672
Fax: +49-521-1066033
eMail: klima@post-uni.bielefeld.de

Antonio Fernandes
R. Constituicao 379 C-2
P-4200 Porto
Portugal
Tel: +351-2-5500044
eMail: antonf@letras.up.pt

Daniel Gijsbers
Cannerweg 117
NL-6213 BA Maastricht
Netherlands
Tel: +31-43-319779
Fax: +31-43-3260392
eMail: gijsbers@mailbox.unimaas.nl

Alan Gilmer
Peat Technology Centre
University College Dublin
Earlsfort Terrace
Dublin 2
Ireland
Tel: +353-1-2959250
Fax: +353-1-2959250
eMail: alan.gilmer@ucd.ie

Reinhard Grünwald
IZT Institut für Zukunftsstudien und
Technologiebewertung
Schopenhauerstr. 16
D-14129 Berlin
Germany
Tel: +49-30-80308817
Fax: +49-30-80308888
eMail: 100726.2352@compuserve.com

Christian Harling
HWWA
Neuer Jungfernstieg 21
D-20347 Hamburg
Germany
Tel: +49-40-3562479
Fax: +49-40-351900
eMail: a.michaelowa@hwwa.uni.hamburg.de

Heddeke Heijnes
Institute for Applied Environmental
Economics (TME)
Grote Marktstraat 24
NL-2511 BJ The Hague
Netherlands
Tel: +31-70-3464422
Fax: +31-70-3623469
eMail: tria@knoware.nl

Anke Herold
Öko-Institut
Postfach 6226
D-79038 Freiburg
Germany
Tel: +49-761-4529526
Fax: +49-761-475437
eMail: herold@freiburg.oeko.de

Falk Huettmann
University of New Brunswick
ACWERN/Fac. Forestry
P.O. Box 44555
Fredericton, N.B. E3B 6C2
Canada
Tel: +1-506-4526033
Fax: +1-506-4533538
eMail: k9wk@unb.ca

Anna Jansson
Ullerakersvägen 74
S-75643 Uppsala
Sweden
Tel: +46-18-710906
Fax: +46-18-696132
eMail: anna.jansson.2284@student.uu.se

Laila Jhaveri
FIELD SOAS
Russel Square 46 - 47
London WC1 B4JP
United Kingdom
Tel: +44-171-6377950
Fax: +44-171-6377951
eMail: W2@SOAS.ac.uk

Michael Kneisz
Ostertor Steinweg 31-33
D-28203 Bremen
Germany
Tel: +49-421-78756

Ewald Korevaar
Department of Science, Technology & Society
Padualaan 14
NL-3584 CH Utrecht
Netherlands
Tel: +31-30-2537647
Fax: +31-30-2537601
eMail: e.korevaar@nwsmail.chem.vuu.nl

Neus La Roca
c/Serpis 15, pta. 7
E-46021 Valencia
Spain
Tel: +34-6-3628538
Fax: +34-6-3864249
eMail: Neus.La.Roca@uv.es

Barbara Limprecht
Wuppertal-Institut für Klima, Umwelt, Energie
Klimapolitik-Abteilung
Doeppersberg 19
D-42103 Wuppertal
Germany

Chunjiang Liu
University of Helsinki
Dept. of Forest Ecology
POB 24
FI-00014 Helsinki
Finland
Tel: +358-9-3745268
Fax: +358-9-1917605
eMail: chunjiang.liu@helsinki.fi

Alexander Maier
Holtenauer Str. 127
D-24118 Kiel
Germany
Tel: +49-431-801065
eMail: Holti127@t-online.de

Ian Paul Milborrow
School of Social Sciences
University of Bath
Bath BA2 7Ay
United Kingdom
Tel: +44-1225-323014
Fax: +44-1225-826381
eMail: hssipm@beth.ac.uk

Ana Monteiro
Institute of Geography
Via Panoramica, s-n
P-4150 Porto
Portugal
Tel: +351-2-6077100
Fax: +351-2-6077153
eMail: anamt@letras.up.pt

Angela Morin
510, North State St
Ann Arbor MI 48104
USA
Fax: +1-313-7634765
eMail: amorin@umich.edu

Peter Moser
Am Turmhügel 14
D-49088 Osnabrück
Germany
Tel: +49-541-187090

Karen Oliveira
University of Kent
Farthings Court 11 A
Canterbury CT2 7UZ
United Kingdom
eMail: kado1@ukc.ac.uk

Giulia Pesaro
Iefe - Università Bocconi
Viale Filippetti, 9
I-20122 Milano
Italy
Tel: +39-2-58363820/1
Fax: +39-2-58363890
eMail: giulia.pesaro@uni-bocconi.it

Bernardo Andrés Pinto
Schaufelderstr. 1
D-30167 Hannover
Germany
Tel: +49-511-713606
Fax: +49-511-713606
eMail: 101675.732@compuserve.com

Bishal Sitaula
Box 5028
N-1432 Aas
Norway
Tel: +47-64-948262
Fax: +47-64-948211
eMail: bishal.situala@ijvf.nlh.no

Reinhard Steurer
Universität Salzburg
Senatsinstitut für Politikwissenschaft
Rudolfskai 42
A-5020 Salzburg
Austria
Tel: +43-662-80446613
eMail: Reinhard.Steurer@mh.sbg.ac.at

Sami Toivanen
LT-Consultants Ltd.
Melkonkatu 9
FIN-00210 Helsinki
Finland
Tel: +358-9-615811
Fax: +358-9-61581430
eMail: sto@ltcon.fi

Willemijn Tuinstra
Tarthorst 1003
6708 JH Wageningen
Netherlands
Tel: +31-317-460502

Marko Ulvila
University of Tampere
Dept of Administrative Science
P.O.Box 607
FIN-33101 Tampere
Finland
Tel: +358-3-2156392
Fax: +358-3-2156020
eMail: ulvila@kaapeli.fi

Jan-Willem van de Ven
EDON International
P.O.Box 519
NL-8000 AM Zwolle
Netherlands
Tel: +31-38-4554936
Fax: +31-38-4554166
eMail: 100733.3021@CompuServe.COM

Frank Vöhringer
Heßbrühlstr. 49 a
D-70550 Stuttgart
Germany
Tel: +49-711-7806131
Fax: +49-711-7803853
eMail: fv@ier.uni-stuttgart.de

Merei Wagenaar
De Vluchtestraat 1-103
NL-7523 BE Enschede
Netherlands
eMail: m.c.f.wagenaar@student.utwente.nl

Edwin Woerdman
Groningen National University
Department of Economics and Public Finance
P.O.Box 716
NL-9700 AS Groningen
Netherlands
Tel: +31-50-3635261
eMail: E.Woerdman@rechten.rug.nl

ZhongXiang Zhang
Department of Economics and Public Finance
University of Groningen
Westerhaven 16 A
NL-9718 AW Groningen
Netherlands
Tel: +31-50-3636882
Fax: +31-50-3637101
eMail: Z.X.Zhang@rechten.RUG.NL

Peter Zapfel
Sternwartestr. 12/34
A-1180 Wien
Austria
eMail: zapfel@ksg.harvard.edu

COLOUR PLATE SECTION

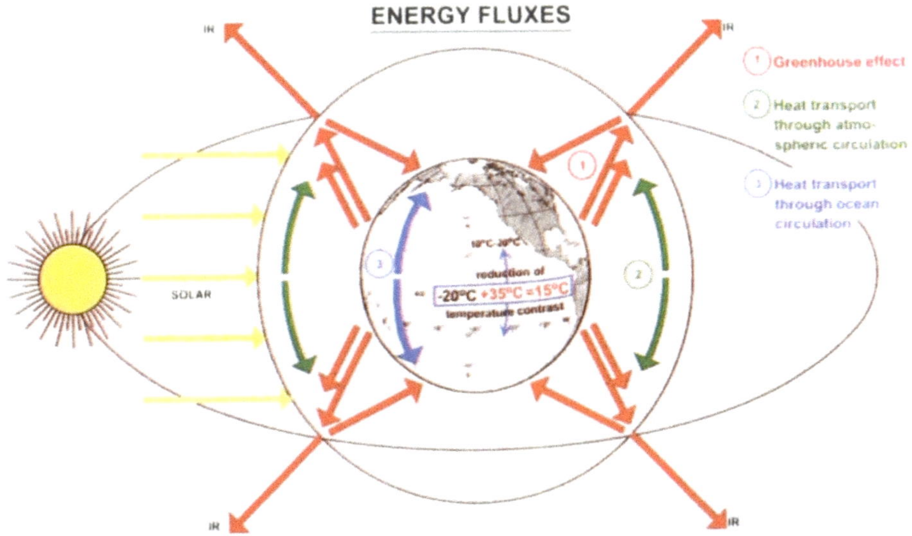

FIGURE A: The energy fluxes in the earth-climate system (Fig. 1, p. 4)

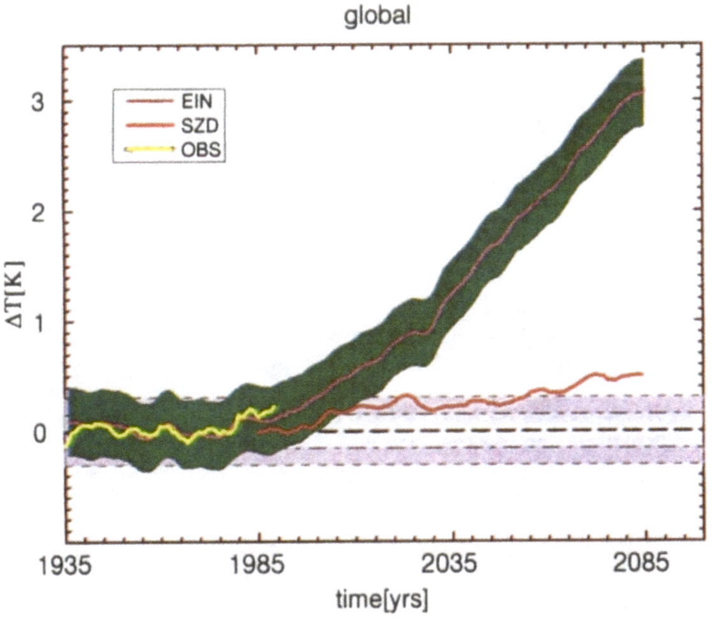

FIGURE B: The global mean temperature change for scenario A (similar to IPCC scenario IS92a) (EIN) and of scenario D (similar to IPCC scenario IS92d) (SZD). The gray and blue belts indicate the model uncertainty, the yellow line the observations (OBS) (after Cubasch et al, 1995) (Fig. 13, p. 16)

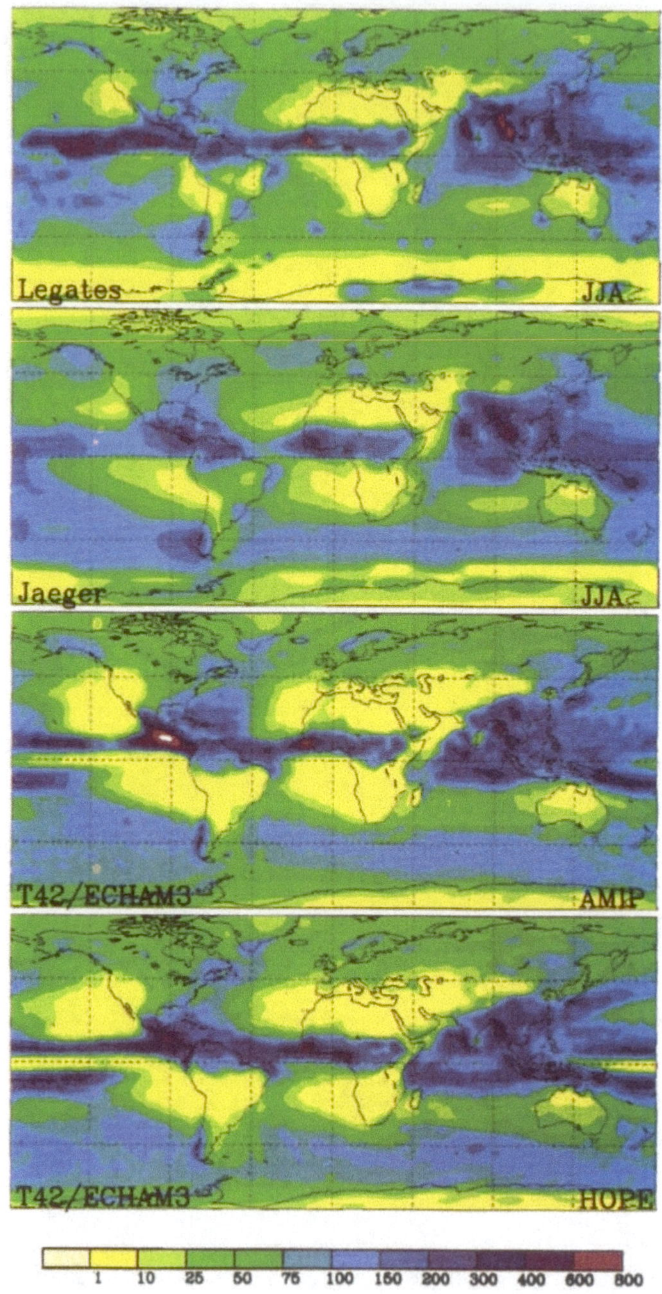

FIGURE C: The observed and the modelled precipitation for summer (JJA). The observations (top two panels) have been compiled by Legatesand Willmott, (1990) and Jäger, one model run was forced with observed sea-surface temperatures (third panel) and one model run has been performed with a fully coupled ocean-atmosphere climate model (bottom panel). (Arpe, pers. com) (Fig. 9a, p. 10)

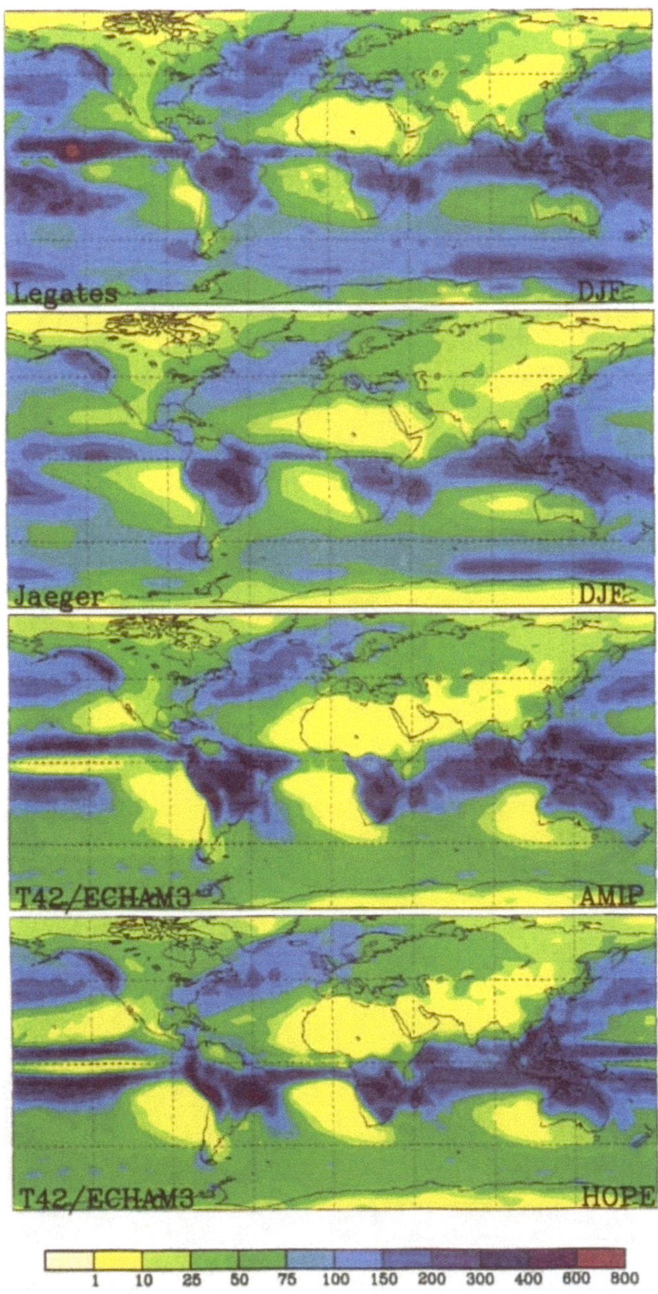

FIGURE D: The observed and the modelled precipitation for winter (DJF). The observations (top two panels) have been compiled by Legatesand Willmott, (1990) and Jäger, one model run was forced with observed sea-surface temperatures (third panel) and one model run has been performed with a fully coupled ocean-atmosphere climate model(bottom panel). (Arpe, pers. com) (Fig. 9 b, p. 11)

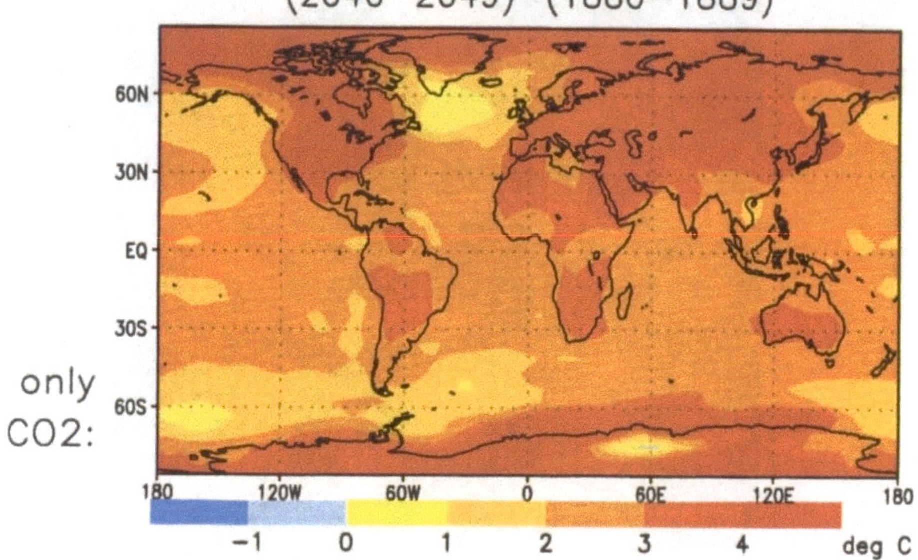

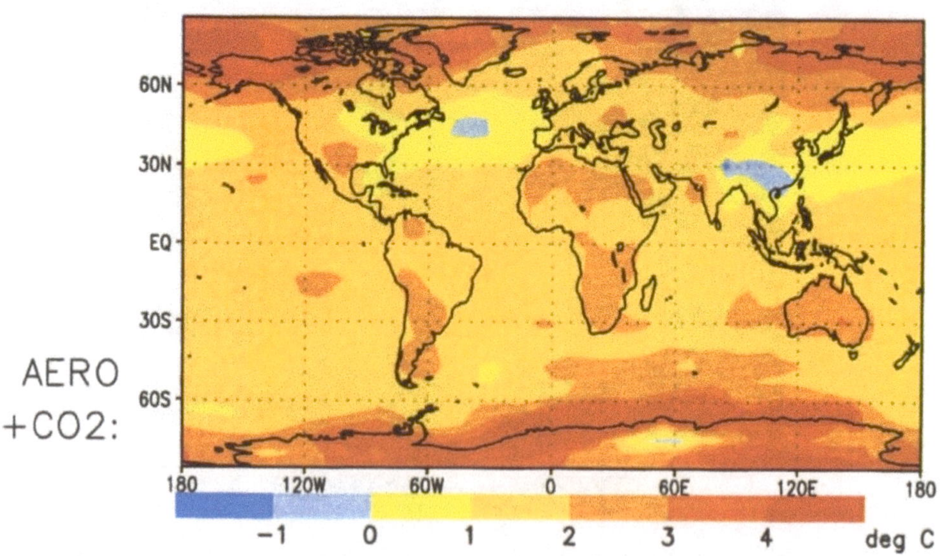

FIGURE E: Change of annual mean near surface temperature (in °C) in the mean over the decade 2041-2049 relative to the initial decade of the simulations (1880-1889) for the greenhouse gas only experiment (top) and the average of two experiments with greenhouse gas-plus-aerosol forcing (bottom) (after Hegerl et al, 1997) (Fig. 16, p. 19)

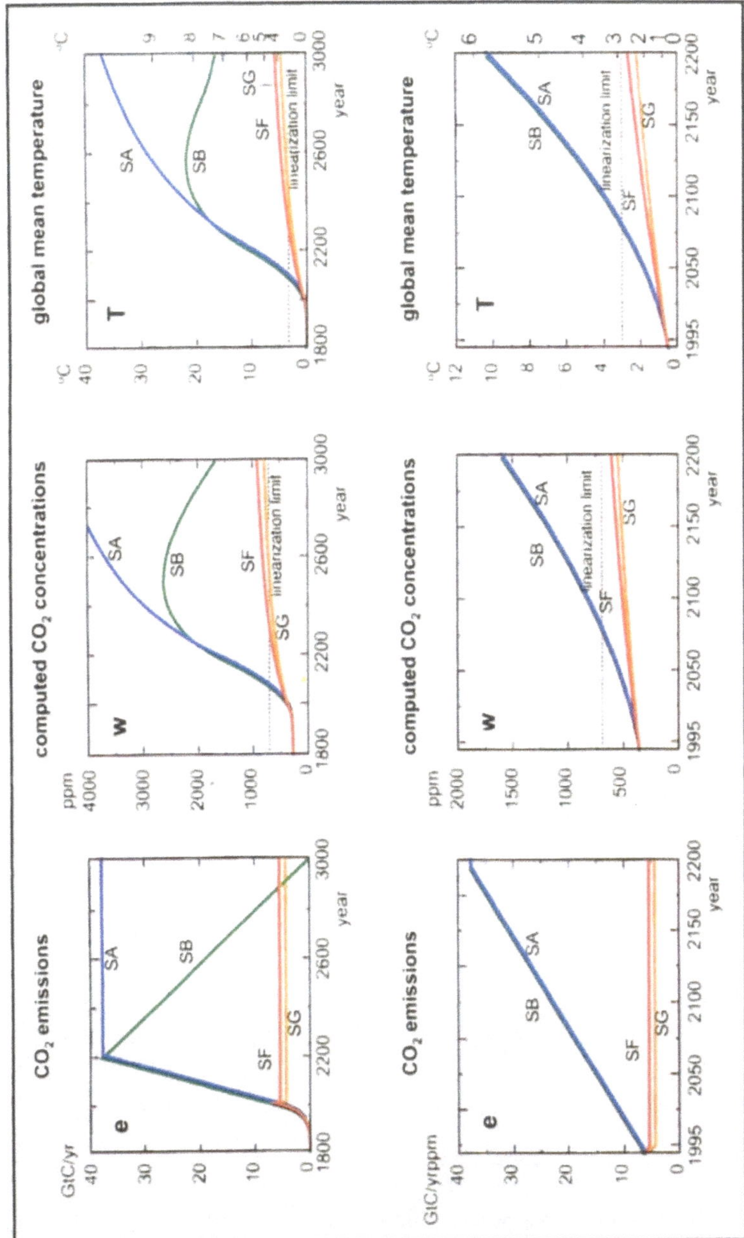

FIGURE F: CO concentrations and global warming (from left to right) for the time periods 1800-3000 (top) and 1995-2200 (bottom) for the "business-as- usual" (BAU) scenario (SA, blue), for the modified BAU scenario (SB, green), frozen emissions at 1990 levels after the year 2000 (SF, red) and 20% reduced emissions relative to the 1990 levels after the year 2000 (SG, orange). The linear model is not applicable above the based levels. (after Hasselmann et al, 1996) (Fig. 20, p. 23)

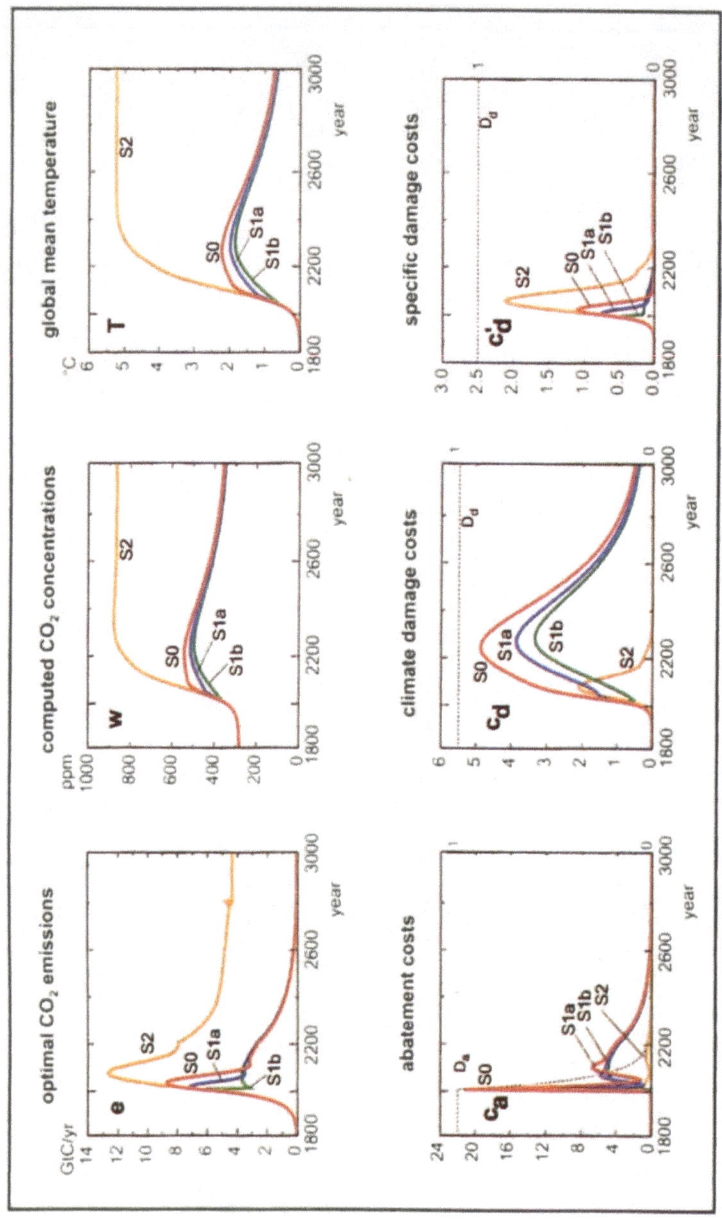

FIGURE G: Evolution over the period 1800 - 3000 of: (top, left to right) CO_2 emissions, CO_2 concentrations, global mean temperature and (bottom, left to right) specific abatement costs (c_a), specific damage costs (c_d) and the contribution to the specific damage cost (c'_d) from the rate of change of temperature for: the baseline reduced-emission scenario S0 (red), with reduced (S1a, blue) and zero initial terms in the abatement cost function (S1b, green) and a modified baseline experiment in which the climate damage costs are assumed to depend only at the rate of the temperature change (S2, orange). Also shown in the lower panel are the exponential abatement and damage cost discount factors D_a and D_d (after Hasselmann et al, 1996) (Fig. 21, p. 24)

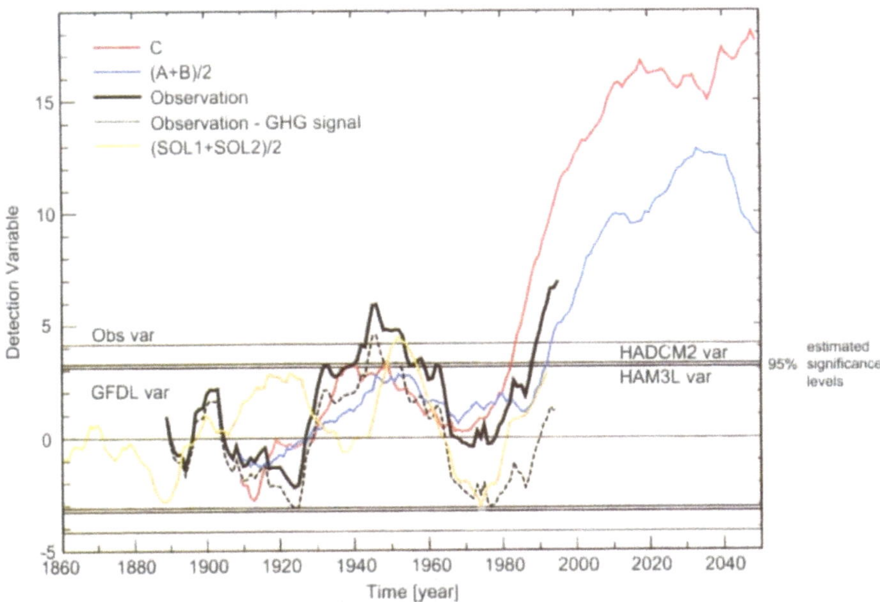

FIGURE H: Evolution of the detection variable computed with an optimal fingerprint for 30-yr. trend patterns from the observations (black), the aerosol experiment ((A + B)/2; blue), the greenhouse gas-only experiment (C; red), an experiment with prescribed solar variability as well as the observations, where the trend induced by the greenhouse gases has been removed (dashed, black). 95% confidence intervals from 4 sets of variability data are also indicated (after Hegerl et al, 1997) (Fig. 17, p. 20)

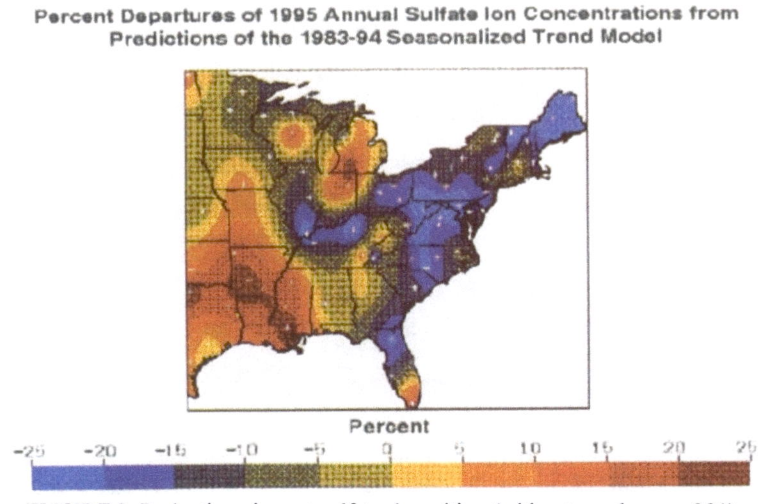

FIGURE I: Reductions in wet sulfate deposition (without number, p. 301)

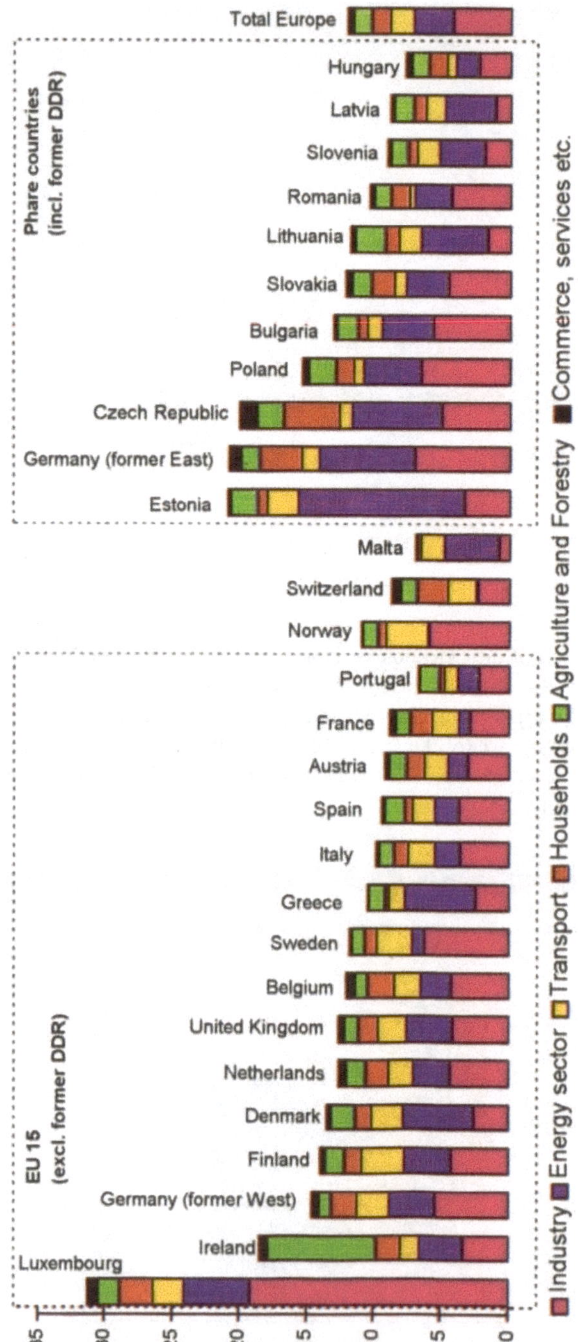

Figure J: Per capita total greenhouse gas emissions in Europe in 1990 by target sector (Mg CO_2 Equiv./capita/year) (Fig. 6. p. 105)

Zeitfracht Medien GmbH
Ferdinand-Jühlke-Straße 7
99095 Erfurt, Deutschland
produktsicherheit@kolibri360.de